Directions in Strong Motion Instrumentation

NATO Science Series

A Series presenting the results of scientific meetings supported under the NATO Science Programme.

The Series is published by IOS Press, Amsterdam, and Kluwer Academic Publishers in conjunction with the NATO Scientific Affairs Division

Sub-Series

I. Life and Behavioural Sciences	IOS Press
II. Mathematics, Physics and Chemistry	Kluwer Academic Publishers
III. Computer and Systems Science	IOS Press
IV. Earth and Environmental Sciences	Kluwer Academic Publishers

The NATO Science Series continues the series of books published formerly as the NATO ASI Series.

The NATO Science Programme offers support for collaboration in civil science between scientists of countries of the Euro-Atlantic Partnership Council. The types of scientific meeting generally supported are "Advanced Study Institutes" and "Advanced Research Workshops", and the NATO Science Series collects together the results of these meetings. The meetings are co-organized bij scientists from NATO countries and scientists from NATO's Partner countries – countries of the CIS and Central and Eastern Europe.

Advanced Study Institutes are high-level tutorial courses offering in-depth study of latest advances in a field.
Advanced Research Workshops are expert meetings aimed at critical assessment of a field, and identification of directions for future action.

As a consequence of the restructuring of the NATO Science Programme in 1999, the NATO Science Series was re-organized to the four sub-series noted above. Please consult the following web sites for information on previous volumes published in the Series.

http://www.nato.int/science
http://www.wkap.nl
http://www.iospress.nl
http://www.wtv-books.de/nato-pco.htm

Series IV: Earth and Environmental Series – Vol. 58

Directions in Strong Motion Instrumentation

edited by

Polat Gülkan
Middle East Technical University,
Ankara, Turkey

and

John G. Anderson
University of Nevada,
Reno, NV, U.S.A.

Springer

Published in cooperation with NATO Public Diplomacy Division

Proceedings of the NATO SFP Workshop on
Future Directions in Instrumentation for Strong Motion and Engineering Seismology, Kusadasi, Izmir
17–21 May 2004

A C.I.P. Catalogue record for this book is available from the Library of Congress.

ISBN-10 1-4020-3782-1 (HB)
ISBN-13 978-1-4020-3782-5 (HB)
ISBN-10 1-4020-3812-7 (e-book)
ISBN-13 978-1-4020-3812-9 (e-book)

Published by Springer,
P.O. Box 17, 3300 AA Dordrecht, The Netherlands.

www.springeronline.com

Printed on acid-free paper

Printed in the Netherlands.

Associate Editors

M. Çelebi (Menlo Park)

H. Sucuoğlu (Ankara)

Principal Contributors

S. Akkar (Ankara)
J.G. Anderson (Reno)
R. Archuleta (Santa Barbara)
J.J. Bommer (London)
D.M. Boore (Menlo Park)
R.D. Borcherdt (Menlo Park)
M. Çelebi (Menlo Park)
M. Erdik (İstanbul)
P. Gülkan (Ankara)
T.C. Hanks (Menlo Park)
S. Kinoshita (Yokohama)
H. Krawinkler (Stanford)
S. Midorikawa (Tokyo)
E. Miranda (Stanford)
J. Schwarz (Weimar)
X. Tao (Harbin)
Y.B. Tsai (Chung-Li)

Co-Contributors

N.A. Abrahamson (San Francisco)
B. Alavi (Wood Dale)
A. Anooshehpour (Reno)
M. Board (Las Vegas)
J.N. Brune (Reno)
C.A. Cornell (Stanford)
C. Dietel (Menlo Park)
E. Durukal (İstanbul)
C. Ende (Weimar)
Y. Fahjan (İstanbul)
G. Glassmoyer (Menlo Park)
D.H. Lang (Weimar)
C.P Lee (Chung-Li)
M.D. Purvance (Reno)
F. Scherbaum (Potsdam)
M. Squibb (Santa Barbara)
J. Steidl (Santa Barbara)
H. Sucuoğlu (Ankara)
H. Wang (Beijing)
R.E. Westerlund (Menlo Park)
A. Yakut (Ankara)
U. Yazgan (Ankara)
F. Zareian (Stanford)
Y. Zhou (Beijing)
Z. Zhou (Beijing)

CONTENTS

Part 3: Arrays and Observations

Preface and Acknowledgments

Earthquakes are a serious social problem. Recent major earthquakes in what should be the best-prepared urban areas, such Kobe, Japan, or Northridge, California, have caused tens to thousands of deaths and several billions of dollars in direct damage. The economic loss due to indirect damage, such as business failures, lost productivity, and business interruptions have similarly been in the range of several billions of dollars, invariably exceeding the direct loss by a substantial margin.

Many cities of the world are not as well prepared. Unlike Japan or the United States, many European and Asian cities have not been built with adequate, and strongly enforced, building codes. Many buildings in cities of former Soviet republics, constructed with untested techniques, have failed during strong shaking. Many structures in cities with theoretically better technical influence, as in Turkey, Greece, and Italy, have also failed. It is probably safe to generalize that a strong earthquake, striking a major urban area in any of the southern European or central Asian countries, has the potential to cause a major disaster. With increasing urbanization, a worst-case scenario is that a major earthquake, striking the major city of a small NATO country or NATO ally, could kill a significant fraction of the total population of the nation. Although the political consequences are outside of our expertise, we have heard suggestions that they could be as dire as completely destabilizing the government. This is a clear situation where greater personal security, and perhaps also national security, can be achieved through the application of scientific knowledge.

The workshop explored the role played by strong ground motion measuring and recording transducers in mitigating the seismic threat. The following report adopted by the workshop participants is a summary of lively discussions that complemented the formally presented contributions.

We owe a debt of gratitude to the Science for Peace Programme of Scientific Affairs Division of NATO that enabled this workshop to be planned and organized under Grant SfP977484, and Dr. Chris DeWispelaere, Programme Director and Dr. Susanne Michaelis, Associate Programme Director, for their continued support and encouragement. We also recognize the Scientific and Technical Research Council of Turkey whose support made it possible to host additional participants who brought energy and focus to the Workshop. This workshop would not have been possible without the support of our own institutions, the Disaster Management Research Center at Middle East Technical University, and the Seismological Laboratory, University of Nevada, Reno. Dr. Ahmet Yakut and İlker Kazaz provided much appreciated help with the assembly and editing of the volume contents.

Polat Gülkan, Ankara
John G. Anderson, Reno
March, 2005

Report Adopted by the Workshop on Future Directions in Instrumentation for Strong Motion and Engineering Seismology May 17-21, 2004, Kuşadası, Turkey

Significance of the Problem

The term "strong" ground motion might be defined in different ways: accelerations that might cause damage, or that exceed some threshold, or that are felt, or that trigger a strong-motion accelerograph. All of these definitions were used in this workshop, which focused on a global need to learn about the ground motions near enough to the fault to impact human activities. This is an area of cooperation between fundamental and applied science. All countries with seismically active regions would benefit tremendously if the scientific community were able to deliver answers to the many questions that have been identified since the first of several meetings on strong motion arrays and their deployment was held a quarter of a century ago. Then, recording and transcribing the motion of the ground used analog technology, and dissemination of the recorded information was through punch cards or nine-track tapes, means that no longer exist today. With the introduction of digital technology of ever decreasing physical size and ever increasing accuracy and capacity, data of hitherto unthinkable amounts can be recorded and saved routinely. The data is retrieved remotely through phone lines, telemetry or over the Internet. Real time acquisition is possible using off-the-shelf technologies. Dissemination is almost exclusively over the Internet through information portals or virtual platforms. Yet, interpretation and conversion of this information into tools of application for the benefit of societies is not all that rapid or automatic because, as we all acknowledge, there is not enough data to even constrain the mean levels of ground motions in many of the most significant cases.

Nature of the Workshop

58 participants met under a workshop sponsored by NATO under Science for Peace Project SfP977484 during May 17-21, 2004. The participants included a selection of leaders in the field of strong motion seismology and instrumentation and response of structures to strong shaking. The state of the art in ground motion observation systems and the future directions in the field were discussed at the workshop.

Specific Issues

The workshop dealt with identifying where we now stand in strong motion seismology, what its current needs are, and where future directions for further research might lie. There exists a global archive consisting of many earthquake seismograms. Most of the large earthquakes in the archive were only recorded on a few analogue instruments. These records might have been recorded by sensors of different design, processed by different means by different organizations, and under conditions that are not well documented. Yet, we can not afford to discard this wealth of information, however impaired it may be, because large earthquakes are rare events and we cannot assume that the handful of recent events that are extensively recorded on digital instruments provide a representative statistical sampling of what might happen. A need exists to address issues associated with lack of uniformity of quality and background information available for older strong motion data in Europe, US, Middle East, Japan, etc. These need to be fully documented, and to the extent possible, processing needs to be standardized. Pairing existing analogue stations with digital sensors would enable to reinterpret old records if simultaneous records could be done during future earthquakes. Older stations were often housed inside buildings, and not in free field conditions as some current criteria insist. So a need also exists to define what is meant by "free field," and generate funding for existing stations to quantify site response issues for strong ground motions. Also, guidance is needed for the conditions under which ground motions recorded in one location can be applied in a different location.

Associated with these issues of the historical strong motion data set are issues of access to strong motion data. Data is of no use if it is not accessible to researchers and engineers. The COSMOS Virtual Database was presented at this workshop as the tool that comes closest to achieving global access to invaluable data. Although this approach has adequate funding for the short term, the history of efforts to make strong motion data accessible is dominated by successful efforts that have lost their funding base in spite of support from the strong motion community. The workshop recognized the great value of the COSMOS database to the global strong-motion community, supports its continuation, and encourages all those who gather strong motion data to provide access either through COSMOS or through some other means.

Near real-time retrieval of data has many attractive advantages: problems with buffer over-flow are minimized, operation (including travel) and manpower costs are reduced, and instrument failures are recognized and repaired sooner. A discussion of when to use near real time retrieval of data is still necessary, though, because of other factors in running such a system: the communications system may not have the required reliability, and immediate access to the data may not be critical. Generation of shake intensity maps only minutes after the occurrence of an earthquake has many advantages for guiding emergency response. Too often response to earthquake emergencies is still driven only by actually reported damages, and not by the likely occurrences of such emergencies predicted from ground motion data. The elimination of these unnecessary delays could save untold numbers of lives by allowing rescue teams to reach the affected area more rapidly. Confidence in the seismological predictions could increase when extrapolation from sparse observations is replaced by a high density of instrumentation that reports ground motions in near real-time. To achieve this, we may need a new generation of less expensive, networked strong motion instruments.

The nature of the ground motion in the near field is of much current interest for earth scientists and engineers. Even though there are a few notable cases where ground motion records were obtained at sites located only a few kilometers from the rupture of large earthquakes (e.g., Kobe, Kocaeli, Chi-Chi), the world-wide collection of such records is still small, and until many more records have been recorded it will remain an open question as to the extent that the few existing records are representative. Yet, we need to include the unique effects of ground motion in the near field on buildings, as well as incorporate these into probabilistic seismic hazard analysis studies. For critical facilities, we also need to understand, and constrain, the extreme ground motions that occur near the fault in very rare events. Because records near the fault are so rare, we need to investigate how well data can be transferred among different physical settings, and conduct further research on existing data. For defining urban hazard and developing seismic regulations in cities located at very short distances to active faults, resolving these issues in near field strong motion are particularly critical. Dense arrays should be deployed in the vicinity of such faults. Setting priorities for faults to be monitored needs the development of criteria, bearing in mind that these are rare events (less frequent than one per year anyplace in the world), so that every time one of these events happens without being recorded it is a critical missed opportunity that delays progress towards well-informed seismic resistant design. For this case also, an order of magnitude reduction in the cost of instruments is necessary to promote the multiple order of magnitude increase of instruments that would be needed to avoid missing rare opportunities to gather the needed data. Even at the current state of knowledge of near-fault ground motions, it would be important to incorporate these effects into seismic resistant design codes.

Determination of site response and basin response is another promising avenue for international cooperation because the physics of the response of near-surface geology depends on material properties but not the latitude and longitude. Efficient methods are needed for determining shallow velocity structure and basin structure. Deployment of new instruments is called for at sites with carefully measured properties and sufficient hazard. A variety of designs measuring basin response, response in vertical profiles, response in areas with a tapering or complicated rock/soil interface, and the relationship of deep borehole data to the response of the urban regions above must be deployed using instruments capable of measuring site response under different amplitudes of the ground disturbance.

Strong motion seismology faces an important challenge in defining ground motion parameters for extremely rare events. At the low-end of the probability scale, statistical tools for defining these parameters are driven exclusively by the distribution of residuals of ground motion prediction equations. These models are derived from available records that only represent a few of the earthquakes of the past 70 years, much shorter than the return periods of these extreme events. The question of how to constrain extreme motions by substituting observations at many points in space for a long duration at a single site requires a careful blending of special instrument deployments, and examination of structures that have survived strong ground motions over geological periods.

The vulnerability of buildings shaken by the ground motions is an area of much current research activity. Arrays that monitor building response are essential for resolving many of the issues related to understanding the way in which complex structural systems respond to ground motions, particularly in the inelastic range. Correlation of low-amplitude building vibration data with response to strong shaking is useful for guiding structural engineers in developing more efficient design tools. Data is required globally for buildings, and for identifying vulnerable buildings. We need to understand why buildings collapse, and many more buildings must be instrumented than is currently is the case to achieve this objective. It is imperative to collect instrumental ground motion data in urban environments where a high density of various types and ages of buildings exist. After a major damaging earthquake these ground motion data will be very valuable for quantitative evaluation of the fragility of existing buildings. This will improve the reliability of earthquake loss assessment of urban areas.

Several of these topics bring up the issue of future instrumentation needs. Future needs can move in two different directions. One is the direction of high resolution (e.g. > 24 bit) instruments that can record both strong and weak motions on the same scale. The weak motions can help to quickly characterize site response and basin response from frequent small earthquakes, and with appropriate methods to be developed, they can help to determine attenuation relations for regions with few strong-motion data. The addition of GPS to these records can give quality control on displacements integrated from the accelerograms, and contribute to development of standard, reliable methods for processing strong motion data, especially at low frequencies. Low frequencies may be significantly underestimated with current processing methods, to the detriment of seismic design codes. GPS can also promote discovery of new phenomena that are in the interface between geodesy and seismic hazard, for example related to silent deformations in the earthquake cycle. The other needed direction is low-cost, high-density strong motion recording for characterizing urban areas, and also allowing instrumentation close to more faults. Although earthquake prediction has not shown notable successes, it is worthwhile to have the capability to deploy dense temporary networks is areas where seismic activity has increased or other anomalous geophysical phenomena are occurring and thus a major earthquake seems to have increased probability.

Earthquake hazard mitigation is a long-term, common concern to many NATO members and partners. Strong motion data play an important role in formulating effective mitigation strategies. However, compared to the standards set in Japan and China, the NATO countries are severely under-instrumented. Therefore, active and concerted instrumentation programs would be highly desirable and beneficial to all.

An important aspect of the conference to many participants was the opportunity for new, younger researchers to interact with those with more experience. A second important aspect was the interaction of seismologists and engineers. These opportunities need to be promoted particularly in countries with high hazards but proportionally fewer experts. Exchange and cooperation with scientists and institutions outside the NATO community will be mutually beneficial and should be encouraged.

Recommendations

Outstanding problems for strong motion instrumentation identified at the Kuşadası workshop include:

- the need for records of strong shaking very close to the causative fault for earthquakes of all mechanisms;
- how to use simple observations of the geology and geophysics at a site to significantly reduce the uncertainties in ground motion prediction;
- to obtain records of structural degradation during a strong earthquake;
- to understand extreme ground motions and to reduce the uncertainties in probabilistic seismic hazard analysis.
- the need for blind predictions to validate methods of ground motion prediction, basin response prediction, and structural response modeling.

The workshop participants strongly supported the interaction of engineers and seismologists to deal jointly with the earthquake problem.

The threat to security due to earthquakes is not being addressed. International cooperation exists at the scientific level. However, an international level organization that commands respect, is able to speak for the seismological and earthquake engineering communities, and has the ear of major national and international governmental organizations does not exist. NATO has the opportunity to take the lead in forming such a committee. The will to respond to the earthquake threat is global, not limited to NATO countries, and the NATO committee should form partners with others throughout the world. NATO can become a catalyst for focusing member country resources, non-public resources, and global resources to address problems. Such focusing is needed to attract the resources that are needed to address the earthquake problem within a useful time frame. This way, the role that NATO can play in enhancing security through scientific cooperation will be emphasized.

Since the threat from earthquakes is global, the solution also should take advantage of opportunities and resources from throughout the world. To this end, future workshops should be encouraged that would continue and expand the global level of collaboration that occurred in the Kuşadası meeting.

Part 1

SEISMIC HAZARD AND EXTREME MOTIONS

DATA NEEDS FOR IMPROVED SEISMIC HAZARD ANALYSIS

John G. Anderson, James N. Brune, Abdolrasool Anooshehpoor and Matthew D. Purvance
Seismological Laboratory and Department of Geological Sciences
University of Nevada
Reno, Nevada 89557

Abstract: Probabilistic seismic hazard analysis (PSHA) attempts to predict the occurrence rates of various ground motion parameters, and is therefore potentially verifiable. At very low probabilities, features that might bound past ground motions, such as surviving precarious rocks, often seem to be inconsistent with the predictions. The PSHA predictions at small probabilities are very sensitive to the way uncertainties are handled. One challenge is to separate uncertainties into their aleatory and epistemic components, which currently are mingled due to making an ergodic assumption when ground motion prediction equations are developed. At Lovejoy Buttes, California, there are several precarious rocks that have not been toppled by about 10,000 years of earthquakes. This paper shows a formal way to evaluate whether a particular probability distribution for peak acceleration from the largest earthquakes, M~8 events on the nearby San Andreas fault, is consistent with existence of those rocks. The paper concludes with suggested data needs from strong motion observations, since the solution to these PSHA issues will ultimately be driven by new data.

Key words: Earthquakes, hazard, seismic hazard analysis, strong-motion, accelerographs, precarious rocks, ground motion prediction, attenuation.

1. INTRODUCTION

Probabilistic seismic hazard analysis (PSHA) is an effort to predict the results of an experiment. The experiment would be to run a strong motion instrument at a fixed site for 10^n years, with *n* determined by the smallest probabilities of interest, and then to compile the statistics of the ground motion parameter of interest, to determine how often it is exceeded. Figures 1 and 2 show an example of such a prediction. Figure 1 shows the location of

P. Gülkan and J.G. Anderson (eds.), Directions in Strong Motion Instrumentation, 1–24.

Lovejoy Buttes, California, where the hazard is dominated by ground motions from earthquakes on the San Andreas fault, such as those that occurred from the earthquake in 1857. The 1857 earthquake had a rupture length of about 340 km, and based on the surface slip the best estimate of the moment magnitude is M_W=7.9 (Sieh, 1978). Figure 2 shows a corresponding hazard curve, giving a prediction for the result of an experiment that would need to be run for ~10^9 years, as it attempts to predict results with probabilities of 10^{-8} per year.

At Lovejoy Buttes, Brune (1999) identified and characterized several rocks that are precarious enough that sufficiently strong earthquake ground motion would cause them to topple. Brune estimates that they have had essentially the same geometry for 10,000 years. At that time, at the end of the last glacial period, the climate of the western US changed. At Lovejoy Buttes, the climate became warmer and drier, and surface erosional processes would have essentially stopped. The seismic hazard for the site, as estimated by the US Geological Survey (Frankel et al., 2002) predicts that a peak acceleration of 0.65g occurs with a probability of 2% in 50 years, corresponding to a mean occurrence rate of 1/2475 per year. This appears to be inconsistent with the survival of these rocks through all the earthquakes that occurred on the San Andreas fault during that time period (Purvance *et al.*, 2002, 2003). The nearest site to Lovejoy Buttes with paleoseismic data is Pallett Creek. At that location, assuming a Poisson process, surface rupturing events have occurred with an average rate of every 135 years (Biasi et al, 2002). Some of those events may not have been as large as the 1857 earthquake, however. Weldon et al (2004) show six earthquakes since about AD850 that, like 1857, ruptured at least from the Carrizo Plain past Pallett Creek, implying a mean occurrence rate about every 200 years. Assuming a mean rate of once every 200 years, one infers that about 50 such events have affected the rocks in their present geometry.

Figure 1. Map of part of southern California. The heavy line shows the approximate extent of rupture of the 1857 Fort Tejon, California, earthquake, Mw=7.9. The large 4-pointed star shows the location of Lovejoy Buttes (~34.60N, 117.85W). The Lovejoy Buttes are significant for the presence of 10,000 year old, semi-precarious rocks that have not been toppled by earthquakes. The inset photo (upper right) shows some of these rocks, with the San Andreas fault in the background. The inset perspective view (lower left) views the region along the direction of rupture propagation in the 1857 earthquake.

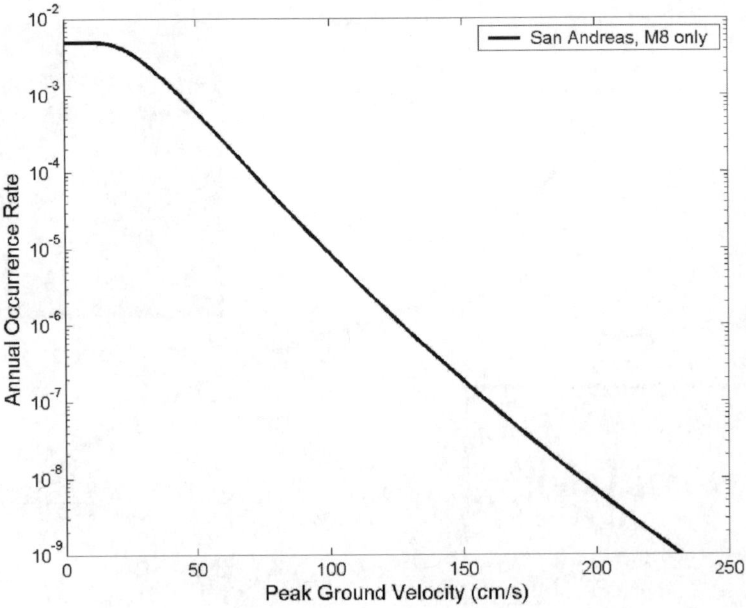

Figure 2. An example of an annual occurrence rate curve for Lovejoy Buttes. This curve is derived from the hazard curve for SA at 1.0 second period, based on the ground motion prediction equations of Sadigh et al (1997) and the assumption that the San Andreas fault has one size of earthquake, with magnitude 8.0, with a mean recurrence rate of 1/200 years. The peak velocity (PGV) is estimated from SA using the conversion equation (Abrahamson, personal communication): ln(PGV)=4.5+0.755 ln(SA), where SA is in units of fraction of the acceleration of gravity, and PGV is in units of cm/s.

This is not the only case where a probabilistic seismic hazard assessment has appeared to overestimate the hazard. Several studies have quantified the ground motions to topple precarious rocks (e.g. Anooshehpoor et al, 2004a; Brune et al, 2004, and references therein) Anderson and Brune (1999b) evaluated a probabilistic seismic hazard analysis for the sites of 26 precarious rocks in Nevada. They equated the probability with which the analysis predicts that ground motion would topple a precarious rock with the confidence level at which the hazard curve should be rejected for that specific site. They found that in all but two or three of those cases the hazard curve obtained by the standard approach would be rejected with more than 95% confidence level.

Another case is the probabilistic seismic hazard analysis for Yucca Mountain, Nevada, which has been designated as the site for the United States high level nuclear waste repository. Stepp et al (2001) carried out what is among the most thorough studies ever published. If correctness of a probabilistic seismic hazard analysis were the inevitable result of following accepted procedure, then there should be no problem in this case. Stepp et al show results to probabilities of 10^{-7}, at which level the analysis predicts peak accelerations of 3.4g. Regulations seek to assure performance to probabilities of 10^{-4} in 10^{4} years, thus corresponding to 10^{-8} probabilities overall. Extended to this level, the analysis predicts peak accelerations of about 11g.

Two comments are in order for the Yucca Mountain study. First, because the ground motions that are predicted at these extremely low probabilities are so far outside of the experience of observational seismology, they have been met with skepticism. Second, they are at probability levels that are far beyond the range of possible experimental verification. There are however proxies, such as the fragility curves for precarious rocks, that can be used to constrain these predictions at low probabilities. Also, consideration of some thought experiments points towards some possible improvements in the analysis methods, and collection of new data has the potential to test some of these ideas.

2. PROBABILISTIC SEISMIC HAZARD ANALYSIS

A probabilistic seismic hazard analysis (PSHA) calculates a mean occurrence rate curve ($\lambda_C(Y)$) giving the average frequency that a ground motion parameter, say *Y*, is equaled or exceeded and a hazard curve ($P_C(Y,T)$) giving the probability that *Y* is equaled or exceeded in time interval *T*. It is important to recognize that $\lambda(Y)$ and $P(Y,T)$ exist and can be measured, at least conceptually if not in practice, by means of the thought experiment described in the introduction. (We drop the subscript to emphasize the difference between the real curve that would result from the experiment and the calculations.) Since strong motion accelerographs were first deployed in the 1930's, the longest instrumental records that are available to estimate empirical hazard curves according to the procedure of the thought experiment are shorter than 100 years, and those are available for only a very few locations. On the other hand, a hazard curve is most useful for sites where new construction is about to occur. Furthermore, hazard curves are interesting for probabilities of significantly less than 10^{-2}. Thus synthetic methods are essential to estimate hazard curves for general locations and for mean occurrence rates and probabilities that are orders of magnitude smaller than 10^{-2}.

Probabilistic seismic hazard analysis as it is currently applied uses an approach that can be traced to Cornell (1968). The basic probabilistic seismic hazard analysis model (e.g. McGuire, 1995; SSHAC, 1997) finds:

$$\lambda_C(Y) = \iint n(M,\mathbf{x})\Phi(y \geq Y \mid \hat{Y}(M,R(\mathbf{x})),\sigma_A)dMd\mathbf{x} \qquad (1)$$

In Equation 1, $n(M,\mathbf{x})$ gives the annual occurrence rate of earthquakes with magnitude M at a location $\mathbf{x}$ in the Earth. The term $R(\mathbf{x})$ is the distance from the earthquake at location $\mathbf{x}$ to the site of interest, and may be defined in different ways for a finite fault. The term $\Phi(y \geq Y \mid \hat{Y}(M,R(\mathbf{x})),\sigma_A)$ is developed from a ground motion prediction equation, and gives the probability that an earthquake with magnitude M at location $\mathbf{x}$ will cause a ground motion that equals or exceeds Y. The probability is conditional on the mean value of a ground motion prediction equation ($\hat{Y}(M,R(\mathbf{x}))$) and its aleatory standard deviation (σ_A). The integrals in Equation 1 are over all relevant magnitudes and over the relevant volume of the Earth. Figure 2 is an example of a mean annual occurrence rate curve for a specialized case where the seismicity is localized to a single source. Equation 1 is a standard equation, from probability theory, for the mean value of a multi-valued distribution.

Assuming that the earthquakes are random, uncorrelated events, then the Poisson model should hold, and we expect that the probability of exceedance in a time interval of duration T is:

$$P_C(Y,T) = 1 - \exp(-\lambda_C(Y)T) \tag{2}$$

Using the expansion:

$$e^x = \sum_{s=0}^{\infty} \frac{x^s}{s!} \tag{3}$$

it is immediately evident that for $\lambda_C(Y)T << 1$, $P_C(Y,T) \approx \lambda_C(Y)T$. The Poisson assumption is widely used, and it is appropriate for uncorrelated, infrequent events.

Virtually all ground motion prediction equations (e.g. Abrahamson and Silva, 1997; Boore et al, 1997; Sadigh et al, 1997) regard the mean value, $\hat{Y}(M,R(\mathbf{x}))$, to be the mean of a lognormal probability distribution with standard deviation σ_A. Under the lognormal assumption, at small probabilities the hazard curve given by Equation 2 is quite sensitive to σ_A. This is illustrated in Figure 3. The figure shows occurrence rate curves calculated for Lovejoy Buttes for a magnitude M_W=8.0 earthquake on the San Andreas fault, 14.5 km from the site. The curve uses $\hat{Y}(M,R(\mathbf{x}))$=0.407g based on Sadigh et al (1997). According to Sadigh et al, σ_A=0.38 (natural logarithm units), but the curves in Figure 3 are computed for σ_A ranging from 0.01 to 0.75. At an occurrence rate of 10^{-8}, the peak acceleration ranges from 0.43 g for σ_A=0.01 to nearly 14g for σ_A=0.75. The extreme accelerations at low occurrence rates, or low probability resulting from large σ_A, are far outside of the range of experience and are likely to seem quite implausible to most experts. Similarly, for the calculations at Yucca Mountain, the extreme ground motions at low probabilities are the result of the high values of uncertainty used in that analysis.

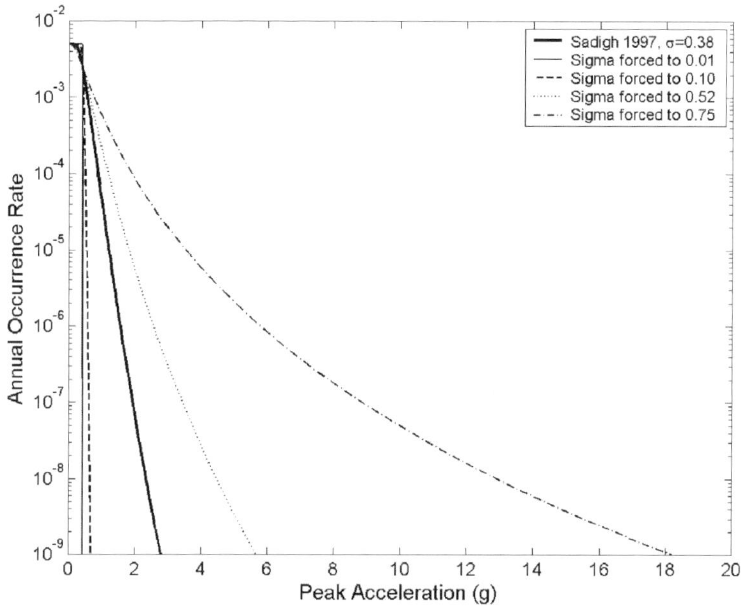

Figure 3. Several models of the predicted annual occurrence rate of peak acceleration at Lovejoy Buttes. All models are derived from the Sadigh et al (1997) ground motion prediction equations, which have a mean peak acceleration of 0.407g at this distance from the fault. The models have differing values of standard deviation, σ_E, however, including the value given by Sadigh et al and other values as indicated. A value of σ_E=0.69 implies that one standard deviation on the uncertainty in the mean value is a factor of two.

The theory for PSHA recognizes different kinds of uncertainties: aleatory and epistemic. Uncertainties in $\hat{Y}(M,R(\mathbf{x}))$ are treated as epistemic, i.e. due to lack of knowledge, in contrast to the uncertainty σ_A which is supposed to be related to the inherent randomness of ground motions from multiple realizations of earthquakes of the same magnitude and distance. Based on the procedure described by SSHAC (1997), the way that epistemic uncertainty is treated is very different from treatment of aleatory uncertainty. Equation 1 shows how aleatory uncertainty is treated. Suppose one identifies alternative estimates for the mean value $\hat{Y}(M,R(\mathbf{x}))$, and assigns relative weights (or probabilities) to each of these models, say w_C, where $\sum w_C = 1$. Then the best estimate of the hazard curve is given by:

$$P_N(Y,T) = \sum_C w_C P_C(Y,T) \tag{4}$$

Figure 4 compares some estimated hazard curves calculated with Equations 2 and 4. These are based on the assumption that the epistemic uncertainty in $\hat{Y}(M,R(\mathbf{x}))$ also obeys a lognormal distribution, with standard deviation σ_E. Furthermore, the total uncertainty σ_T is constrained to be constant by the assumption:

$$\sigma_T^2 = \sigma_A^2 + \sigma_E^2. \tag{5}$$

The four frames in Figure 4 are defined for four different time intervals *T*, selected to be 0.1, 1, 10, and 100 times the mean repeat time of the M_W=8.0 earthquake on the San Andreas fault. Each of the frames compares the two cases where σ_A=0.38 and where σ_T=0.38 while σ_A=0.10, with σ_E adjusted accordingly. The frames also show the most probable hazard curve, i.e. the curve with the highest value of w_C. This curve has used $\hat{Y}(M,R(\mathbf{x}))$=0.407g and σ_A=0.10.

The frames in Figure 4 show that the division of uncertainties into aleatory and epistemic components has an effect on the estimated hazard curve when the time interval is long compared to the mean repeat time of the earthquakes, i.e. when repetition of the same earthquake dominates the hazard. This is because the uncertainty in the mean value will be resolved, at least in part, with the first earthquake occurrence. The very large tails at low probabilities remain, however, compared to the most probable curve, since until the first event has happened the extremely high mean values have not been ruled out.

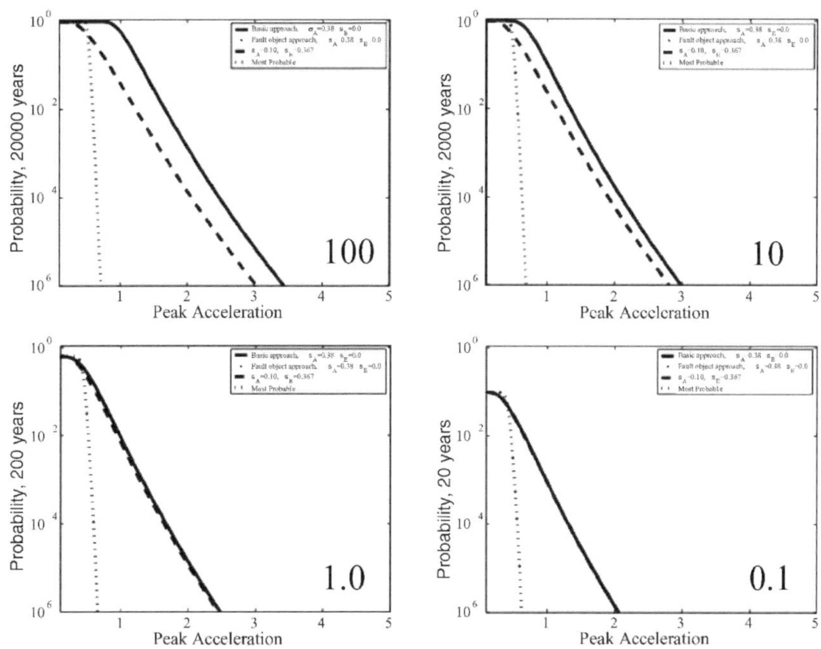

Figure 4. Numerical experiments with alternative ways to estimate the probabilistic hazard for peak acceleration at Lovejoy Buttes. The hazard is estimated for time intervals of 100, 10, 1, and 0.1 (upper left, upper right, lower left, and lower right, respectively) times the mean repeat time for 1857-like earthquakes on the San Andreas fault. Each frame has three curves. The solid line, reproduced by Field et al's fault object approach (dots, not visible because they overlay the solid line) gives the probability of exceedance of the peak acceleration in the given time interval, assuming all of the uncertainty is aleatory. The other lines are based on the arbitrary assumption that σ_A =0.10, and the rest of the uncertainty is epistemic. In this case, the most probable hazard curve (dotted) is the one where the mean is as given by the ground motion prediction equation. The dashed line is obtained by a linear averaging of a suite of alternative hazard curves with σ_A =0.10, but with different mean values defined by the epistemic uncertainty.

The very large difference between the most probable curve and the curves based on Equations 1 and 4 provides motivation to carefully examine the understanding of the two kinds of uncertainty, to examine whether the mathematical treatment as given in Equations 1-5 is really the correct approach for handling these uncertainties in the hazard analysis, and to examine whether any of the hazard curves can be ruled out based on observations.

Figure 5 shows the equivalent of Figure 4, but for a different choice of probabilities. In all cases σ_T =0.75. In one case it is all aleatory, while in the case where much of the uncertainty is considered epistemic, σ_A =0.52, giving σ_E =0.54. These uncertainties are closer to those that appear in Stepp et al (2002).

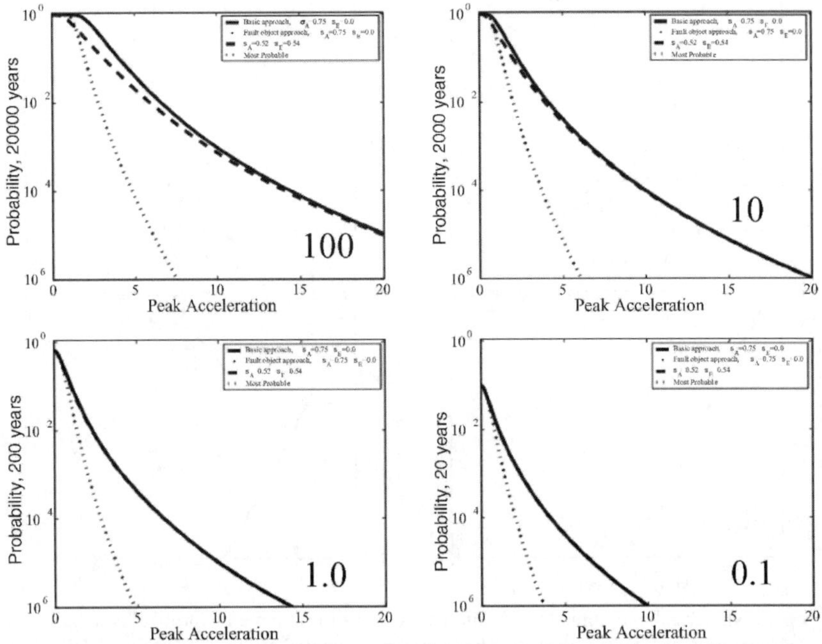

Figure 5. Equivalent to Figure 4, except that the aleatory, epistemic, and total uncertainty are larger, as given in the legends. The total uncertainty σ_T is always 0.75, while σ_A=0.75 for the solid line, and σ_A =0.52 for the other curves

3. THE ERGODIC ASSUMPTION

The division of the uncertainty into aleatory and epistemic parts is not a straightforward task. Anderson and Brune (1998b) point out that an assumption, that they refer to as an "ergodic" assumption, is almost always made, and may be incorrect. Figure 6a illustrates a geometry where several strong motion stations observe the ground motions on a "racetrack" at a common distance from a vertical strike-slip fault. This situation has been reproduced using synthetic seismograms by Zeng (personal communication). The calculated peak accelerations for the racetrack of stations at several distances are shown in Figure 6b. Figure 6c shows the standard deviation determined at each distance, based on the lognormal assumption about the distribution of peak accelerations. The magnitude of the standard deviation in Figure 6c, between 0.25 and 0.35, will be discussed later.

It is clear that the standard deviation in Figure 6c is due entirely to a spatial distribution of ground motions. The synthetic accelerograms are all computed for the same flat-layered half space, and thus there is no variability in the wave propagation to increase the variability in the ground motions. The variability is the result of differences in the slip function on the fault, radiation pattern, and directivity in the synthetic seismogram model.

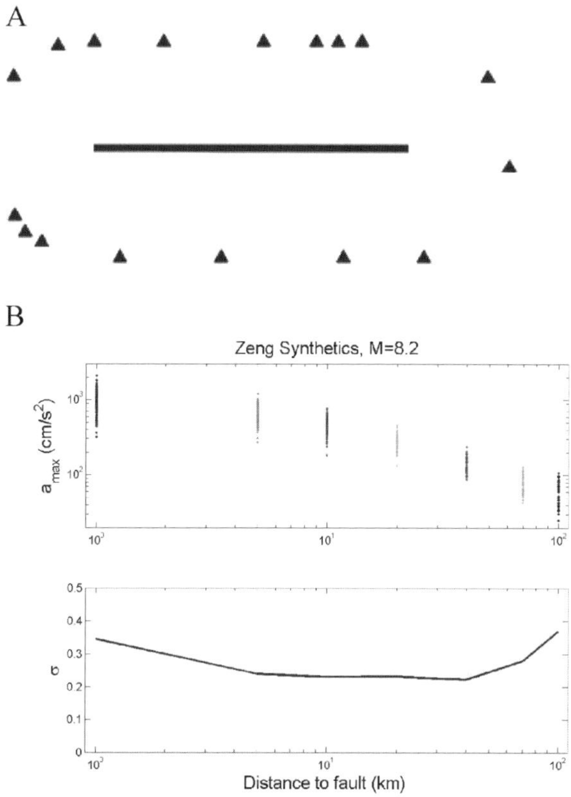

Figure 6. A) Illustration of a geometry where several stations all record the ground motion at the same distance from a vertical strike-slip fault.
B) Peak acceleration results of a numerical experiment by Yuehua Zeng (personal communication) in which the experiment in part A is implemented using a numerical scheme to calculate synthetic seismograms (Zeng et al, 1994). The experiment was performed at several distances, as given by the horizontal axis.
C) Estimated standard deviation at each distance.

Figure 7 shows, for the 5 km distance, the lognormal distribution of ground motions over all stations determined by the synthetics. It also shows the distribution at a single station in the forward directivity direction. At the single station, the standard deviation is schematically reduced to 0.10, which is consistent with values obtained by more than one synthetic model for this controlled geometry. The variability at a single station is much smaller than the total variability. The total variability in Figure 6c is largely controlled by the distribution of motions from individual sites at different azimuths. We also note that even if the variability at a single station is lognormal, it is not true that the sum of lognormal distributions is also lognormal. The distribution determined over all stations has a much larger tail than the tail from the distribution at a single station in the forward directivity direction. This gives one explanation for why the tails of the distribution of values obtained from the ground motion prediction equations might be larger than reality.

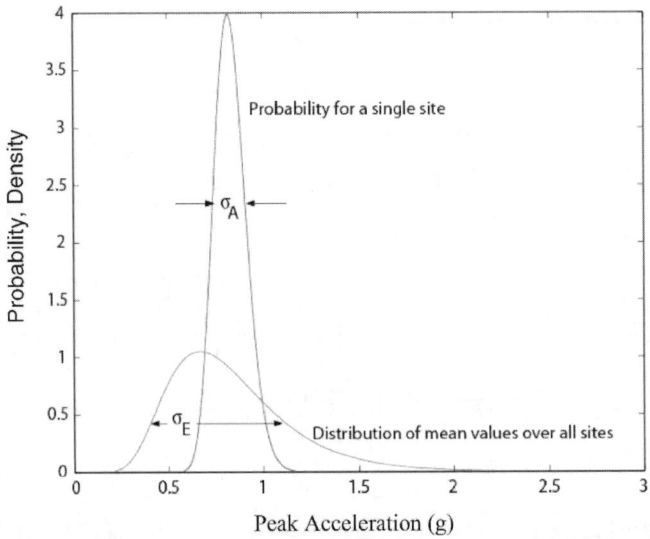

Figure 7. Illustration of the difference between epistemic and alcatory uncertainty, based on the numerical experiment in Figure 6. The green line shows the distribution of mean values of peak acceleration over the set of equidistant stations from the fault. The blue line shows the probability distribution for repeated recordings of the ground motion at a single site.

4. HOW IS THE TOTAL STANDARD DEVIATION DISTRIBUTED AMONG VARIOUS EFFECTS, ASSUMING THE LOGNORMAL DISTRIBUTION?

Typical values of the standard deviation from the ground motion models are higher than those shown in Figure 6. For instance the models identified above have the following standard deviations: Abrahamson and Silva, 0.43; Boore et al, 0.47; Sadigh et al, 0.38. These are all greater than estimates from the synthetics. Variability in ground motions introduced by variability in wave propagation and site characteristics could easily be responsible for these differences.

Cua and Heaton (2004) found that data recorded on Trinet, after careful correction for site conditions, could be modeled using a ground motion prediction equation with sigma = 0.243. This is in the range predicted in Figure 6, for stations at distances of 10 km or greater from the fault.

The similarity of the Cua and Heaton estimate and the results in Figure 6c may be just a coincidence. However, if we speculate that the synthetic and empirical approach have converged to a standard deviation that is approximately realistic, we can come up with Table 1, a speculative model for the distribution of the standard deviation in ground motion prediction equations.

Table 1. A Speculative Model for the Total Standard Deviation in Ground Motion Prediction Equations

	Total sigma	Contribution from this phenomenon
Single fault-station geometry, due to variability at the source, in repeated ruptures of the same fault plane with the same sense of directivity.	0.1	σ_A=0.1
Uncertainty in a flat-layered half space due to radiation pattern effects.	0.25	σ_{space}=0.23
Uncertainty due to variability of site effects.	0.39-0.47	σ_{site}=0.3-0.4

An important data need for probabilistic seismic hazard analysis is to determine carefully how the total standard deviation, as revealed by ground motion prediction equations, should be divided among the various factors that contribute.

5. SOME IDEAS ON HOW TO CONSTRAIN THE STANDARD DEVIATION

Old geological indicators or monuments might help to constrain the standard deviation. We present an example using some precariously balanced rocks in southern California. Figure 1 illustrates a semi-precarious rock near the San Andreas fault in southern California. Purvance (2004) and Purvance et al (2004) carefully studied the rocking problem. This included testing the response of the rock in Figure 1 to a suite of accelerograms, all of which have frequency content similar to what is expected from a magnitude ~8.0 earthquake on the San Andreas fault. Figure 8a shows what might be called the "fragility curve", f(a), for the precarious rock that results from these numerical tests.

Whether or not a rock topples in a given strong motion record depends on a highly nonlinear physical process which depends on the phase of the ground motion and is also rather sensitive to the detailed geometry and the assumed physics at the base of the rock. The curve f(a) appears relatively smooth and monotonic in Figure 8a, but that smoothness comes from averaging over several accelerograms. Above the threshold for toppling from a constant acceleration, whether the rock remains standing is very sensitive to small changes in the seismogram. Since it is highly unlikely that successive ruptures on the San Andreas fault will be identical in the phase at high frequencies, this smoothing process used to derive f(a) is appropriate.

Figure 8a also shows the derivative of a hazard curve for the site, $-d\Phi/dy$, where y is peak acceleration in this case, developed based on the Sadigh et al regression (mean = 0.407g, σ_A= 0.38). The product of the two curves in Figure 8a gives the probability density of toppling as a function of peak acceleration. This probability density is shown in Figure 8b, along with the integral to get the overall probability of toppling due to all accelerations

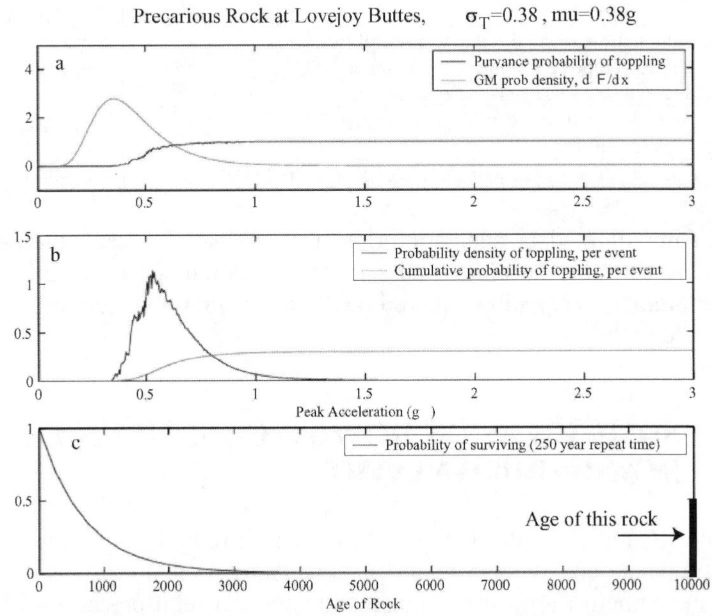

Figure 8. A) Fragility curve for toppling a balanced rock with parameters typical of Lovejoy Butte. The monotonic increasing curve gives the probability that the rock will topple as a function of increasing peak acceleration in a suite of records with frequency content appropriate to a magnitude 8 earthquake. This is a highly nonlinear problem, where the outcome for any individual record is either toppling or not, and switches between the two domains as peak acceleration increases. This curve is smoothed and approximately monotonic due to summing the curves for a large suite of records. The peaked curve on this plot is the probability density distribution for the peak acceleration based on the Sadigh et al (1997) ground motion prediction equation, conditional on the earthquake happening.
B) The peaked line gives the probability density for the rock to topple, conditional on the earthquake happening, obtained as the product of the two curves in part A. The monotonic increasing line is the cumulative probability of toppling per event. Under the assumptions used here, the overall probability of toppling is about 0.3 in each characteristic earthquake.
C) Probability of the rock surviving as a function of it's age. With a probability of toppling of about 50% per earthquake, the half life of an individual rock of this geometry is about 400 years, or two earthquake cycles. The bold vertical line shows the age of the rock.

conditional on the occurrence of an earthquake. Since the earthquake has a significant probability that it will not topple the rock, the integrated probability of toppling is substantially less than 1, in fact about p_0=0.35.

If the rock is exposed to n earthquakes, its chance of surviving is

$$p_s = 1 - (1 - p_0)^n \tag{6}$$

Figure 8c presumes that events of this size occur on the San Andreas fault every 200 years to calculate *n* as a function of the time interval, and then applies Equation 6 to determine its probability of survival. The chance of surviving two earthquakes, or 400 years, is about 50%, so one might designate 400 years as the expected half-life of the rock. An alternative way to look at Figure 8c is to select the estimated age of the rock (10^4 years in this case), and determine the probability of it surviving for that duration of time. In this case, the probability for it to survive so long is much less than 1%, so following the suggestion of Anderson and Brune (1998b), we would reject the Sadigh combination of mean and aleatory uncertainty for this earthquake-site pair with a confidence level of well over 99%.

Figures 9 and 10 explore the combinations of mean and aleatory uncertainty that are not rejected by this fragility curve. Figure 9 contours the half-life of the rock as a function of the mean and aleatory uncertainty of the ground motion prediction equation for this earthquake. Figure 10 contours the probability with which a particular value of the mean and the aleatory uncertainty would be rejected. It is evident that the mean and aleatory uncertainty for these three ground motion prediction equations (Abrahamson and Silva, 1997; Boore et al, 1997; Campbell, 1997) would be rejected.

O'Connell and LaForge (2004) suggest that through application of the Boore et al. regression for very hard rock, the precarious rock is not inconsistent with the PSHA results at a probability of 2% in 50 years. If they had extended the hazard curve to 0.5% in 50 years (or about 10^{-4} / year), they would have found a contradiction. But still, they also recognized that with an appropriate combination of mean and standard deviation, contradictions of the probabilistic seismic hazard analysis with the existence of the semi-precarious rocks can be resolved.

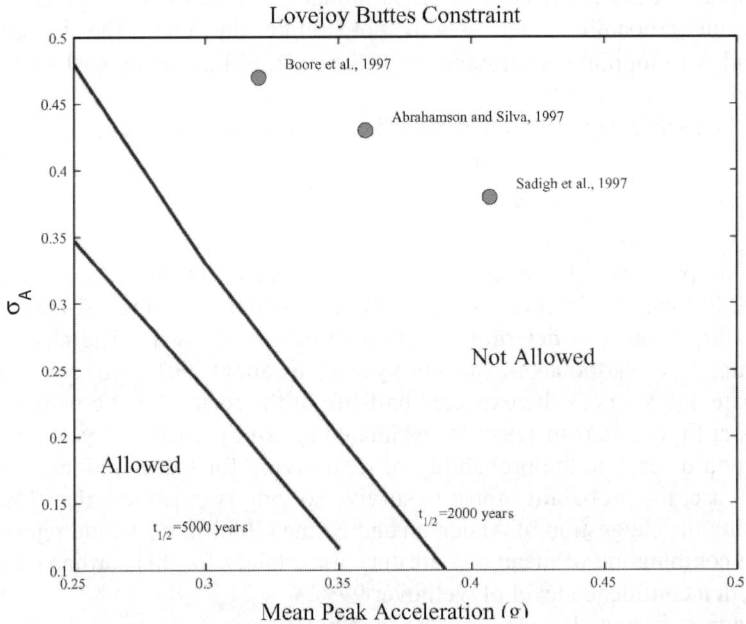

Figure 9. One way of evaluating ground motion prediction equations for the specific case of Lovejoy Buttes. The contours show two half-live lines in the peak acceleration – aleatory uncertainty plane. Since the rocks at Lovejoy Buttes are over 10,000 years old, ground motion prediction equations substantially to the right of the 2000 year half-life line are inconsistent.

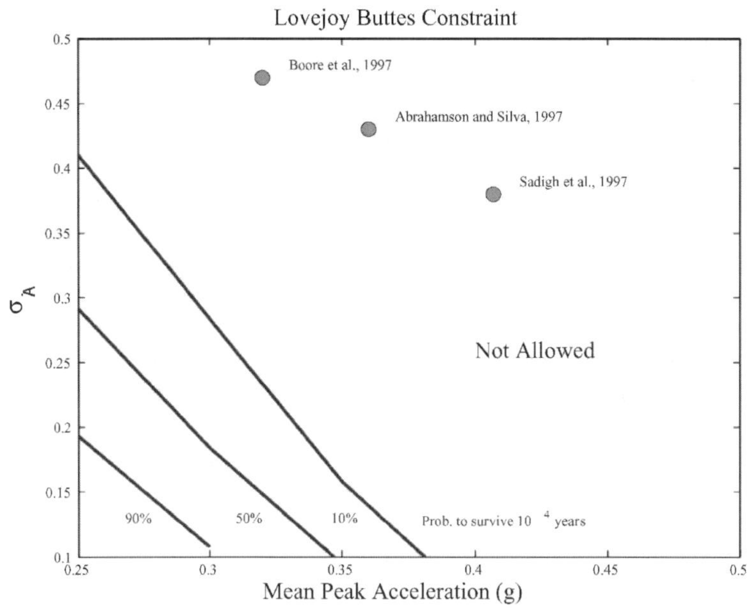

Figure 10. An alternative way of evaluating ground motion prediction equations. The contours on the peak acceleration – aleatory uncertainty plane give the probability for a rock with a certain geometry to survive 10^4 years.

6. SUMMARY

In some cases, probabilistic seismic hazard analyses yield ground motions that appear to be too high compared with past experience or with geological indicators. There are three possible explanations, any or all of which might contribute in varying amounts. These possibilities are that the mean ground motion estimates are too high, that the uncertainty is too large, or that uncertainties are not being handled in the best way.

The results of a probabilistic seismic hazard analysis are very sensitive to σ_A, especially at small probabilities. Furthermore, for time periods that are long compared to the repeat times of a dominant earthquake in the analysis, the results are sensitive to the way the total uncertainty is divided between σ_A and σ_E. The ergodic assumption appears to erroneously transfer uncertainties from σ_E into σ_A, resulting in an overestimate of σ_A. The problem of how to separate σ_A and σ_E is however a very difficult one, and needs concentrated research.

Precarious rocks constrain the mean and standard deviation for ground motions from specific earthquakes. This paper suggests that the mean and standard deviation of some current ground motion prediction equations for a magnitude 8 earthquake on the San Andreas fault are outside the range allowed by the rocks at Lovejoy Buttes, California. If σ_A has been significantly overestimated, the discrepancy will be much smaller in all cases, and may completely go away in some. Since the ergodic assumption has been applied, the standard deviations are likely to be overestimated. If this conclusion is correct, the ergodic assumption is having an important effect on probabilistic seismic hazard estimates.

It is important to avoid generalizing too much beyond what the data allows. The suggestion that ground motions are overestimated in this one case, for a very mature and active fault, should not be taken to imply that the mean values of ground motions are overestimated by the current generation of ground motion prediction equations in all cases. Indeed, Anderson et al (2002) identified several physical mechanisms that might lead to smaller ground motions from these large-magnitude earthquakes on the most active faults.

7. DATA NEEDS FOR SEISMIC HAZARD ANALYSIS

The discussion above shows that for seismic hazard analysis at very low probabilities, a good knowledge of the standard deviation can have a greater impact than improvements in the mean. This introduces very significant challenges, since it is easier to determine the mean of a distribution than the standard deviation. For large magnitude earthquakes at short distances, where there is insufficient data to even determine the mean, this might seem to be an impossible task.

The following are our ideas on data and studies that can have a significant impact on hazard analysis.

7.1. More Instruments Close to Major Faults

The global seismicity rate is about one earthquake with magnitude 8.0 or larger per year, and about 17 earthquakes with magnitudes between 7.0 and 7.9 per year. Table 2 gives a very preliminary estimate on the rates of different types of events, constrained by a review of the Harvard moment tensor catalog and the catalog of surface rupturing events given by Wells and Coppersmith (1994).

Table 2. Preliminary Estimate of the Rates of Continental Earthquakes With Magnitude 7.0 or Larger.

MW	Global Total	Subduction, Deep, Other	Continental Thrust	Continental Strike Slip	Continental Normal
8.0	1	~1	<<1 (~0.02?)	<<1 (~0.02?)	<0.01
7.0	17	15	~0.3-1	~0.5-1	~0.1

Considering Table 2, it is clear that large continental earthquakes, of the size where our uncertainties in prediction of ground motions are largest, are a very scarce resource. Thus, it is very important to have sufficient instruments out to gather data from these events when they do occur.

As mentioned above, the uncertainty on the mean of ground motions close to the fault from such events is currently modeled as a lognormal distribution with standard deviation of about a factor of two. Since these uncertainties get multiplied by the aleatory uncertainty, reduction of the uncertainty in the mean will have a significant impact at low probabilities.

7.2. Precarious Rocks, Monuments, Other Geological Indicators

This paper has proposed an approach for the use of precarious rocks, and by extension, ancient monuments and structures, to place constraints on historical ground motions in the past 1000 to 10000 years. To understand the constraints better, we need to operate strong motion stations at these features, so that uncertainties due to site effects is eliminated. We also need to continue to improve our models of the response of these structures to ground motions. We need to think creatively about making maximum use of all kinds of geological indicators.

7.3. Determine the Aleatory Uncertainty Defined as the Uncertainty in Ground Motions Caused by Repeated Earthquakes on the Same Source, Recorded at the Same Station

Unfortunately, given the repeat times of large earthquakes, this thought experiment is not practical. There are however some proxies that might inform our judgment.

Schaff and Richards (2004) report that among 14000 earthquakes in China, about 10% appear at regional distances to be doublets with locations no more than 1 km apart. These are identified by filtering the seismograms between 0.5 Hz and 5 Hz. Taking these as approximately rupturing the same fault, the differences between the seismograms at higher frequencies could be a measure of the aleatory uncertainty at those higher frequencies. A study of this nature would need to take into account that the doublet pairs generally are not believed to rupture exactly the same patch of the fault, and they may differ in magnitude. The difference in location may be unimportant when the observation point is several hundred kilometers away. Differences in magnitude could correspond to differences in corner frequency, which would automatically cause differences in the high frequencies of the ground motions.

Care would be needed to manage this variable, as by definition, repeating earthquake used to determine the aleatory uncertainty should have the same moment. Several other studies of earthquake doublets and multiplets show that such events are present in the seismic catalogs in many other locations (e.g. Nadeau and McEvilly, 1999; Wiens and Snyder, 2001; Schaff et al, 2002; Shearer et al, 2003; Alparone and Gambino, 2003; Igarashi et al., 2003; Scarfi et al, 2003).

A second approach to understanding the aleatory uncertainty is to consider modeled earthquakes in the laboratory. The simulated earthquakes on foam rubber have several advantages for such studies. The wavelengths of body waves in foam are short compared to the size of the foam samples, thus allowing bulk properties of the medium to be investigated. The Poisson's ratio of foam (about 0.3) is about the same as it is for rock. Stick-slip events in foam show many of the same properties as earthquakes, including nucleation at a point and directivity as the rupture spreads over the surface of the fault. The waveforms from such events can be modeled using earthquake dynamic codes designed for application to crustal earthquakes. Preliminary studies (e.g. Anooshehpoor et al, 2004b) indicate that the ergodic assumption would be wrong to apply to prediction of ground motions from these stick-slip events. Uncertainties at specific stations for repeated ruptures all initiating at the same location thus might be helpful in providing an analogous estimate for aleatory uncertainty.

Additional studies such as those by Cua and Heaton (2004) could also help constrain the aleatory uncertainty. As mentioned above, their study which carefully removes the site response achieves a very large reduction in the standard deviation. This type of analysis could be applied to more strong motion data sets, such as data from Mexico or Japan where there are large numbers of strong motion records per station. Beyond that, remaining residuals should be studied for focal mechanism and directivity effects.

Another approach might be to search the strong motion database for repeated ruptures of large earthquakes recorded on the same station. The nearest example we can think of is the pair of records from Imperial Valley recorded in El Centro in 1940 and 1979. Fault slip on the surface was nearly identical on the part of the fault nearest El Centro. However, the location of the 1940 station was on a massive foundation (Brune and Anooshehpoor, 1991) while the nearest station recording the 1979 event was a couple hundred meters away, in a more nearly free-field setting. Besides that, the fault ruptured in opposite directions in the two events. Another complication is the difficulty encountered in digitizing the 1940 accelerogram, due to overlapping traces. It would be worthwhile to search other strong motion sites that have been occupied for long times to see if better examples can be found.

7.4. Dense Instrumentation to Get Better Sampling of Spatial Uncertainty and Extremes

There are many sources of uncertainties in ground motions, including focal mechanism, directivity, wave propagation, and site effects. Of these the statistical effects of focal mechanism, for instance, is understood and yields a probability distribution with a finite range of amplitudes, i.e. without an unbounded tail. The same could be said of directivity (for a band-limited signal) and site response. The wave propagation effect is perhaps one of the most difficult to bound using theoretical models, since the statistics of focusing are not well known in detail. Dense sampling of ground motions would give an empirical estimate of the spatial uncertainty associated with the combination of wave propagation and site response. A careful investigation of the distribution of residuals in this case could help constrain the shape of the "true" residuals relative to the mean of a ground motion prediction equation.

7.5. Consistent and Improved Data on Wave-Propagation Properties of the Near-Surface Geology for Strong Motion Stations

There is an extensive literature on the effect of near-surface geology on strong motion. Since this introduces a significant amount of variability into the ground motions, it has a significant effect on the uncertainty. A vision for the future is that when the near-surface is characterized sufficiently well, this uncertainty is significantly reduced.

7.6. More Consideration of How Sigma is Used in the Seismic Hazard Analysis

Anderson and Brune (1999a) found that the thought experiment about Lovejoy Buttes was informative about how to use aleatory and epistemic uncertainty. However, a deeper understanding of uncertainties is still needed. The large difference between the most probable model and the hazard curve in Figures 4 and 5 would seem to be an indicator that the methodology in putting material together needs to be examined very carefully.

7.7. The Strong Motion Community Needs to Develop a New Paradigm for Strong Motion Network Operations

It is unlikely for government agencies to have the willingness to support the number and density of instruments needed to achieve these goals. The way they can be achieved is through a one to two order of magnitude reduction of the cost of deploying and maintaining instruments. This must probably be accompanied through development and application of a mechanism through which the private sector becomes motivated to bear the cost but the data is freely shared with those who are interested to use it.

ACKNOWLEDGEMENTS

The authors thank Yuehua Zeng for sharing results on synthetic seismogram generation. James Scott reviewed the manuscript. Part of this research was completed while the author was on a sabbatical leave as a visiting scientist at the Lamont-Doherty Earth Observatory. The support and warm welcome from the staff at Lamont is gratefully acknowledged. This research was carried out with partial support of the Southern California Earthquake Center. This material is also based in part upon work supported by the National Science Foundation under Grant No. 0000050, which partly supports the Guerrero accelerograph network and associated research on data from that network. The Guerrero accelerograph network also receives support from the Instituto de Ingenieria at Universidad Nacional Autonoma de Mexico, Mexico City. The author also wishes to acknowledge support of the Office of Repository Development for support of the Yucca Mountain Seismic Monitoring. That work was supported by contract DE-FC28-04RW12232 administered by the Harry Reid Center for Environmental Studies, Las Vegas, Nevada.

REFERENCES

1. Abrahamson, N. A. and W. J. Silva (1997). Empirical response spectral attenuation relations for shallow crustal earthquakes, Seismological Research Letters 68, 94-127.
2. Alparone, S; Gambino, S. 2003. High precision locations of multiplets on south-eastern flank of Mt. Etna (Italy): reconstruction of fault plane geometry. Physics of the Earth and Planetary Interiors 135 (4): 281-289.
3. Anderson, J. G. and J. N. Brune (1999a). Probabilistic seismic hazard analysis without the ergodic assumption, Seismological Research Letters 70, 19-28.
4. Anderson, J. G. and J. N. Brune (1999b). Methodology for using precarious rocks in Nevada to test seismic hazard models, Bull. Seism. Soc. Am. 89, 456-467.
5. Anderson, J.G., J. N. Brune, S. G. Wesnousky (2002) Physical Phenomena Controlling High-Frequency Seismic Wave Generation In Earthquakes in Proceedings of the Seventh

US National Conference of Earthquake Engineering, July 21-25, Boston, Massachusetts, CD Rom, Paper # 00272, Earthquake Engineering Research Institute, Palo Alto, California.

6. Anooshehpoor, A., J. N. Brune and Y. Zeng (2004a). Methodology for obtaining constraints on ground motion from precariously balanced rocks, Bull. Seism. Soc. Am. 94, 285-303.
7. Anooshehpoor, A., S. H. Gonzalez, S. M. Day, and J. N. Brune (2004b). Foam rubber and numerical simulation of rupture dynamics, Seism. Res. Lett., 75, p. 279.
8. Biasi, GP; Weldon, RJ; Fumal, TE; Seitz, GG. 2002. Paleoseismic event dating and the conditional probability of large earthquakes on the southern San Andreas fault, California. Bull. Seism. Soc. Am. 92 (7): 2761-2781.
9. Boore, D. M., W. B. Joyner and T. E. Fumal (1997). Equations for estimating horizontal response spectra and peak acceleration from western North American earthquakes: a summary of recent work, Seismological Research Letters 68, 128-153.
10. Brune, J. N. (1999). Precarious rocks along the Mojave section of the San Andreas fault, California: constraints on ground motion from great earthquakes, Seis. Res. Lett. 70,
11. Brune, J. N. and A. Anooshehpoor (1991), Foam rubber modeling of the El Centro terminal substation building, EERI: Earthquake Spectra, 7, 45-79.
12. Brune, J. N., A. Anooshehpoor, B. Shi and Y. Zeng (2004). Precarious rock and overturned transformer evidence for ground shaking in the MS=7.6 Kern County earthquake: an analog for disastrous shaking from a major thrust fault in the Los Angeles basin, Bull. Seism. Soc. Am. (in press).
13. Campbell, K. W. (1997). Empirical near-source attenuation relationships for horizontal and vertical components of peak ground acceleration, peak ground velocity, and pseudo-absolute acceleration response spectra, Seismological Research Letters 68, 154-179
14. Cornell, C. A. (1968). Engineering seismic risk analysis. Bull. Seism. Soc. Am. 58, 1583-1606.
15. Cua, G. B. and T. H. Heaton (2004). Characterizing average properties of southern California earthquake ground motions over a wide range of magnitudes, distances, and site conditions (abstract), Seismological Research Letters 74, 247.
16. Frankel, A. D., M. D. Petersen, C. S. Mueller, K. M. Haller, R. L. Wheeler, E. V. Leyendecker, R. L. Wesson, S. C. Harmsen, C. H. Cramer, D. M. Perkins and K. S. Rukstales (2002). Documentation for the 2002 Update of the National Seismic Hazard Maps, U.S. Geological Survey Open-File Report 02-420.
17. Igarashi, T; Matsuzawa, T; Hasegawa, A. 2003. Repeating earthquakes and interplate aseismic slip in the northeastern Japan subduction zone. Journal of Geophysical Research-Solid Earth 108 (B5): art. no.-2249.
18. McGuire, R. K. (1995). Probabilistic seismic hazard analysis and design earthquakes: closing the loop, Bull. Seism. Soc. Am. 85, 1275-1284.
19. Nadeau, RM; McEvilly, TV. 1999. Fault slip rates at depth from recurrence intervals of repeating microearthquakes. SCIENCE 285 (5428): 718-721.
20. O'Connell, D. R. H. and R. LaForge (2004). Balanced rocks demonstrate that site-specific PSHA works (abstract), Seismological Research Letters 74, 284.
21. Purvance, M. D. (2004) parameters contributing to the toppling of precariously balanced rocks, (Abstract), Annual Meeting, Southern California Earthquake Center.
22. Purvance, M. D., R. Anooshehpoor, and J. N. Brune (2002). Sensitivity of near San Andreas precariously balanced rocks to strong ground motions (Abstract), Annual Meeting, Southern California Earthquake Center.

23. Purvance, M. D., R. Anooshehpoor, and J. N. Brune (2003). Precarious rock constraints on strong ground motions, (Abstract), Annual Meeting, Southern California Earthquake Center.
24. Purvance, M. D., J. Brune, R. Anooshehpoor, H. K. Thio, and P. Somerville (2004). Precariously balanced rock toppling constraints and vector-valued hazard, (Abstract), Annual Meeting, Southern California Earthquake Center.
25. Sadigh, K., C.-Y. Chang, J. A. Egan, F. Makdisi and R. R. Youngs (1997). Attenuation relationships for shallow crustal earthquakes based on California strong motion data, Seismological Research Letters 68, 180-198.
26. Scarfi, L; Langer, H; Gresta, S. 2003. High-precision relative locations of two microearthquake clusters in southeastern Sicily, Italy. Bull. Seism. Soc. Am. 93 (4): 1479-1497.
27. Schaff, DP; Bokelmann, GHR; Beroza, GC; Waldhauser, F; Ellsworth, WL. 2002. High-resolution image of Calaveras Fault seismicity. Journal of Geophysical Research-Solid Earth 107 (B9): art. no.-2186.
28. Schaff, DP; Richards, PG. 2004. Repeating seismic events in China. SCIENCE 303 (5661): 1176-1178.
29. Senior Seismic Hazard Analysis Committee (SSHAC) (1997). Recommendations for Probabilistic Seismic Hazard Analysis: Guidance on Uncertainty and Use of Experts, NUREG/CR-6372, U. S. Nuclear Regulatory Commission.
30. Shearer, PM; Hardebeck, JL; Astiz, L; Richards-Dinger, KB. 2003. Analysis of similar event clusters in aftershocks of the 1994 Northridge, California, earthquake. Journal of Geophysical Research-Solid Earth 108 (B1): art. no.-2035.
31. Sieh, K. E. 1978. Slip along San-Andreas Fault associated with great 1857 earthquake. *Bulletin of the Seismological Society of America* 68 (5): 1421-&.
32. Stepp, J. C., I. Wong, J. Whitney, R. Quittmeyer, N. Abrahamson, G. Toro, R. Youngs, K. Coppersmith, J. Savy, and T. Sullivan (2001). Probabilistic Seismic Hazard Analyses for Ground Motions and Fault Displacement at Yucca Mountain, Nevada, Earthquake Spectra 17, 113-151.
33. Weldon, R., K. Scharer, T. Fumal and G. Biasi (2004). Wrightwood and the earthquake cycle: what a long recurrence record tells us about how a fault works, GSA Today 14, No. 9, 4-10.
34. Wells, D. L. and K. J. Coppersmith (1994). New empirical relationships among magnitude, rupture length, rupture width, rupture area, and surface displacement, Bull. Seism. Soc. Am. 84, 974-1002.
35. Wiens, DA; Snider, N.O. (2001). Repeating deep earthquakes: Evidence for fault reactivation at great depth. SCIENCE 293 (5534): 1463-1466.

CAPTURING AND LIMITING GROUND-MOTION UNCERTAINTY IN SEISMIC HAZARD ASSESSMENT

Julian J. Bommer[1] and Frank Scherbaum[2]

[1]Department of Civil & Environmental Engineering, Imperial College London, South Kensington campus, London SW7 2AZ, U.K.; [2]Institut für Geowissenschaften, Universität Potsdam, D-14415 Potsdam, Germany

Abstract: Seismic hazard assessments provide quantitative evaluations of the nature of ground shaking at a specified location that could be induced by future earthquakes. Such evaluations serve to inform engineering decisions about the location and design of new projects and the safety of existing structures. In order to provide the engineers and planners with complete information on which to base their decisions, the assessments must identify and, to the extent that is possible, quantify the associated uncertainties. There are major uncertainties associated with both the seismicity model and the ground-motion model in any seismic hazard assessment, but the uncertainties associated with the latter will generally have the larger impact on the results. Uncertainties in ground-motion prediction equations can be characterized as aleatory variability and epistemic uncertainty; the former can be directly integrated into the hazard calculations, although for very low annual exceedance frequencies it can become necessary to impose physical limits on the distribution of residuals. Epistemic uncertainty can be handled using logic-tree formulations. Combining several ground-motion prediction equations in a logic tree often requires adjustments to be made to compensate for the use of different parameter definitions; without these adjustments, the epistemic uncertainty can be grossly over- or underestimated. However, the adjustments themselves, which are often empirically derived, carry their own uncertainty and this must be included in the analyses.

Key words: Seismic hazard assessment; ground-motion prediction equations; aleatory variability; epistemic uncertainty; logic trees.

1. INTRODUCTION

The earthquake-resistant design of critical engineering projects involves decision making regarding the balance between investing in the provision of

P. Gülkan and J.G. Anderson (eds.), Directions in Strong Motion Instrumentation, 25–40.

additional seismic resistance and the consequences, in terms of human, social, environmental and economic losses, that may be incurred if the resistance is insufficient. The decision-making process must also consider the balance between the possibility that during its useful life the structure may not experience any significant ground shaking and the fact that should a major earthquake occur close to the site it is likely to represent the most severe loading case to which the structure will be subjected. Therefore, during the design life of the structure, the possibilities range from no earthquake to the most severe earthquake scenario realizable at this location. The decision makers are faced with the challenge of not knowing where their project is situated within this hazard space, as a result of all the uncertainties in the underlying processes of earthquake occurrence and ground-motion generation. Hence a key element of any seismic hazard assessment is to identify and quantify, as far as is possible, these uncertainties in order to define the range of possible ground shaking scenarios that engineers and decision makers should consider in the design process.

In deterministic seismic hazard assessment (DSHA) the uncertainty is dealt with, according to the claims of its proponents (e.g. Krinitzsky, 1998), by effectively enveloping the hazard space through expert (or arbitrary, depending upon one's philosophical viewpoint) judgements that are frequently classified by the ambiguous and misleading adjective 'conservative'. One of the weaknesses in this approach is that it does not, despite the claims made, define the worst-case scenario: the 'conservatism' is usually applied to the magnitude and/or location of the scenario earthquake and then ground-motion amplitudes are calculated using, at most, the 84-percentile levels from ground-motion prediction equations (Krinitzsky, 2002a). For the magnitude and distance ranges currently well covered by data, ground motion amplitudes two or three logarithmic standard deviations above the logarithmic mean predicted by equations are common (e.g. Abrahamson, 2000; Bommer et al., 2004a) and for a typical equation for the prediction of PGA, the increase in amplitude due to adding just one standard deviation is equivalent to increasing the magnitude by one complete unit or to halving a source-to-site distance of 15 km. A related problem with DSHA is caused by the fact that the probability density functions for ground motion amplitudes as expressed in empirical ground motion models ("attenuation relations") are in the strict sense conditional probability functions given the occurrence of the scenario earthquake. As a result of the usual practice to ignore the actual frequency of occurrence of the scenario earthquake in DSHA, design ground motions derived from DSHA can relate to vastly differing absolute probabilities of exceedances, even if the same percentile levels were chosen on the ground motion models.

Whatever the relative merits and disadvantages between deterministic and probabilistic approaches to seismic hazard assessment, there is no doubt that PSHA is better suited to capturing and quantifying the uncertainty in the hazard estimates. Although it could be argued that the way design return periods are currently selected is another example of arbitrary selection (Bommer, 2002, 2004), PSHA does provide a framework in which decision-makers can be presented with a quantified range of possibilities in terms of ground motions at the site of interest. PSHA achieves this by considering the uncertainties in the models for earthquake occurrence (e.g. estimates of the *b*-value in the recurrence relationship; the shape of the recurrence relationship for large magnitudes; the maximum feasible magnitude) and for ground-motion generation. A major problem arises, when low annual frequencies of exceedance (or long return periods) are considered because the scatter associated with ground-motion prediction equations is generally modelled by an untruncated lognormal distribution (e.g. Restrepo-Vélez and Bommer, 2003). Arbitrary truncations of the distribution do not constitute a solution to this problem: physical constraints on the maximum possible levels of ground motion are required. On this point, there is a convergence, or at least a shared problem, between PSHA and DSHA, since physical estimates of the upper bounds can be considered a missing piece from the current formulations of both methods.

The view of the authors is that uncertainties should be identified and their distributions (centre, body and range) determined so that they can be incorporated in a quantitative and transparent manner in the hazard calculations. Any 'conservatism' should be applied by the decision maker (who is neither the seismologist nor the engineer, even though the decision-making process should involve interactions with these experts) only at the end of the process and on a clear basis of considerations of acceptable risks and costs.

In this paper we focus on the uncertainties in ground-motion prediction models – which are generally the source of the largest uncertainties in seismic hazard estimates – and present different procedures for dealing with them in seismic hazard assessment.

2. UNCERTAINTY IN GROUND-MOTION MODELS

Within SHA, uncertainty is now usually classified as aleatory or epistemic. Although these terms, which replace randomness and uncertainty whose meaning has become blurred by being used interchangeably (Bommer, 2003), sometimes cause confusion – and occasionally even displeasure (Krinitzsky, 2002b) – their importance lies in the clarification of the distinction between natural variability (aleatory) and uncertainty as a result of incomplete knowledge (epistemic). Whilst this vocabulary has been adopted to avoid ambiguity in discussions of uncertainty, it has to be stressed that this distinction is not merely semantic since only the aleatory part of ground motion "uncertainty" is part of the kernel of the hazard integral. Therefore, this distinction should not be ignored. Informative discussions of the concepts of epistemic uncertainty and aleatory variability in ground-motion predictions can be found in Anderson and Brune (1999) and Toro et al. (1997).

2.1. Aleatory Variability in Ground-motion Models

The simplest example of aleatory uncertainty is the result of rolling a dice (*alea* in Latin). The outcome of such an experiment can only be described in a meaningful way by the probability distribution for the natural variability of the dicing process but not by an individual single number. Aleatory variability in ground-motion prediction equations is the apparent randomness of measured values of a given parameter with respect to the median predictions. When regressions are performed on the logarithms of the ground-motion parameter, the residuals are generally found to follow a normal distribution, at least within the range of ±2 standard deviations (sigmas) from the mean. The use of ever larger strong-motion datasets and the inclusion of more explanatory variables in the predictive equations have had little impact on the standard deviations associated with predictive equations (Douglas, 2003). The lowest published values of sigma for $\log_{10}$(PGA) are of the order of 0.18 and such low values seem sometimes to be achieved through careful selection of the data (Fig. 1); higher values are found for peak ground velocity (PGV) and for response spectral ordinates at intermediate and long periods.

There appears to be little scope for any major reduction of sigma values in the foreseeable future. Given the complexity of the physical process being modeled, and the crudeness of the models employed, this is not surprising. However, it should also be noted that based on the analysis of precarious rocks that have survived several large earthquakes in California, it has been suggested that the aleatory uncertainty in empirical ground-motion models could also be biased on the high side in the cases where the earthquakes are characteristic events instead of samples drawn from a Gutenberg-Richter relationship (Anderson and Brune, 1999). The case is made for performing PSHA without making the ergodic assumption whereby spatial variation (from recordings of earthquakes at different locations) is used to model temporal variation of ground motion at one location.

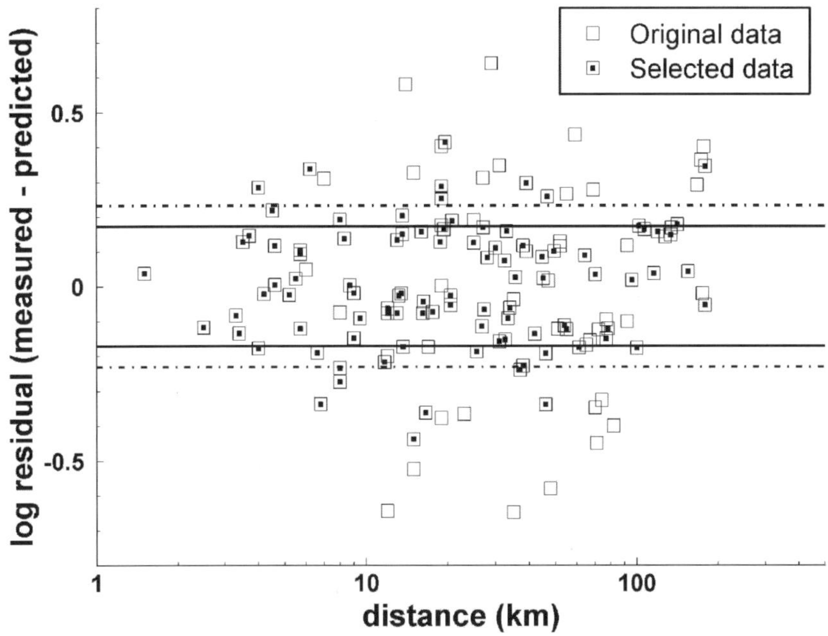

Figure 1. Residuals of PGA from the equation of Sabetta and Pugliese (1996); the solid lines show the $\pm\sigma$ levels obtained from the selected records used to derive the equation, whereas the dashed lines are obtained using all of the available records from the same earthquakes.

Most ground-motion prediction equations include a homoskedastic model of the aleatory variability but a few studies have identified a variation of the sigma with magnitude, with smaller variability at higher magnitudes (Youngs et al., 1995; Abrahamson and Silva, 1997). Campbell (1997) identified a reduction of the scatter with both increasing magnitude and increasing ground-motion amplitude; the latter effect was first identified by Donovan and Bornstein (1978) who attributed the phenomenon in part to the non-linear response of soil sites. Midorikawa and Ohtake (2004), on the other hand, studied the variance of PGA and PGV in Japanese strong-motion data and found decreasing variance with increasing magnitude and amplitude but also an even stronger decrease with decreasing distance, concluding that the apparent reduction with increasing ground-motion amplitude is the combined effect of the magnitude and distance dependencies.

2.2. Epistemic Uncertainty in Ground-motion Models

The aleatory variability in ground-motion models is directly incorporated into the calculations of PSHA to determine annual exceedance frequencies of different amplitudes of ground motion. Direct integration over the distribution representing the aleatory variability does not imply that all of the uncertainty in the ground-motion predictions is being included in the hazard estimates; this would only be the case if the empirical equation were derived from a very large strong-motion dataset obtained from dense and uniformly distributed accelerograph networks triggered by earthquakes whose source characteristics reflected the full range of possible parameter variations. Clearly no such strong-motion dataset, and hence ground-motion prediction equations, exist and SHA thus employs equations based on datasets that are generally biased both in terms of the source characteristics of the earthquakes and the spatial (distance and azimuth) sampling of the recordings.

The derivation of empirical equations from sparse and biased strong-motion datasets often leads to the publication of several predictive models for those regions that possess a large number of indigenous recordings. The differences between the median predictions from these equations, which are equally valid in terms of their derivation and empirical basis, are epistemic uncertainty. A simple example of epistemic uncertainty is the result of measuring the length of a table using different measuring instruments. In contrast to aleatory uncertainty, the outcome of such an experiment can well be described by an individual single number, although we may not be able to determine it with any of the measuring devices we are using. This contrasts with making repeated measurements of a table with the same instrument (the experiment through which Gauss obtained the distribution that bears his name), where there will be aleatory variability in the measurements. The latter variation could be interpreted to represent the inherent scatter in recorded ground motions, whereas the use of different measure instruments could be considered equivalent to using different functional forms for the regression on the data.

Epistemic uncertainty in ground motion is not captured in seismic hazard assessment by making use of any one of the available empirical equations alone. For seismic hazard assessment in regions with limited indigenous strong-motion data, the use of several predictive equations becomes even more important. In many regions there are only accelerograms generated by small magnitude earthquakes with few or no records from the magnitude-distance combinations that generally control the hazard at the return periods commonly considered for design of buildings and infrastructure. In such cases the epistemic uncertainty is even higher because of the uncertainty associated with extrapolating ground-motion predictions from small to large magnitude earthquakes. In regions with no locally recorded accelerograms, there is effectively no constraint and hence many ground-motions models will need to be considered in order to ensure that the epistemic uncertainty is captured. The tool that allows a number of ground-motion prediction equations to be considered simultaneously in a seismic hazard assessment is the logic tree, which is dealt with in detail in the following section. At this point it is worthwhile pointing out that even in regions with large strong-motion datasets, using equations from elsewhere may be deemed necessary. Since the earthquakes and the sampling of the resulting ground shaking may both be biased towards lower or higher values, the differences amongst the median predictions from equations derived for, say, California, may not represent the full extent of the epistemic uncertainty. This consideration may lead an analyst performing a PSHA for a site in California, to include in the logic-tree formulation branches for predictive equations derived in other parts of the world. Although this may seem like a poorly justified proposal to some, it is already the case that some of the predictive equations used in California, such as Abrahamson and Silva (1997), are derived from datasets that include accelerograms from Europe, the Middle East and Japan, added to improve the coverage in magnitude-distance space. Seismic hazard assessments in regions of extensional tectonics, including the Basin and Range Provinces in the USA, routinely employ the equations of Spudich et al. (1999) that were derived from a global database including a large number of records from Europe.

3. LOGIC TREES FOR GROUND-MOTION MODELS

Although logic trees are widely used in the practice of seismic hazard assessment, there is little published guidance on the details of their deployment and the potential pitfalls in their use.

3.1. Branches and Weights

The branches of the ground-motion section of a logic tree carry ground-motion prediction equations that the analyst considers to have the potential for predicting future earthquake ground-motions in the region under study (Cotton et al., 2004). The weights applied to the branches reflect the relative confidence in each of the models providing the best estimate of future ground-motions in the region. However, further refinement is possible, particularly if the weights are allowed to vary across magnitude-distance bins. A convenient way to separate the influences that may be considered in assigning rankings to the candidate equations is to group them into two categories: intrinsic and application-specific characteristics. The intrinsic factors are those related to the confidence of the user in a particular equation by virtue of how it was derived, regardless of where the model is actually being applied. The grading for these factors should be individually applied to different bins of magnitude and distance, as well as to different response frequencies. This will enable maximum advantage to be taken of the strengths of each equation, whilst reducing the influence of any given equation for cases where it is weakly constrained. The primary basis for assigning the grading for intrinsic characteristics of an equation will be the distribution and quality of the strong-motion data used for the derivation of the equation. The distribution of the data within each magnitude-distance bin defined for the weighting will be the basic feature for assigning the grading, but other factors may also be considered.

The application-specific characteristics can be divided into those related to the conversions adopted for the hazard calculations (section 3.2) and those related to the specific environment for which the hazard assessment is performed (section 3.3). The applicability of an equation to a particular region can be quantitatively assessed using strong-motion recordings from that region. Scherbaum et al. (2004b) have proposed the use of recorded response spectra to judge the match or mismatch of ground-motion models to a particular target region. This method is based on the evaluation of how well a particular ("modified") ground-motion model "predicts" observed reference data in a quantitative way.

Provided at least four ground-motion prediction equations are included in the logic tree, Sabetta et al. (2004) find that the median hazard estimates, at least at annual frequencies of exceedance down to 10^{-3}, are rather insensitive to the branch weights.

3.2. Conversions for Parameter Compatibility

A wide variety of definitions are encountered in the literature for the basic variables included in ground-motion prediction equations, such as the horizontal component of motion, the magnitude scale and the measure of source-to-site distance. If the equations combined in a logic tree employ different definitions for any of these parameters, then suitable adjustments <u>must</u> be made in order to capture the genuine epistemic uncertainty. Conversions for different component definitions and different magnitude scales are presented by Bommer et al. (2004b).

The parameter incompatibility that has the greatest impact on the hazard estimates is the use of different measures of distance; Scherbaum et al. (2004a) provide conversions between all of the commonly used distance metrics.

Another potential source of incompatibility is when the logic tree includes equations that include a predictor variable that is not present in others, which in principle can be addressed by adding aleatory variability for those parameters not included in a particular ground motion model (e. g. Campbell, 2003). The most commonly encountered example of an additional parameter appearing in some of the selected equations is the style-of-faulting, for which a procedure to achieve compatibility of the median motions – the effect on the scatter is not considered – is presented by Bommer et al. (2003). Another important incompatibility can arise because site classes, such as 'soft soil' or 'rock', may be defined differently in the equations combined in a logic tree. A clear example of this is the rock conditions in western North America, corresponding to shear-wave velocities of the order of 620 m/s (Boore et al.,

1997), and those in eastern North America that have velocities of 2,800 m/s (Atkinson and Boore, 1997). However, because of differences in near-surface attenuation, it is not sufficient in such cases to simply correct for the differences in V_s values (Cotton et al., 2004).

3.3. Adjustments for Regional Applicability

The conversions to achieve compatibility amongst the equations used in a logic tree must be applied for it to make sense to combine them in the hazard calculations. The adjustments made to the equations may also include conversions to account for differences between the region for which the equation was derived and the region for which the SHA is being performed; this then would lead to a hybrid ground-motion model in the sense of Campbell (2003). Practical aspects of such host-to-target conversion are discussed in Cotton et al. (2004). One major issue in this context is the fact that the generating data sets for empirical ground-motion models do not usually come from a single tectonic region, even if from the same geographical area: the equations of Sabetta and Pugliese (1996) are derived from Italian strong-motion data but this includes recordings from reverse faulting events in the Friuli region and from normal events in the regions of extensional tectonics such as the Apennines. Therefore, the set of stochastic parameters to be used for the hybrid method corresponds to a virtual host region and has to be derived from the empirical ground motion models themselves, e.g. by inversion of synthetic data as demonstrated by Scherbaum et al. (2004c).

3.4. Uncertainties in Adjustments

Unless appropriate adjustments are applied to achieve compatibility amongst the ground-motion prediction equations, and indeed between each equation and the seismicity model employed in the hazard calculations, the epistemic uncertainty will not be captured correctly, as indicated previously. However, since the adjustments are generally made using empirical relationships, which carry their own associated uncertainty, there is a price to pay in their application in terms of increased uncertainty associated with each ground-motion model to which an adjustment is made. There will often be two or more empirical relations available to the analyst for each adjustment, without anyone of them being unambiguously the most appropriate to the particular application under consideration. Furthermore, being derived from regressions on empirical data, the conversions have an associated random variability represented in the scatter of the data about the regression line. The analyst will need to decide in each case whether both the epistemic uncertainty (choice of empirical relationship) and the aleatory variability (scatter associated with each relationship) need to be considered in the error propagation. For example, Bommer et al. (2004b) concluded that mean ratios of horizontal motions obtained using different treatments of the two horizontal components, derived from a large strong-motion databank, are very stable and hence it is not necessary to consider alternative models. Furthermore, the scatter associated with the ratios between different measures of the horizontal motions (such as the larger component and the geometric mean) is sufficiently small to be ignored.

For the case of magnitude conversions, the situation is somewhat different. Let us assume that the logic tree is combining ground-motion prediction equations defined in terms of M_w and others in terms of M_s, and that the hazard calculations will be based on M_w: as the hazard calculations are performed, each M_w value needs to be converted to M_s for those equations that are based on the latter scale. There is no uncertainty at all in the M_w values: in PSHA, the scenarios are defined exactly and the location and size of these 'virtual' earthquakes are defined unambiguously and without uncertainty. However, since the relationship between moment magnitude and surface-wave magnitude is not unique, the converted M_s values will carry uncertainty. For hazard assessments in southern Europe, for example, one could use the M_s-M_w relationship of Ambraseys and Free (1997) or that of Bungum et al. (2002). Branches could be added to the logic tree to use both equations, weighted according to the analyst's degree of confidence in each being the more reliable. However, the differences between the M_s values obtained from the two equations are very small, so to use branches for this conversion would probably not be justified in terms of the additional computational effort hence a pragmatic approach would be to select the relationship considered more appropriate by the analyst. However, whichever of the two empirical relations is chosen, it will carry an appreciable aleatory variability: this means that in various realizations of a virtual scenario of M_w 6.5, say, the M_s value will vary according to the distribution of the residuals in the empirical relation and this variability will result in an increase of the aleatory variability of the ground-motion model.

4. UNCERTAINTY IN SEISMIC HAZARD ASSESSMENT

Incorporating uncertainties into seismic hazard assessments involves identifying and quantifying the uncertainties, and then estimating the impact both on the results of the hazard calculations and on the uncertainty in the hazard estimates.

4.1. Sigma Values and Upper Bounds

The aleatory variability in ground-motion models is treated in a very simple manner in DSHA, whereby the median or 84-percentile level of motion from the selected ground-motion model is calculated for the magnitude-distance combination of the design earthquake scenario. A weakness in current DSHA practice is that it provides no mechanism for dealing with the possibility of larger ground motions: Krinitzsky (2003) claims that "*the 84 percentile is judged to be reasonable and safe for engineering purposes*". If the lognormal distribution is valid, then in one in six realizations of the design earthquake scenario, the 84-percentile motions will be exceeded and this possibility is simply discounted.

If ignoring the possibility of the estimated ground motions being exceeded is the Achille's heel of DSHA, a potential weakness of current PSHA practice is the opposite problem of estimating ground motions that may not be physically possible. PSHA involves integrating over all possible combinations of magnitude, M, distance, D, and percentiles of the ground-motion amplitude for each M-D pair, usually expressed in term of the number of logarithmic standard deviations, ε. The range of values of M and D are limited, the former by the recurrence relationship and the latter by the geographical boundaries of the seismic source zones; by contrast, the value of ε is not limited when the variability is modeled as a Gaussian distribution. As a result, the hazard estimates do not saturate but grow indefinitely with decreasing annual frequency of exceedance unless some truncation is applied. This is generally not an important issue except for those projects of sufficient importance to warrant consideration of ground motions with very low annual exceedance frequencies, such as 10^{-7} in the PEGASOS project (Abrahamson et al., 2002) and 10^{-8} in the study for the nuclear waste repository at Yucca Mountain (Stepp et al., 2001). In such cases, the hazard can be driven by the tails of the lognormal distribution of the residuals – at levels assumed by extrapolation of the model rather than constrained by the data – and may exceed physically realizable levels (Bommer et al., 2004a). The challenge facing engineering seismology is to find physical limits on earthquake ground motions, which will undoubtedly be in excess of the largest motions recorded to date and which cannot be inferred from statistical analysis of the empirical data.

The feasibility of fixing upper bounds on ground motions in the near future is not certain because it is likely to rely on robust and fully validated dynamic source models, which are still under development. A possible outcome of this work could be only to confirm that certain high levels of motion cannot be proven to be unphysical; impossible motions may prove impossible to identify hence the solution may be lie in defining the levels of motion with extremely low, but finite, probabilities of exceedance. This still requires research to identify the true nature of the upper tails of the residuals.

One way or another, the challenge must be addressed if very long-term hazard estimates are to be made. Arbitrary truncations of the scatter are a tool of convenience but they do not constitute a solution to the problem. As mentioned previously, there seems to be little scope for major reductions of the aleatory variability in ground-motion prediction equations in the near future. Sigbjörnsson and Ambraseys (2003) obtain seismic hazard curves that are not strongly influenced by the aleatory variability at low annual exceedance frequencies but this is achieved by modeling the scatter only as function of the variability of a few selected parameters that constitute their theoretical ground-motion model. Whilst this produces results that appear to have solved, or circumvented, the upper bounds problem, it is convenient rather than robust: the aleatory variability is primarily the result of the numerous parameters that are not included in the predictive modes – and in many cases not yet even identified (Bommer et al., 2004c).

A final comment on the treatment of aleatory variability in DSHA may be in order. A more rational approach than the current practices of either neglecting it or taking a single standard deviation above the mean in all cases, would be to select a level of exceedance that together with the estimated recurrence interval of the design earthquake scenario (obtained by discretizing the recurrence relationship) yields a return period that the client or regulator can accept as sufficiently low. For a high seismicity region where earthquakes of the magnitude close to that of the design event may occur every few decades, it may be appropriate, for important structures, to use 97.7-percentile motions (two standard deviations above the mean), whereas for regions where large earthquakes have recurrence intervals of several centuries, it may be appropriate to use the median ground motions. However, caution is required because by definition there is less data in low seismicity regions and hence the epistemic uncertainty in both the earthquake recurrence frequency and the ground-motion model are likely to be high.

4.2. Planting and Harvesting Logic Tree

The logic-tree formulation is a very useful tool for handling epistemic uncertainty in seismic hazard analysis. From general practice, and from the literature, it would appear that logic trees are exclusive to PSHA; this is reinforced by the fact that the criticisms of PSHA by the proponents of DSHA have included arguments against the use of logic trees (Krinitzsky, 1995). Some of the opinions presented are valid arguments not against logic trees *per se* but rather against the way they are often used, which stems from the interpretation of the branch weights as being probabilities (which would be sufficient to turn those who favor DSHA against the use of logic trees). Although the weights at a given node sum to unity, giving them the appearance of probabilities, they are not statistical probabilities since the branches rarely, if ever, contain an exhaustive set of possibilities and the weights to do not represent the relative rate at which the alternative models occur in Nature (Abrahamson and Bommer, 2004). There is, in fact, absolutely no reason why logic trees should not be used in DSHA – without undermining any of the principles that 'determinists' hold dear, apart from that of presenting the engineer with a single design value and thus relieving the project owner from making informed choices regarding risks in the face of uncertainty. And here lies the key issue: the fruits of the logic tree – apart from the mean hazard, which would only have a straightforward statistical meaning if the weights could be considered probabilities (Abrahamson and Bommer, 2004) – are hazard estimates with associated confidence levels, which permit owners and regulators to make informed decisions on risk management. The owner must decide the appropriate design level – be it the deterministic scenario or the ground motion with a specified return period – and then the logic tree allows the design motions to be chosen according to the degree of confidence that is desired with respect to whether or not the ground motions at this level have been captured by the analysis.

5. CONCLUDING REMARKS

Uncertainty in ground-motion prediction models is large and will remain so for many years to come. An essential component of seismic hazard assessment is therefore to identify, quantify and capture this uncertainty, and to demonstrate its impact on the results in order to inform risk-based decisions regarding earthquake-resistant design. The distinction between aleatory variability and epistemic uncertainty, although at times ambiguous, is needed for rational treatment of the uncertainties: aleatory variability can be measured and must be included explicitly in the hazard integral. Epistemic uncertainty on the other hand is not part of the hazard integral kernel and can be managed through logic trees for both probabilistic and deterministic hazard assessments.

ACKNOWLEDGMENTS

The first author extends his thanks to Polat Gülkan and to John Anderson for the invitation to participate in the Kuşadasi Workshop and for the opportunity to discuss the ideas presented herein with the other participants. These ideas have been developed, and continue to evolve, through lively and prolonged discussions with Hilmar Bungum, Fabrice Cotton and Fabio Sabetta, who together with the authors formed the Expert Panel on Ground Motions in the PEGASOS project for the assessment of seismic hazard at nuclear power plant sites in Switzerland (Abrahamson et al., 2002). As a group we are indebted to Norm Abrahamson, our Technical Facilitator

Integrator in the PEGASOS project, for his guidance and instruction, and for continuously challenging our ideas and assumptions. We thank the organizers of the PEGASOS project for providing a stimulating environment and continuous support for the development of the ideas presented herein. Thanks are also due to Jonathan Hancock for assistance with the figure and to Juliet Bird for her helpful review of the manuscript.

REFERENCES

1. Abrahamson, N. A., 2000, State of the practice of seismic hazard evaluation, *Proc. GeoEng 2000*, Melbourne, Australia. Vol. 1: 659-685.
2. Abrahamson, N. A. and Bommer, J. J., 2004, Probability and uncertainty in seismic hazard analysis, *Submitted to Earthqu. Spectra.*
3. Abrahamson, N. A. and Silva, W. J., 1997, Empirical response spectral attenuation relations for shallow crustal earthquakes, *Seism.. Res. Letts.* **68** (1): 94-127.
4. Abrahamson, N. A., Birkhauser, P., Koller, M., Mayer-Rosa, D., Smit, P., Sprecher, C., Tinic, S. and Graf, S., 2002, PEGASOS – a comprehensive probabilistic seismic hazard assessment for nuclear power plants in Switzerland, *Proc. 12th Eur. Conf. Earthqu. Engrg*, London, Paper no. 633.
5. Ambraseys, N.N. and Free, M.W., 1997, Surface-wave magnitude calibration for European region earthquakes, *J. Earthqu.. Engrg* **1**(1)**:** 1-22.
6. Ambraseys, N. N., Simpson, K. A. and Bommer, J. J., 1996, Prediction of horizontal response spectra in Europe, *Earthqu. Engrg Struct. Dyn.* **25**: 401-412.
7. Anderson, J. G. and Brune, J. N., 1999, Probabilistic seismic hazard analysis without the ergodic assumption, *Seism.. Res. Letts.* **70** (1): 19-28.
8. Atkinson, G. M. and Boore, D. M., 1997, Some comparisons between recent ground-motion relations, *Seis. Res. Letts* **68**(1): 24-40.
9. Bommer, J. J., 2002, Deterministic vs. probabilistic seismic hazard assessment: an exaggerated and obstructive dichotomy, *J. Earthqu.. Engrg* **6**(spec. iss. 1)**:** 43-73.
10. Bommer, J. J., 2003, Uncertainty about the uncertainty in seismic hazard analysis, *Engrg Geol.* **30:** 165-168.
11. Bommer, J. J., 2004, Earthquake actions in seismic design codes: can current formats meet the needs of PBSD?, *Proc. Int. Workshop on Performance-Based Seismic Design*, Bled, Slovenia, Pacific Earthquake Engineering Research Center.
12. Bommer, J. J., Abrahamson, N. A., Strasser, F. O., Pecker, A., Bard, P-Y., Bungum, H., Cotton, F., Fäh, D., Sabetta, F., Scherbaum, F. and Studer, J., 2004a, The challenge of defining upper bounds on earthquake ground motions, *Seism. Res. Letts* **75**(1): 82-95.
13. Bommer, J. J., Douglas, J. and Strasser, F. O., 2003, Style-of-faulting in ground-motion prediction equations, *Bull. Earthqu. Engrg* **1**(2): 171-203.
14. Bommer, J. J., Sabetta, F., Scherbaum, F., Bungum, H., Cotton, F. and Abrahamson, N. A., 2004b, On the use of logic trees for ground-motion prediction equations for PSHA, *Submitted to Bull. Seism. Soc. Am.*
15. Bommer, J. J., Scherbaum, F., Cotton, F., Bungum, H. and Sabetta, F., 2004c, Discussion of "Uncertainty analysis of strong-motion and seismic hazard" by R. Sigbjörnsson and N.N. Ambraseys, *Bull. Earthqu. Engrg* **2**(2), *in press*.

16. Boore, D. M., Joyner, W. B. and Fumal, T. E., 1997, Equations for estimating horizontal response spectra and peak accelerations from western North American earthquakes: a summary of recent work, *Seis. Res. Letts* **68**(1): 128-153.

17. Bungum, H., Lindholm, C.D. and Dahle, A., 2003, Long-period ground-motions for large European earthquakes, 1905-1992, and comparisons with stochastic predictions, *Jrnl Seismology* **7**(3): 377-396.

18. Campbell, K. W., 1997, Empirical near-source attenuation relationships for horizontal and vertical components of peak ground acceleration, peak ground velocity, and pseudo-absolute acceleration response spectra, *Seism.. Res. Letts.* **68** (1): 154-179.

19. Campbell, K. W., 2003, Prediction of strong ground motion using the hybrid empirical method and its use in the development of ground-motion (attenuation) relations in Eastern North America, *Bull. Seism. Soc. Am.* **93**(3): 1012-1033.

20. Cotton, F., Scherbaum, F., Bommer, J. J., Bungum, H. and Sabetta, F., 2004, Criteria for selecting and adapting ground-motion models for specific target regions, *In preparation*.

21. Donovan, N. C. and Bornstein, A. E., 1978, Uncertainties in seismic risk procedures, *J. Geotech. Div. ACSE* **104**(7): 869-887.

22. Douglas, J., 2003, Earthquake ground motion estimation using strong-motion records: a review of equations for the estimation of peak ground acceleration and response spectral ordinates, *Earth Science Reviews* **61**: 43-104.

23. Krinitzsky, E. L., 1995, Problems with logic trees in earthquake hazard evaluation, *Engrg Geol.* **39**: 1-3.

24. Krinitzsky, E. L., 1998, The hazard in using probabilistic seismic hazard assessment for engineering, *Env. Engrg Geosci.* **4**(4): 425-443.

25. Krinitzsky, E. L., 2002a, How to obtain earthquake ground motions for engineering design, *Engrg Geol.* **65**: 1-16.

26. Krinitzsky, E. L., 2002b, Epistematic and aleatory uncertainty: a new shtick for probabilistic seismic hazard analysis, *Engrg Geol.* **66**: 157-159.

27. Midorikawa, S. and Ohtake, Y., 2004, Variance of peak ground acceleration and velocity in attenuation relationships, *Proc. 13th World Conf. Earthqu. Engrg*, Vancouver, Paper no. 325.

28. Restrepo-Velez, L. F. and Bommer, J. J., 2003, An exploration of the nature of the scatter in ground-motion prediction equations and the implications for seismic hazard assessment, *J. Earthqu.. Engrg* **7**(spec. iss. 1): 171-199.

29. Sabetta, F., Lucantoni, A., Bungum, H and Bommer, J. J., 2004, Sensitivity of PSHA results to ground-motion prediction relations and logic-tree weights, *Submitted to Soil Dyn. Earthqu. Engrg*.

30. Sabetta, F. and Pugliese, A., 1996, Estimation of response spectra and simulation of nonstationary earthquake ground motions, *Bull. Seis. Soc. Am.* **86**(2): 337-352.

31. Scherbaum, F., Cotton, F. and Smit, P., 2004b, On the use of response spectral reference data for the selection and ranking of ground motion models for seismic hazard analysis in regions of low to moderate seismicity, *Bull. Seism. Soc. Am., in press*.

32. Scherbaum, F., Cotton, F. and Steadtke, H., 2004c, The determination of equivalent stochastic models from empirical ground-motion prediction equations (abstract), *Seism. Res. Letts* **74**(2): 287.

33. Scherbaum, F., Schmedes, J. and Cotton, F., 2004a, On the conversion of source-to-site distance measures for extended earthquake source models, *Bull. Seism. Soc. Am.* **94**(3): 1053-1069.

34. Sigbjörnsson, R. and Ambraseys, N. N., 2003, Uncertainty analysis of strong-motion and seismic hazard, *Bull. Earthqu. Engrg* **1**(3): 321-347.

35. Spudich, P., Joyner, W. B., Lindh, A. G., Boore, D. M., Margaris, B. M. and Fletcher, J. B., 1999, SEA99: a revised ground motion prediction relation for use in extensional tectonic regimes, *Bull. Seis. Soc. Am.* **89**(5): 1156-1170.

36. Stepp, J. C, Wong, I. Whitney, J., Quittemeyer, R., Abrahamson, N., Toro, G., Youngs, R., Coppersmith, K., Savy, J., Sullivan, T. and Yucca Mountain PSHA Project Members, 2001, Probabilistic seismic hazard analyses for ground motions and fault displacements at Yucca Mountain, Nevada, *Earthqu. Spectra* **17**(1): 113-151.

37. Toro, G. R., Abrahamson, N. A. and Schneider, J. F., 1997, Model of strong ground motions from earthquakes in Central and Eastern North America: best estimates and uncertainties, *Seism.. Res. Letts.* **68** (1): 41-57.

38. Youngs, R. R., Abrahamson, N. A., Makdisi, F. I. and Sadigh, K., 1995, Magnitude-dependent variance of peak ground acceleration, *Bull. Seism. Soc. Am.* **85**(4): 1161-1176.

LONG-PERIOD GROUND MOTIONS FROM DIGITAL ACCELERATION RECORDINGS: A NEW ERA IN ENGINEERING SEISMOLOGY

David M. Boore

U.S. Geological Survey, MS 977, 345 Middlefield Road, Menlo Park, CA 94025

Abstract: Digital strong-motion instruments allow routine recovery of ground motions at periods much longer than possible with older analog instruments, and under favorable conditions it may be possible to recover the residual displacements near earthquakes, particularly if both the rotational and translational components of motion are recorded (currently only the translational components are recorded). In practice, however, digital recordings are commonly plagued by drifts in the velocity and displacement traces obtained by integrating the recorded acceleration traces. Various baseline-correction schemes can be designed that give the appearance of removing these drifts. Although comparisons with independent measures of residual displacements, such as from GPS or InSAR measurements, show that such schemes can work, in general the sources of the drifts are such as to prevent routine corrections for the baseline problems; removal of low-frequencies by filtering is then required for many recordings. That is the bad news. The good news is that the filter corners can be so low that little of engineering interest is lost. Recent data from several earthquakes (e.g., 1999 Hector Mine and 2002 Denali) shows that for displacement response spectra, the transition from increasing to constant spectral levels occurs at significantly longer periods than in Eurocode 8; the transition period is in good agreement, however, with the recent 2003 NEHRP code.

Key words: Long-period, displacements, displacement response spectra, digital accelerographs

1. INTRODUCTION

There are many benefits of digital recordings over the older analog recordings. The most obvious of these advantages is that the data are available for processing much more quickly than before. But another

P. Gülkan and J.G. Anderson (eds.), Directions in Strong Motion Instrumentation, 41–54.

important advantage is that the resolution of the recordings is usually higher for digital than for analog recordings. The typical digitization of an analog record has a maximum resolution of 2 cm/s^2, whereas digital instruments with ±2g full scale recording have resolutions from about 1 cm/s^2 for 12-bit analog-to-digital conversion (ADC) to 0.0002 cm/s^2 for 24-bit ADC. The Fourier acceleration spectrum decreases at long period, and thus the higher resolution should allow data at longer periods to be obtained than from analog recordings. These longer period data are useful in defining residual displacements, long-period site response and basin waves, as well as the long-period behavior of displacement response spectra and the strains within structures. An example of a digital recording of acceleration and the velocity and displacement obtained by straightforward integration of the acceleration (after filtering to remove long-period drifts) is given in Figure 1. Note the low amplitudes and the clear presence of basin waves (due to propagation across part of the Los Angeles basin).

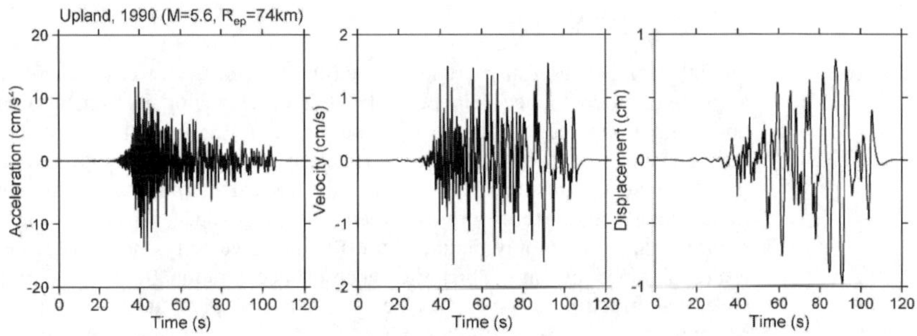

Figure 1. Acceleration, velocity, and displacement time series for one horizontal component of a 16-bit recording of the 1990 Upland earthquake, with a 0.1 Hz acausal low-cut filter. The waves traveling to the station have traversed part of the Los Angeles basin, and therefore large surface waves are present on the displacement time series. (From Boore, 2004a.)

2. POTENTIAL AND REALITY: LONG-PERIOD "NOISE"

Under the most ideal circumstances, with no instrumental or physical sources of long-period noise, the potential for obtaining long-period information is controlled by the number of bits used in the analog-to-digital converter (ADC) and the signal amplitude (assuming that the transducer is not the limiting factor). One measure of the potential is given by the accumulated error in the displacement trace from random digitizing errors in the acceleration trace. The expected error in the final displacement derived from double integration of a suite of digitizations will be zero, but as in a random-walk process, any single realization will show a drift in the displacement. The standard deviation of the final displacement after double integration for a length of time is a measure of the error in the displacement due to digitizing the acceleration time series. Using the equations in Boore et al. (2002), the standard deviation of the final displacement after double integration of a 200 sps acceleration time series for 100 s would be 11 cm, 1 cm, and 0.003 cm for 12-bit, 16-bit, and 24-bit ADC of recordings with a full scale of ±2g, respectively. The actual drifts in displacements obtained by double integration are often much larger than this (e.g., Boore, 2001, Boore et al., 2002, Wang et al., 2003). This indicates that the theoretical potential is usually not realized because a number of other sources of long-period noise may be present in digital recordings. By "noise" I mean undesired energy in the time series, whether it be due to instrumental instabilities, the effect of ADC on the baseline (Boore, 2003), or the effect of ground tilt and rotation (thereby putting a small component of gravitational acceleration onto the horizontal sensor, as discussed, for example, by Graizer, 1989, and Trifunac and Todorovska, 2001). Even with the presence of long-period noise, however, digital recordings are a significant improvement over analog recordings.

2.1. Removing Long-period "Noise"

The long-period noise can be removed in several ways: making baseline corrections, filtering, or a combination of both.

2.1.1. Baseline corrections

A number of methods for correcting for baseline variations have been used in the processing of strong-motion recordings. These usually have little physical basis and largely amount to fitting low-order polynomials to the velocity time series. Another method that is motivated by physical causes for baseline offsets was proposed by Iwan et al. (1985), with modifications by Boore (2001). But as shown in Figure 2, the residual displacements can be sensitive to a parameter t_2 in the baseline correction. The displacement response spectrum, however, is relatively insensitive to the parameter t_2 at periods of usual engineering concern (Figure 3).

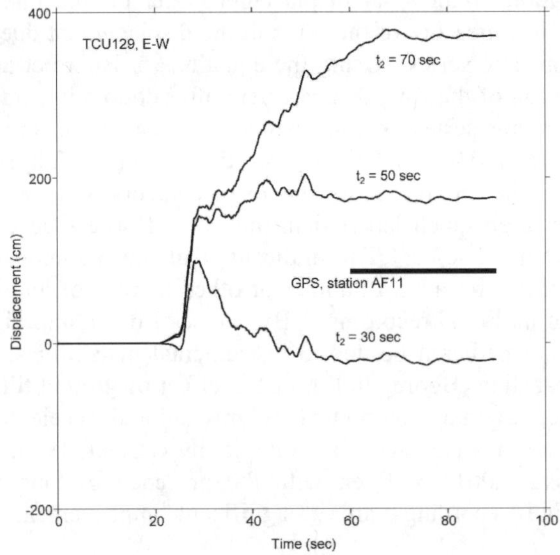

Figure 2. Displacements obtained by double integration of the EW component of acceleration recorded at TCU129 and modified using three values of T_2 to correct for baseline shifts. The GPS level was obtained at a station 2.3 km from TCU129, on the foot wall side of the fault (as is TCU129). (Modified from Boore, 2001.)

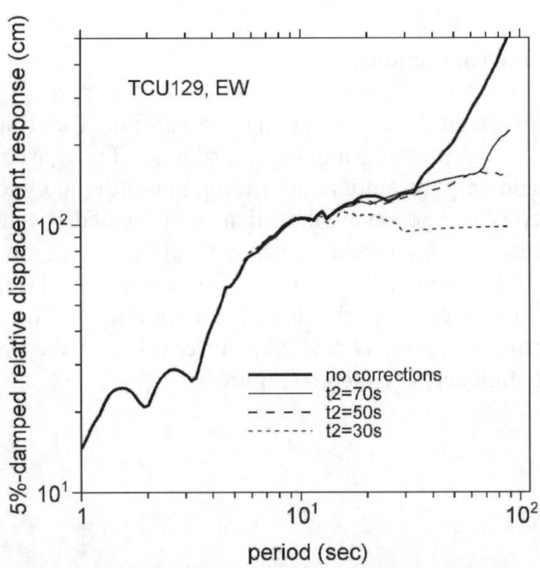

Figure 3. Displacement response spectra corresponding to processing shown in the previous figure.

2.1.2. Low-cut (high-pass) Filtering

Although baseline corrections result in removal of some long-period noise, often low-cut filtering is needed as well. In many cases, filtering is all that is needed, and filtering has the advantage that the effect in the frequency domain is more apparent than when baseline corrections are used. As discussed by Boore and Akkar (2003), acausal (phaseless) filtering is preferred to causal filtering. An example of the velocity and displacement time series obtained by integration of an acceleration time series that was filtered using different filter corners is given in Figure 4. The displacement waveforms and peak displacements are sensitive to the filter corners, but as shown in Figure 5, the displacement response spectrum is insensitive to the filter corners at periods of engineering interest when the amplitude of the signal at long periods allows the use of relatively low-frequency filter cutoffs.

The increased resolution of digital instruments allows the low-cut filter corners to be chosen at lower frequencies, in general, than for analog recordings. This is shown in Figure 6 (plotted in terms of period rather than frequency).

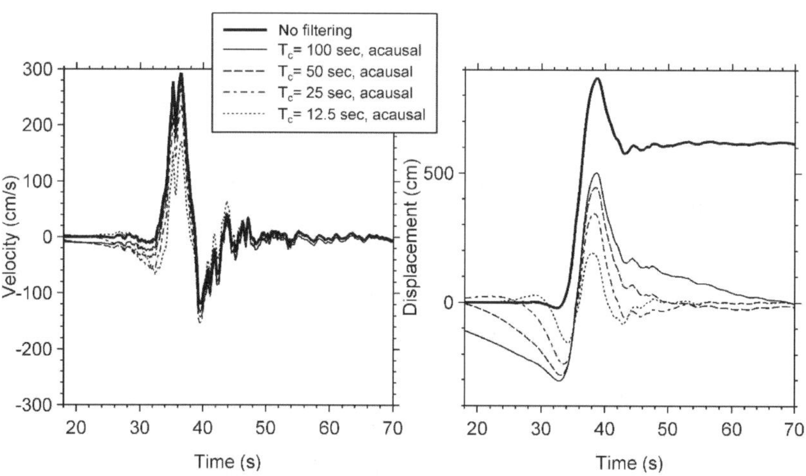

Figure 4. The velocity and displacement time series derived from the TCU068 recording. In all cases a simple baseline correction (the "v0" correction of Boore, 2001) was used, with and without filtering. The filters were low-cut acausal Butterworth filters having filter corner periods of 12.5, 25, 50, and 100 sec.

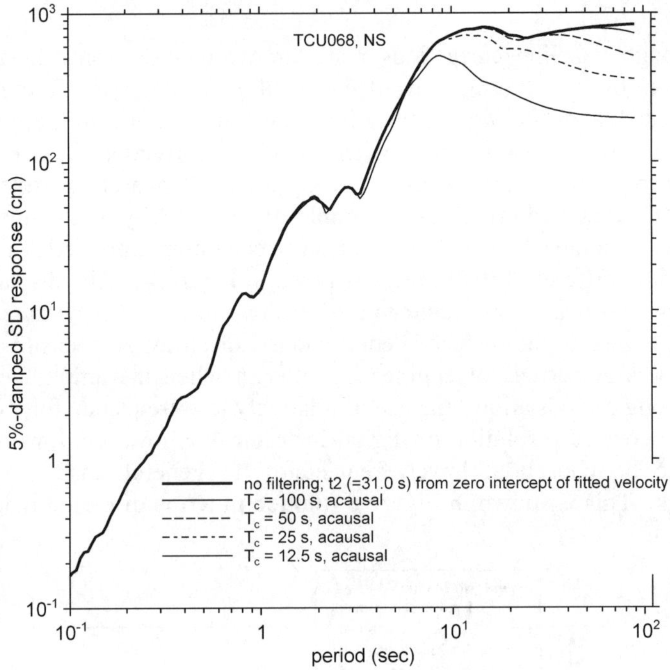

Figure 5. 5%-damped relative displacement response spectra for the time series processed using acausal filters.

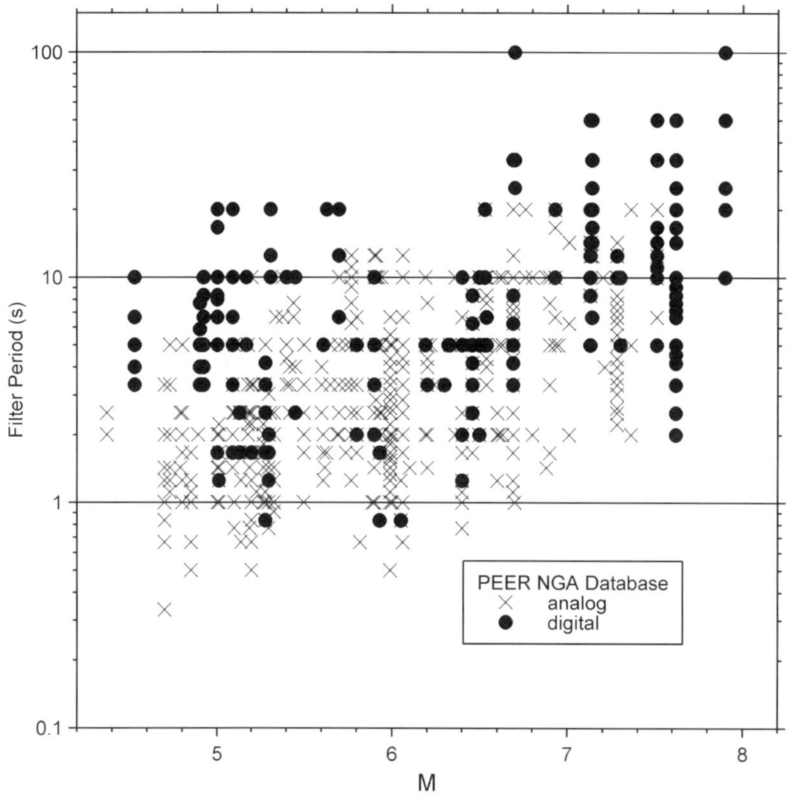

Figure 6. Filter period vs. moment magnitude, with data from analog and digital instruments indicated by the symbols. An effort has been made to determine whether each recording was made on an analog or a digital instrument, but there are undoubtedly a few errors. The filter for which the period is given removes long-period energy from the record, and the filter periods are those in a database being developed for use in the Pacific Earthquake Engineering Research Center (PEER) New Generation Attenuation (NGA) project. An attempt has been made in developing the PEER NGA database to apply uniform processing to all records. Some of the recordings, such as the analog recordings of the M 7.4 Tabas, Iran earthquake, may have had significant amounts of long-period noise removed by the use of baseline corrections, in which case the filter period is not a good measure of the long-period content of the record. For this reason the Tabas data are not shown here. In some other cases, such as the M 7.6 1999 Chi-Chi, Taiwan earthquake, longer filter periods could have been chosen.

3. SPATIAL SIMILARITY OF DISPLACEMENT WAVEFORMS: DEVELOPING CONFIDENCE IN LONG-PERIOD MOTIONS

As shown by Hanks (1975), a good way of judging the adequacy of the processing at long periods is to compare the displacement waveforms for stations within small distances of one another. The idea here is that long-period motions have wavelengths so large that the motions will be insensitive to site-to-site variations in geology (and thus site response). An example from the 1999 Hector Mine earthquake is shown in Figure 7. Note that the variations in the velocity waveforms are larger than in the displacement waveforms, as would be expected because the high frequencies in velocity waveforms are more sensitive to variations in near-surface geology.

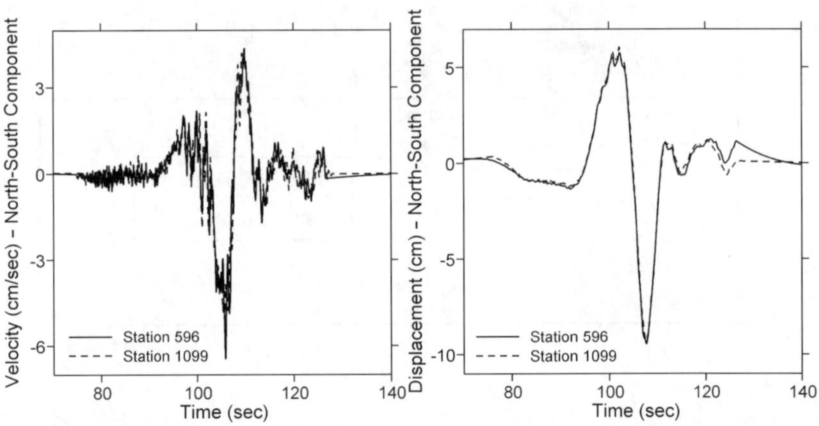

Figure 7. Comparison of a) velocities and b) displacements for two closely spaced (1.6 km) stations that recorded the 1999 Hector Mine, California, earthquake. The recorded accelerations from which the velocities and displacements are derived were processed using a constrained quadratic fit to velocity, followed by acausal filtering with a corner frequency of 0.02 Hz. The traces are aligned using time shifts derived by visually aligning the large pulse to the south on the displacement trace. (From Boore et al., 2002, but using acausal rather than causal filtering.)

4. DISPLACEMENT RESPONSE SPECTRA (SD) AND BUILDING CODES

The ability to determine motions at longer periods than before has an important application in displacement-based design (Bommer and Elnashai, 1999; Faccioli et al., 2004). An important consideration in seismic design codes is the period at which the displacement response spectrum (SD) no longer increases. As seen in Figure 8, this period can exceed 10 s. Theoretical seismic source scaling can be used as a guide to the period at which SD flattens. A single corner period model predicts the transition period as a function of moment magnitude shown in Figure 9, but Atkinson and Boore (1998) show that this model is not in agreement with the relation between surface wave magnitude and moment magnitude for large earthquakes. In better agreement with the magnitudes are two-corner-period models, such as the model of Atkinson and Silva (2000) shown in Figure 9. If the smaller of the two corner periods (corresponding to the line labeled f_b in Figure 9) is taken as the start of the transition of SD to a constant value, then Figure 9 shows that digitally-recorded data should be able to capture this transition period.

Figure 10 compares SD from several recent codes with that from the M 7.9 Denali Fault, Alaska, earthquake. Clearly the comparison with the recent 2003 NEHRP code is much better than with either the 2000 NEHRP code, for which SD increases without limit, or Eurocode 8, for which the transition to a constant SD occurs at too short a period. The transition period in the 2003 NEHRP code is taken from maps, and the values in the maps are based on the modal magnitude from dissaggregation of 2%-in-50-year hazard maps. Given the modal magnitude, the transition corner period is based on a combination of theoretical and observational considerations (the Denali Fault earthquake data were not used in deriving the transition periods).

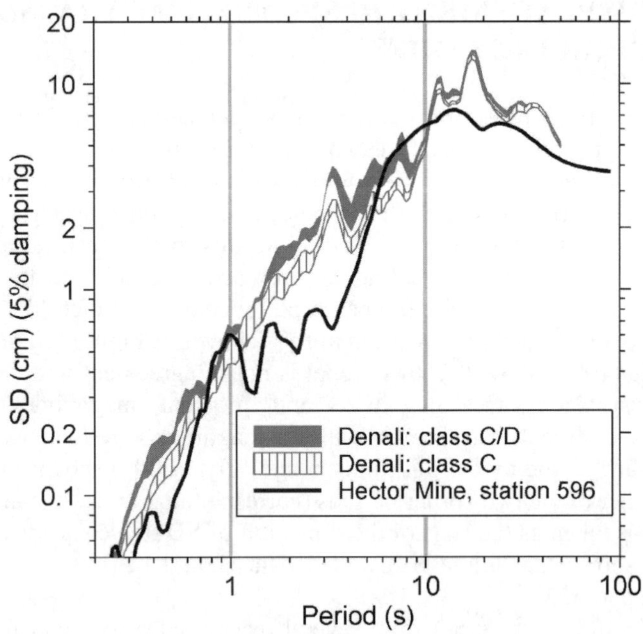

Figure 8. Displacement response spectrum (SD) for the 2002 Denali Fault, Alaska, earthquake recordings and for the transverse component of the 1999 Hector Mine, California, earthquake (after dividing the Hector Mine spectrum by 2.6 to correct for the different distances to the earthquakes). All records were low-cut filtered with a low-order acausal Butterworth filter with a corner at 0.02 Hz (the filter response goes as f^4 for low frequencies). (From Boore, 2004b).

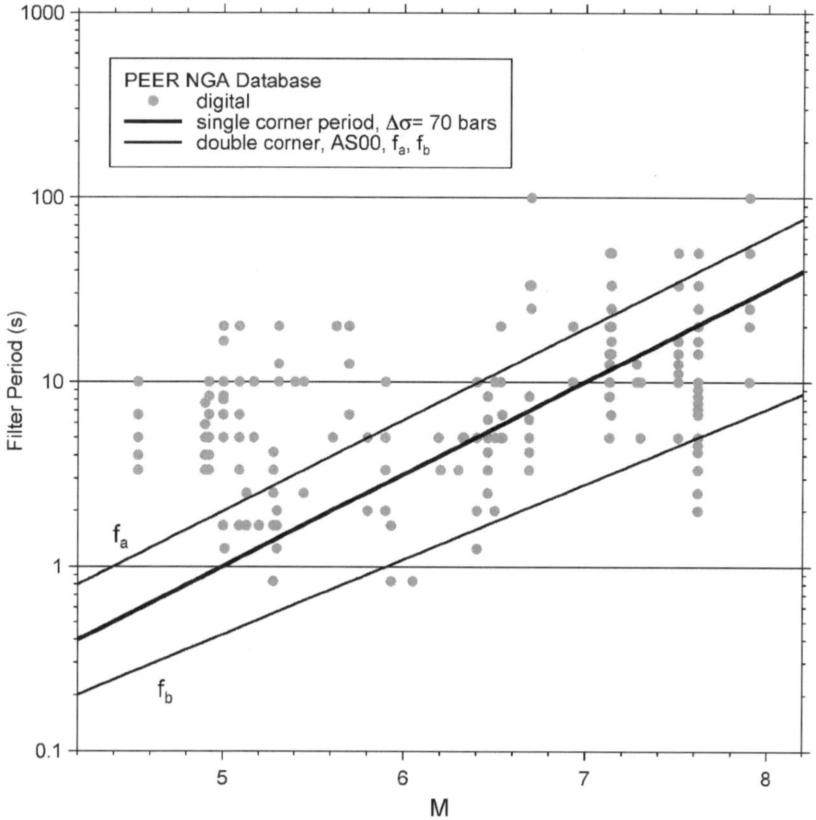

Figure 9. Corner periods from two source-scaling models, compared to the filter periods used in processing strong-motion data (AS00 = Atkinson and Silva, 2000).

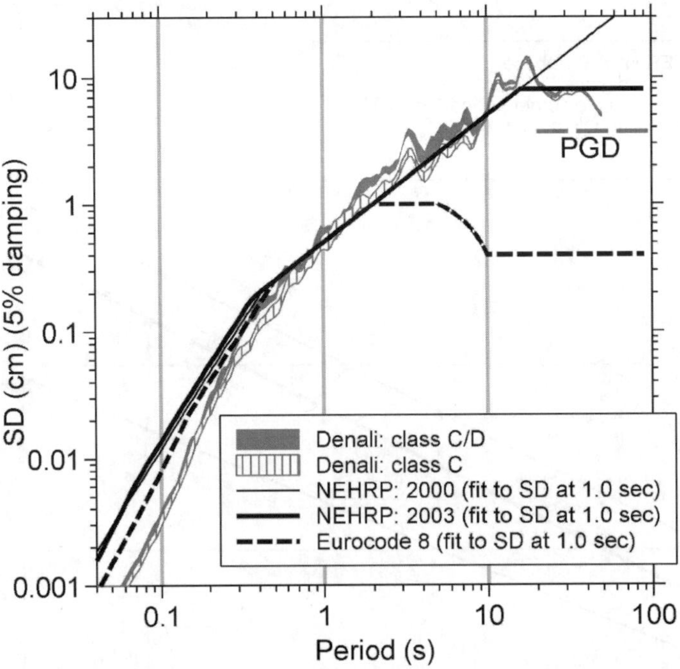

Figure 10. Comparison of observed displacement response spectrum (SD) for class C/D and C sites in Anchorage to the spectra constructed following the procedures for the following building codes: 2000 NEHRP (BSSC, 2001), 2003 NEHRP (BSSC, 2004), and Eurocode 8 (CEN, 1998). Also shown is the peak ground displacement (PGD) for the Anchorage motions; the displacement response spectrum will by asymptotic to this value at sufficiently long periods. For NEHRP, the mapped values of spectral response at 0.2 and 1.0 s were used, but to emphasize the shape, the spectra for the three codes have been normalized to a representative value of the data at 1 s. (From Boore, 2004b.)

5. SUMMARY

Digital strong-motion instruments allow routine recovery of ground motions at periods much longer than possible with the older analog instruments.

In spite of the theoretical possibility of recovering even the residual displacements in the vicinity of earthquakes, in practice digital recordings are often plagued by drifts in the velocity and displacement traces obtained by integrating the recorded acceleration traces. Baseline-correction schemes may not give stable results, and removal of low-frequencies by filtering is often required. The filter corners can be so low, however, that little of engineering interest is lost.

Recent data from several earthquakes (e. g. 1999 Hector Mine and 2002 Denali) shows that for displacement response spectra, the transition from increasing to constant spectral levels occurs at significantly longer periods than in Eurocode 8; the transition period is in good agreement, however, with the recent 2003 NEHRP code.

ACKNOWLEDGMENTS

Many people helped with the research discussed here, but before acknowledging specific individuals, I thank Polat Gülkan for the invitation to the Workshop and NATO for sponsoring the Workshop. Among those that helped with the paper, I wish to thank in particular Julian Bommer for help both cerebral and practical, Sinan Akkar for answering numerous inquiries about Turkish strong-motion data, Basil Margaris for doing the same for earthquakes in Greece, E. V. Leyendecker for information regarding the 2003 NEHRP code, and Tom Hanks and Chris Stephens for reviewing the paper.

REFERENCES

1. Atkinson, G. M. and Boore, D. M., 1998. Evaluation of models for earthquake source spectra in eastern North America, *Bull. Seism. Soc. Am.* **88:** 917—934.
2. Atkinson, G. M. and Silva, W., 2000, Stochastic modeling of California ground motions, *Bull. Seism. Soc. Am.* **90:** 255—274.
3. Bommer, J. J. and Elnashai, A. S., 1999, Displacement spectra for seismic design, *J. Earthquake Eng.* **3:** 1—32.
4. Boore, D. M., 2001, Effect of baseline corrections on displacements and response spectra for several recordings of the 1999 Chi-Chi, Taiwan, earthquake, *Bull. Seism. Soc. Am.* **91:** 1199—1211.
5. Boore, D. M., 2003, Analog-to-digital conversion as a source of drifts in displacements derived from digital recordings of ground acceleration, *Bull. Seism. Soc. Am.* **93**: 2017—2024.
6. Boore, D. M., 2004a, Can site response be predicted?, *J. Earthquake Eng.* **8:** Special Issue 1, 1-- 41.
7. Boore, D. M., 2004b, Ground motion in Anchorage, Alaska, from the 2002 Denali fault earthquake: Site response and displacement pulses, *Bull. Seism. Soc. Am.* **94:** (in press).
8. Boore, D. M. and Akkar S., 2003, Effect of causal and acausal filters on elastic and inelastic response spectra, *Earthq. Eng. And Struct. Dyn.*, **32:** 1729—1748.
9. Boore, D. M., Stephens, C. D., and Joyner, W. B., 2002, Comments on baseline correction of digital strong-motion data: Examples from the 1999 Hector Mine, California, earthquake, *Bull. Seism. Soc. Am.* **92:** 1543—1560.
10. BSSC, 2001, NEHRP recommended provisions for seismic regulations for new buildings and other structures, 2000 Edition, Part 1: Provisions, prepared by the Building Seismic Safety Council for the Federal Emergency Management Agency (Report No. FEMA 368), Washington, D.C. (available from http://www.bssconline.org/NEHRP2000/comments/provisions/).
11. BSSC, 2004, NEHRP Recommended Provisions for Seismic Regulations for New Buildings and other Structures, 2003 Edition, Part 1 - Provisions, Part 2 - Commentary,

prepared by the Building Seismic Safety Council for the Federal Emergency Management Agency (Report No. FEMA 450), Washington, D.C. (available from http://www.bssconline.org/).

12. CEN, 2002, Eurocode 8: Design of structures for earthquake resistance, Draft No. 5, CEN (European Committee for Standardization) Doc CEN/TC250/SC8/N317 (prEN 1998-1:200X), Brussels.
13. Faccioli, E., Paolucci, R., and Rey, J., 2004, Displacement spectra for long periods, *Earthquake Spectra* **20:** 347--376.
14. Graizer, V. M., 1989, On inertial seismometry, Izvestiya, Physics of the Solid Earth **25:** 26--29.
15. Hanks, T. C., 1975, Strong ground motion of the San Fernando, California, earthquake: ground displacements, *Bull. Seism. Soc. Am.* **65:** 193—225.
16. Iwan, W. D., Moser, M. A., and Peng, C.-Y., 1985, Some observations on strong-motion earthquake measurement using a digital accelerograph, *Bull. Seism. Soc. Am.* 75: 1225--1246.
17. Trifunac, M. D. and Todorovska, M. I., 2001, A note on the useable dynamic range of accelerographs recording translation, Soil Dynamics and Structural Eng. **21:** 275—286.
18. Wang, G.-Q., Boore, D. M., Igel, H., and Zhou, X.-Y., 2003, Some observations on colocated and closely-spaced strong ground motion records of the 1999, Chi-Chi, Taiwan earthquake, *Bull. Seism. Soc. Am.* **93:** 674—693.

OBSERVED GROUND MOTIONS, EXTREME GROUND MOTIONS, AND PHYSICAL LIMITS TO GROUND MOTIONS

T. C. Hanks[1], N.A. Abrahamson[2], M. Board[3], D.M. Boore[4], J.N. Brune[5], C.A. Cornell[6]

[1]U. S. Geological Survey, 345 Middlefield Road, Menlo Park, CA 94025, thanks@usgs.gov; [2]Pacific Gas & Electric Company, SanFrancisco, CA 94611, naa2@pge.com; [3]Mark_Board@ymp.gov; [4]U.S. Geological Survey, 345 Middlefield Road, Menlo Park 94025, boore@usgs.gov; [5]Seismological Laboratory, University of Nevada, Reno, NV 89577, brune@seismo.unr.edu; [6]Department of Civil and Environmental Engineering, Stanford University, Stanford, CA, 94305, cornell@stanford.edu

Abstract: This paper examines whether the extremely high ground motion values that are calculated for probabilistic seismic hazard analyses in critical facilities are physically attainable.

Keywords: probabilistic seismic hazard analysis, nuclear waste repository, truncation, bounding value of ground motion, asperity

The probabilistic seismic hazard analysis (PSHA) concluded in 1998 for Yucca Mountain, the designated site of the underground repository for the United States' high-level radioactive waste, was the most complicated and complex PSHA ever undertaken at the time. The procedures, methods, and results of this PSHA are described in Stepp *et al.* (2001), mostly in the context of a probability of exceedance (hazard) of 10^{-4}/yr for ground motion at Site A, a hypothetical, reference rock outcrop site at the elevation of the emplacement drifts within the mountain.

When extended to lower hazard levels, the resulting ground motions become progressive higher: at mean-value hazard levels of 10^{-6}/yr, 10^{-7}/yr, and 10^{-8}/yr, the resulting peak ground accelerations (PGA) and peak ground velocities (PGV) are 3 g, 6 g, and 11 g and 3.5 m/sec, 7 m/sec, and 13 m/sec, respectively. We refer to these as *extreme ground motions*, the very large-amplitude ground motions that will arise in any PSHA at very low probabilities, given untruncated ground-motion distribution functions.

P. Gülkan and J.G. Anderson (eds.), Directions in Strong Motion Instrumentation, 55–59.

Needless to say, such large-amplitude ground motions have generated considerable concern if not consternation in the scientific, engineering, and regulatory communities, for such PGV's and PGA's have never been recorded for earthquakes. Indeed, not a small minority of informed opinion believe such ground motions are "physically unrealizable", which is to say that there are physical limits to earthquake ground motions. Such physical limits would provide the basis for truncating the high-end tail of the ground-motion distribution. A vertical truncation would mean that any ground motion greater than the truncation value has zero probability of occurrence, but physical limits will surely come with uncertainty, meaning that one ground-motion distribution function will be replaced by another, with lower ground motions at the same mean-value hazard.

The PEGASOS project is a PSHA for nuclear power plants in Switzerland, comparable in scope to the 1998 Yucca Mountain PSHA (Abrahamson *et al.*, 2002). According to Bommer *et al.*(2004), because of the Yucca Mountain results at low hazard levels, "the PEGASOS project has required experts … to specify bounding values on the ground motions." Bommer *et al.* (2004) does not tell us how this happened nor what the results are, but it does provide a useful summary of thinking on upper-bound ground motions, mostly from the perspectives of strong-motion seismology and geotechnical engineering as they have developed over the last 40 years.

Most of the participants at this meeting will know seismologists to be a noisy lot, at least the American variety of the species, so it may come as some surprise as to how silent seismologists have been on this matter of "upper bounds' and "physical limits". Part of this reticence is surely born of the specific upper bounds articulated by engineering colleagues in the 1960's and 1970's that Nature contradicted in short order, most notably with the Parkfield (June 28, 1966; **M** = 6.0) and San Fernando (Feb.9, 1971; **M** = 6.6) earthquakes in California. More generally, however, most earth scientists are philosophically indisposed to say things like "No, the Earth just can't do this or that, like have a mantle that convects." We have been wrong too many times.

The Science and Technology Program of the Office of Civilian Radioactive Waste Management in the U.S. Department of Energy have asked the six authors of this paper to take stock of the extreme ground motions at Yucca Mountain and to recommend further research that can speak to physical limits, however uncertain they may be. Our approach to this problem will include several scientific and intellectual dimensions, reaching well beyond the perspectives of engineering seismology because they do not take us very far.

For example, it is fairly easy to show from the data shown graphically in Figure 9 of Stepp *et al.* (2001) that the 3 g PGA at 10^{-6}/yr hazard is mostly due to the experts' epistemic uncertainty on the median PGA at close distances. By consequence, most of the hazard arises solely from the Solitario Canyon fault. As revealed by paleoseismic data, **M** 6 1/2 to 7 earthquakes occur on the Solitario Canyon fault with a recurrence interval of ~50,000 years, so the 1 in 20 ground motion record at close distances suffices for a hazard level of 10^{-6}/yr. Although some debate attends distance metrics at such close distances, available strong-motion data may suffice to conclude that the 3g and 3.5 m/sec ground motions at the 10^{-6}/yr hazard level are too high, especially when they are normalized for the hard-rock conditions of the reference Site A. Such data at close distances are too few, however, to allow us to say much at hazard levels of 10^{-7}/yr and 10^{-8}/yr, for which we need more extensive databases and/or longer-term records.

Seismologists interested in the earthquake source have come to understand that ground motion generated by a fault or fault patch is proportional to the stress drop on that fault or fault patch, commonly referred to as an asperity. Large-amplitude ground motion can only be generated by faults or fault patches with large stress drops, a kilobar or more in the case of the extreme ground motions detailed above. For such earthquakes, body-wave magnitude (m_b) will be large relative to its surface-wave magnitude (M_S) compared to earthquakes with average stress drop of several tens of bars. Such m_b–M_S data were long ago recognized as a first-order discriminant between underground nuclear explosions and crustal earthquakes (*e.g.*, Liebermann and Pomeroy, 1969) and should easily discriminate earthquakes with unusually large stress drops from those with conventional values. Since the advent of the Worldwide Standardized Seismographic Network (WWSSN) in the early 1960's, 50,000, perhaps even 100,000 m_b–M_S pairs have been collected for **M** > 5 crustal earthquakes.

Earthquake stress drops have been determined for thousands of crustal earthquakes by now, and they also speak to the frequency of kilobar stress drop earthquakes, although most of these determinations have been made for **M** < 5 events. There is substantial scatter in these data, with at least a few events possessing stress drops ~1 kilobar. The rms-acceleration stress drop (Hanks and McGuire, 1981) or the "stress parameter" of Boore (1983) shows far less variability. While fewer events have been analyzed according to this methodolgy, the reason for this difference in variability, if real, is not known.

We can also reach back to Geology 101 and the Principle of Uniformitarianism, the idea that we can understand the past by understanding processes observable in the present. In summary form, "The present is the key to the past." A variation on this theme, first articulated by David Schwartz in the context of paleoseismology, is that "The past is the key to the future". This is to say that anything that the Earth can do in the future, it has already done in the past. We need only find one example of something that has happened in the past to conclude that it can happen again. The converse proposition, that if something has never happened in the past, for example one level or another of extreme ground motions, then it can never happen in the future is much more difficult to deal with, given a geologic record that is incomplete in both space and time.

Yucca Mountain lies within the extensional regime of the Basin and Range Province, in the southwestern Nevada volcanic field. It is built from a sequence of volcanic tuffs, the product of volcanism active from 14 to 11 million years ago. These units have been offset ~300 m by down-to-the-west normal faulting along the Solitario Canyon fault, which forms the western escarpment of the back-tilted (to the east) Yucca Mountain block. The volcanic stratigraphy includes the so-called lithophysal units, lithophysae being voids in the rock caused by exsolved gases as the tuffs cooled. These voids weaken the rock considerably, but these units show no evidence of fracturing or damage. This means that PGV in excess of several m/sec, where "several" is yet to be defined as 3 or 4 or maybe 5 with corresponding uncertainty, have not passed through these rocks in the 10^7 years since they formed.

Similarly, precipitous talus slopes, precarious rocks, undamaged speleothems, and intact archeological excavations all speak to ground-motion levels that have not occurred in the time intervals for which these features have existed. By consequence, they also speak to probability of exceedances of ground-motion levels that would destroy them. Moreover, we are confident that single-event fault scarps of ~10 m or more would be identifiable as such for a million years or more in and around Yucca Mountain, should such large-displacement faulting events be necessary to excite the extreme ground motion at Yucca Mountain. Such single-event faulting displacements are unknown in the Yucca Mountain area or, for that matter anywhere else in the Basin and Range Province.

We are persuaded that, if physical limits to earthquake ground motion exist in meaningful forms, it will be because they are rooted in nonlinear dissipative mechanisms. Linear elasticity has been the workhorse of seismology for more than a century, and it has been the basic tool for earthquake ground-motion estimation over the past 50 years. But we will have to move beyond linear elasticity to deal with extreme ground motions in any fundamental way. There is much that we have yet to learn about the nonlinear response of earth materials to extreme ground motions, and it is certain that any physical limits that we may derive from them will come with significant uncertainty.

Numerous underground nuclear explosions were detonated on the Nevada Test Site over a period of almost forty years in volcanic rocks not dissimilar from the Yucca Mountain tuffs, and many seismograms exist for these explosions. Beginning with an energy density vastly greater than any known crustal earthquake, the explosions first vaporize rock, then melt it, and fracture it to distances that depend on the energy release. At and beyond the "elastic radius", progressive wavetrains develop, the amplitudes of which are determined by the maximum stresses that these rocks can sustain. These observations should also provide constraints on ground-motion amplitudes at Yucca Mountain, although we must take into account the variability of the data (which translates into uncertainty in the constraint) and whatever differences may exist in the rock properties in the vicinity of the explosion and those in Yucca Mountain.

Confining pressure increases with depth in the earth due to the weight of the overlying rocks, and it is possible that strong excitation in the source region at typical earthquake depths of 10 to 15 km would still be invisible to us at the surface above the source. Deep seismometers will tell us what happens between the source and surface, and separate analysis of upgoing and downgoing radiation should reveal how strong the source excitation can be. Considerable work has already been done along the second line of research. Analysis of teleseismic signals generally yields earthquake source parameters consistent with those determined from local and regional recordings. At the higher frequencies that determine the radiated energy, however, teleseismic amplitudes are *lower* than expected, even for "expected" levels of excitation let alone something stronger (J.L. Boatwright, personal communication).

Exploring the physical limits to the radiated field of earthquakes, either in the near-surface environment but especially in the source region, presents unusual challenges not only to seismologists but also to much of the community of earthquake scientists. Among these will be overcoming our historical reticence to dealing with physical limits. Another will be learning to live in a "beyond-linear-elasticity" world in which these physical limits will be defined. As is so often the case, other seismologists have preceded us here as well, most notably those working in the then largely-classified world of nuclear-explosion seismology (*e.g.*, Rodean ,1971).

REFERENCES

1. Abrahamson, N.A. and seven others (2002). PEGASOS: A comprehensive probabilistic seismic hazard assessement for nuclear power plants in Switzerland, in *Proceedings of the Twelfth European Conference on Earthquake Engineering*, Elsevier, Paper 663, London.
2. Bommer, J. J. and ten others (2004). The challenge of defining upper bounds on earthquake ground motions, *Seismological Research Letters* **75**, 82-95
3. Boore, D.M. (1983). Stochastic simulation of high-frequency ground motions based on seismological models of the radiated spectra, *Bulletin of the Seismological Society of America* **73,** 1865-1894.
4. Hanks, T.C. and R.K. McGuire (1981). The character of high-frequency strong ground motion, *Bulletin of the Seismological Society of America* **71**, 2071-2095.
5. Liebermann, R.C. and P.W. Pomeroy (1969). Relative excitation of surface waves by earthquakes and underground explosions, *Journal of Geophysical Research* **74**, 1575-1590.
6. Rodean, H.C. (1971). Nuclear-explosion seismology, *U.S. Atomic Energy Commission Division of Technical Information.*
7. Stepp, J.C. and nine others (2001). Probabilistic seismic hazard analyses for ground motions and fault displacement at Yucca Mountain, Nevada, *Earthquake Spectra* **17**, 113-151.

Part 2

ENGINEERING USES OF STRONG MOTION SEISMOGRAMS

RAISED DRIFT DEMANDS FOR FRAMED BUILDINGS DURING NEAR-FIELD EARTHQUAKES

P. Gülkan and U. Yazgan
Disaster Management Research Center and Department of Civil Engineering / Middle East Technical University, Ankara 06531, Turkey

Abstract: The improvement of understanding structural capacity against large displacement demand of near-fault conditions has become a subject of research interest for the last decade. Parallel to the developments in performance-based seismic design (PBSD) this issue has attracted researcher interest further, as assessment of structural displacement capacity has become one of the main issues for the procedures employed in this new concept. This study focuses on the near-fault ground motion demand on framed structures. We have used soil site, near-field records from various M > 6.5 events including the 1999 Turkey and Taiwan earthquakes. The spectral quantities were computed using a ground motion prediction relationship that is based partly on these near-fault ground motion records. The spectral quantities were evaluated for the global displacement demand definition of such ground motions. We employ this global demand definition to calculate the distance and magnitude dependent inter-story drift demand limits for frame-type structures. Comparison of these preliminary findings with code provisions is encouraging.

Keywords: Near-field earthquake, drift, nonlinear drift calculation, peak velocity, ground motion prediction

1. INTRODUCTION

Unexpected structural damage observed in framed buildings situated in the near field during the 1994 Northridge, and 1995 Kobe earthquakes were forceful reminders of what some seismologists had speculated might distinguish distant and close earthquake effects (Bolt, 1996). It appeared that the large displacement demand of near-fault ground motions due to pulse-like velocity waveforms could not be met by structures proportioned to fulfill requirements of traditional force-based design (FBD). The 1999 Turkey and Chi-Chi, Taiwan, earthquakes displayed cases that seemed to support these observations, directions of stronger polarization do not always match

P. Gülkan and J.G. Anderson (eds.), Directions in Strong Motion Instrumentation, 61–81.

theoretical expectations, (e.g., Akkar and Gülkan, 2002; Akkar et al., 2004). These near-fault events have motivated a considerable amount of research for design against effects of near-by ruptures. Some of these have integrated their results with the PBSD concept in which the structural performance is described by the earthquake-induced displacements through the implementation of displacement-based design (DBD) procedures.

Performance based procedures require that, both for assessment and for design purposes a means should be at hand to estimate structural displacements in structures situated at different distances to the earthquake-generating fault. An attenuation relationship (appropriate, in this case, for near-fault situations) then needs to be utilized to convert the ground motion descriptors into some estimate for the structural drift. A good deal of research has been conducted on strong ground motion records to predict what some desired index of the likely ground motion as a function of magnitude, distance, site geology or type of faulting. These relations have also been derived for spectral quantities so that design spectra can be constructed to provide more direct guidance to structural engineers. Widespread use of prediction curves commonly ignores a basic caveat of the trade: recording instrument characteristics, pre-processing techniques applied to the records, any filter characteristics and other features all leave subtle but important marks on the records, and more importantly, on calculated quantities for which they serve as the source. Even if we ignore these spurious effects features unique to each earthquake and tectonic environment make it difficult to develop an all-seeing set of prediction curves valid globally. Indeed, with the increase in the number of instruments deployed in many different parts of the world, and their increasing sensitivity attenuation relations undergo constant updating and revision. There exists no convincing evidence for the transportability of ground motion relations, and yet they are often used for design purposes in environments quite different from those under which they have been generated. Even curves that have been developed for a given region have been updated over time as new data has been collected. This builds up the case that in urban centers close to or traversed by known active faults strong motion arrays should be deployed in carefully designed patterns so that seismological features can be embedded into design rules with enhanced confidence.

A test of the hypothesis on the cross utilization of ground motion prediction relations is attempted in this article by using a suite of near-field records that include ground motions from the 1999 Turkey and Taiwan earthquakes in calculating the exact drifts in framed structures. An attenuation relationship calculated from a number of records from Turkey is employed in computing the magnitude- and site-dependent spectral values for periods up to 2 s. These results are used for describing the amount of near-source roof-level displacement demand that is a critical measure for near-field excitations. The adequacy of near-fault factors implemented by the UBC97 (ICBO, 1997) provisions is discussed in view of the relationship based on a statistical study of records made in Turkey. Based on this new prediction relationship, the drift limits for frame type structures will be calculated as a function of magnitude and distance to account for the local displacement demand of near-source ground motions. These will then be compared with the drift limits specified by UBC97 and NEHRP (FEMA, 1997) provisions. The results presented are expected to clarify some important aspects of near-fault ground motions for engineering studies that aim to improve the DBD procedures employed in PBSD.

2. NEAR-FAULT GROUND MOTION DATA BASE

The near-fault database employed in this study consists of horizontal components of firm soil records from shallow crustal earthquakes. The ground motions are either from free field stations or from instruments housed in shelters and ground levels of mostly one-story structures. We have used the shortest distance from the station to fault rupture as the distance measure. The records selected have peak ground velocity (PGV) greater than 40 cm/s at least in one component. Our PGV preference is attributed to the results of near-fault related studies that anticipated the structural damage of such excitations to their large amplitude and long period velocity pulses (Hall et al., 1995). Almost all of the records pertain to moment magnitude (M_w) 6.5 or greater events. The record that we used from the 1986 Palm Springs earthquake ($M_w = 6.2$) contravenes this criterion but we have chosen it due to its large velocity (PGV > 70 cm/s). Table 1 lists the basic features of these components for the selected records. Soil characterization is according to UBC (1997) and FEMA-302 criteria.

Studies in strong motion seismology confirm that forward directivity (the alignment of rupture direction and wave propagation) is important for near-fault ground motions, and causes higher demands in fault normal components as these comprise a larger amount of accumulated energy in their pulse-like signals (Somerville et al., 1997). This observation suggests that fault-normal components would be a major consideration in near-fault related predictions relationships. However, it is not always straightforward to obtain the parameters that help in defining the fault-normal components. Akkar and Gülkan (2001, 2002) observed evidence that the maximum-velocity-direction component is as critical as the corresponding fault-normal component not only for records with forward directivity but also for cases with backward directivity effects (i.e. rupture propagating away from the recording site). This component appears to mask the essential characteristics (i.e. large velocities and/or pulse-like waveforms) that cause each near-fault case potentially critical for structural damage.

Table 1. Near-fault records - maximum velocity direction components

Earthquake	Record description	M_W	d** (km)	Site*	PGA (cm/s^2)	PGV (cm/s)	PGD (cm)
San Fernando 02/09/71	Pacoima dam (CDMG 279)	6.6	2.8	S_C	1277.7	118.9	33.9
Tabas, Iran 09/16/78	Tabas	7.4	3.0	S_D	839.1	122.3	91.8
Imperial Valley 10/15/79	Meloland Overp ff (CDMG 5155)	6.5	0.5	S_D	370.9	115.1	40.2
Imperial Valley 10/15/79	El Centro Array #7 (USGS 5028)	6.5	0.6	S_D	453.4	109.5	43.7
Imperial Valley 10/15/79	El Centro Array #5 (USGS 952)	6.5	1.0	S_D	361.9	92.9	57.0
Imperial Valley 10/15/79	El Centro Array #6 (CDMG 942)	6.5	1.0	S_D	424.9	117.4	68.2
Imperial Valley 10/15/79	El Centro Array #4 (USGS 955)	6.5	4.2	S_D	337.8	81.9	55.4
Imperial Valley 10/15/79	El Centro Diff Array (USGS 5165)	6.5	5.3	S_D	339.8	71.3	45.8
Imperial Valley 10/15/79	EC CO Center ff (CDMG 5154)	6.5	7.6	S_D	229.9	68.8	39.5
Coalinga 05/02/83	Pleasent Valley PP – Yard (USGS 1162)	6.5	8.5	S_D	581.8	60.2	8.7
Palm Springs 07/08/86	N Palm Spr P.O. (USGS 5070)	6.2	8.2	S_C	627.6	73.9	11.8
Superstition Hills (B) 11/24/87	(USGS 5051)	6.6	0.7	S_C	489.7	116.0	54.3
Loma Prieta 10/18/89	Saratoga W Valley Coll (CDMG 58235)	6.9	13.7	S_C	408.2	74.8	28.9
Erzincan 03/13/92	Erzincan Station (ERD)	6.8	2.0	S_D	466.2	139.8	58.4
Cape Mendocino 04/25/92	Petrolia (CDMG 89156)	7.1	9.5	S_D	695.2	98.3	34.6
Northridge 01/17/94	Sylmar – Hospital (CDMG 24514)	6.7	6.4	S_D	816.7	132.0	33.9
Northridge 01/17/94	Newhall (CDMG 24279)	6.7	7.1	S_D	722.5	121.4	33.3
Northridge 01/17/94	Pacoima Kagel Canyon (CDMG 24088)	6.7	8.2	S_C	511.2	56.5	10.9
Kobe 01/16/95	KJM	6.9	0.6	S_C	836.6	95.8	24.4
Kocaeli 08/17/99	Yarimca Station (BU)	7.4	3.28	S_D	328.6	90.8	65.5
Bolu-Düzce 11/12/99	Düzce Station (ERD)	7.1	8.23	S_D	439.5	93.9	118.1
ChiChi, 09/21/99	TCU067	7.6	1.11	S_D	555.3	108.4	203.5
ChiChi, 09/21/99	TCU129	7.6	2.21	S_D	829.6	78.7	128.2
ChiChi, 09/21/99	TCU065	7.6	2.49	S_D	768.2	147.4	244.3
ChiChi, 09/21/99	TCU076	7.6	3.17	S_D	417.7	87.6	128.7
ChiChi, 09/21/99	TCU075	7.6	3.38	S_D	324.4	116.2	172.2
ChiChi, 09/21/99	TCU071	7.6	4.88	S_D	551.28	85.04	303.74
ChiChi, 09/21/99	TCU072	7.6	7.87	S_D	465.09	85.56	209.90
ChiChi, 09/21/99	TCU078	7.6	8.27	S_D	470.65	45.75	175.61
ChiChi, 09/21/99	TCU089	7.6	8.33	S_C	343.72	43.38	159.81
ChiChi, 09/21/99	TCU122	7.6	9.22	S_D	225.15	50.16	76.75
ChiChi, 09/21/99	CHY024	7.6	9.26	S_D	300.88	62.78	97.52
ChiChi, 09/21/99	TCU120	7.6	9.87	S_C	219.07	63.16	110.95
ChiChi, 09/21/99	TCU079	7.6	10.95	S_D	561.34	69.24	244.13
ChiChi, 09/21/99	TCU138	7.6	11.29	S_D	197.75	40.47	51.71
ChiChi, 09/21/99	TCU084	7.6	11.40	S_C	914.99	127.75	355.73
ChiChi, 09/21/99	TCU074	7.6	13.75	S_D	605.69	76.08	225.98

* Site Conditions are determined according to UBC97 and NEHRP: S_C : 360 m/s – 750 m/s, S_D : 180 m/s – 360 m/s, S_E : < 180 m/s

** Shortest distance from the station to rupture surface.

4. GROUND MOTION PREDICTION RELATIONSHIP

For purposes of this study we adopt the form used by Joyner and Boore (1997) for the ground motion prediction relationship:

$$lnY = b_1 + b_2 (M - 6) + b_3 (M - 6)^2 + b_5 \ln r + b_V \ln (V_S / V_A) \quad (1)$$

$$r = (r_{cl}^2 + h^2)^{1/2} \quad (2)$$

Here Y is the ground motion parameter (peak horizontal acceleration (PGA) or pseudo spectral acceleration (PSA) in g for 5 percent damping, M is the (moment) magnitude, r_{cl} is the closest horizontal distance (or Joyner-Boore distance) from the station to a site of interest in km, V_S is the characteristic shear wave velocity for the station in m/s, b_1, b_2, b_3, b_5, h, b_V, and V_A are the parameters to be determined. In the expression h is a fictitious depth, and V_A is a fictitious velocity that is determined by regression. This form has been adopted by Gülkan and Kalkan (2002) in deriving a prediction equation on the basis of 47 records recorded prior to 1999 during earthquakes in Turkey. Table 2 lists the revised array of regression constants and the corresponding error terms for up to T = 2 s. The damping is 5 percent of critical, and the mean of two horizontal components is represented. The values for V_s assumed for soft soil, soil and rock sites were 200, 350, and 700 m/s, respectively. This simplification follows from the poorly known site geology for most strong motion stations in Turkey, but it is not wholly in agreement with the site descriptions of Table 1.

Table 2. Coefficients for ground motion prediction relation adopted from Gülkan and Kalkan (2002) (mean horizontal PGA and 5 percent damped S_a)

Period, s	b_1	b_2	b_3	b_5	b_V	V_A	h	$\sigma_{Ln(Y)}$
PGA	-0.682	0.253	0.036	-0.562	-0.297	1381	4.48	0.562
0.1	-0.139	0.2	-0.003	-0.553	-0.167	1063	3.76	0.621
0.11	0.031	0.235	-0.007	-0.573	-0.181	1413	3.89	0.618
0.12	0.123	0.228	-0.031	-0.586	-0.208	1501	4.72	0.615
0.13	0.138	0.216	-0.007	-0.59	-0.237	1591	5.46	0.634
0.14	0.1	0.186	0.014	-0.585	-0.249	1833	4.98	0.635
0.15	0.09	0.21	-0.013	-0.549	-0.196	1810	2.77	0.62
0.16	-0.128	0.214	0.007	-0.519	-0.224	2193	1.32	0.627
0.17	-0.107	0.187	0.037	-0.535	-0.243	2433	1.67	0.621
0.18	0.045	0.168	0.043	-0.556	-0.256	2041	2.44	0.599
0.19	0.053	0.18	0.063	-0.57	-0.288	2086	2.97	0.601
0.2	0.127	0.192	0.065	-0.597	-0.303	2238	3.48	0.611
0.22	-0.081	0.214	0.006	-0.532	-0.319	2198	1.98	0.584
0.24	-0.167	0.265	-0.035	-0.531	-0.382	2198	2.55	0.569
0.26	-0.129	0.345	-0.039	-0.552	-0.395	2160	3.45	0.549
0.28	0.14	0.428	-0.096	-0.616	-0.369	2179	4.95	0.53
0.3	0.296	0.471	-0.14	-0.642	-0.346	2149	6.11	0.54
0.32	0.454	0.476	-0.168	-0.653	-0.29	2144	7.38	0.555
0.34	0.422	0.471	-0.152	-0.651	-0.3	2083	8.3	0.562
0.36	0.554	0.509	-0.114	-0.692	-0.287	2043	9.18	0.563
0.38	0.254	0.499	-0.105	-0.645	-0.341	2009	9.92	0.562
0.4	0.231	0.497	-0.105	-0.647	-0.333	1968	9.92	0.604
0.42	0.12	0.518	-0.135	-0.612	-0.313	1905	9.09	0.634
0.44	0.035	0.544	-0.142	-0.583	-0.286	1899	9.25	0.627
0.46	-0.077	0.58	-0.147	-0.563	-0.285	1863	8.98	0.642
0.48	-0.154	0.611	-0.154	-0.552	-0.293	1801	8.96	0.653
0.5	-0.078	0.638	-0.161	-0.565	-0.259	1768	9.06	0.679
0.55	-0.169	0.707	-0.179	-0.539	-0.216	1724	8.29	0.71
0.6	-0.387	0.698	-0.187	-0.506	-0.259	1629	8.24	0.707
0.65	-0.583	0.689	-0.159	-0.5	-0.304	1607	7.64	0.736
0.7	-0.681	0.698	-0.143	-0.517	-0.36	1530	7.76	0.743
0.75	-0.717	0.73	-0.143	-0.516	-0.331	1492	7.12	0.74
0.8	-0.763	0.757	-0.113	-0.525	-0.302	1491	6.98	0.742
0.85	-0.778	0.81	-0.123	-0.529	-0.283	1438	6.57	0.758
0.9	-0.837	0.856	-0.13	-0.512	-0.252	1446	7.25	0.754
0.95	-0.957	0.87	-0.127	-0.472	-0.163	1384	7.24	0.752
1	-1.112	0.904	-0.169	-0.443	-0.2	1391	6.63	0.756
1.1	-1.459	0.898	-0.147	-0.414	-0.252	1380	6.21	0.792
1.2	-1.437	0.962	-0.156	-0.463	-0.267	1415	7.17	0.802
1.3	-1.321	1	-0.147	-0.517	-0.219	1429	7.66	0.796
1.4	-1.212	1	-0.088	-0.584	-0.178	1454	9.1	0.79
1.5	-1.468	0.995	-0.048	-0.582	-0.25	1479	10	0.788
1.6	-1.518	0.997	-0.047	-0.588	-0.24	1524	10.11	0.787
1.7	-1.668	0.991	-0.04	-0.58	-0.245	1644	9.79	0.789
1.8	-1.75	0.989	-0.039	-0.578	-0.231	1645	9.61	0.827
1.9	-1.793	0.995	-0.037	-0.589	-0.22	1714	9.74	0.864
2	-1.82	1.015	-0.062	-0.61	-0.23	1795	10	0.895

5. COMPARISON WITH RECENT GROUND MOTION PREDICTION EQUATIONS

The prediction relations given in Equation 1 with the coefficients in Table 2 was compared to those recently developed by Ambraseys et al. (1996), Boore et al. (1997), Campbell (1997), Sadigh et al. (1997), and finally Spudich et al. (1999). There exist some interpretation difficulties in transcribing condition among competing relations. The equations in Boore et al. (1997) and Ambraseys et al. (1996) divide site classes into four groups according to their shear wave velocities. The Campbell (1997) equations refer to alluvium (or firm soil), soft rock and hard rock. Sadigh et al. (1997) and Spudich et al. (1999) state that their equations are applicable for rock and soil sites. Since the backbone of the equation we use here is the equivalence of that of Boore et al. (1997), Figure 1 gives the comparison between this model and that adapted for Turkey for soft soil site condition. The parametric surfaces shown in this figure give insight by setting the magnitude unconstrained. These figures exhibit more rapid fall off of Boore et al. (1997) along both magnitude and distance axes in contrast to that utilized here. The attenuation of PGA for M_w = 7.4 earthquake for rock and soil sites is compared next with recent models in Figure 2. The measured data points from the 1999 Kocaeli event are marked on these figures to show how predictions fit with the observations. As it is inferred from the comparisons, ground motion amplitudes estimated by this model are generally consistent with the measured data points-not unexpected because it is based on those very points. The best estimate curves in these figures correspond to mean values. Plus and minus sigma curves of the model are also drawn to show the prediction band corresponding to the 84 percentile probability. The peak acceleration estimate given here has a standard deviation of 0.562, which is within an acceptable band when the scatter in the data about the regression line is considered. Comparing the predictions of different models, we observe that our ground motion predictions at short distances fall below the others. However, at longer distances this effect is reversed, resulting in higher predictions. Until additional records are obtained this situation can not be remedied.

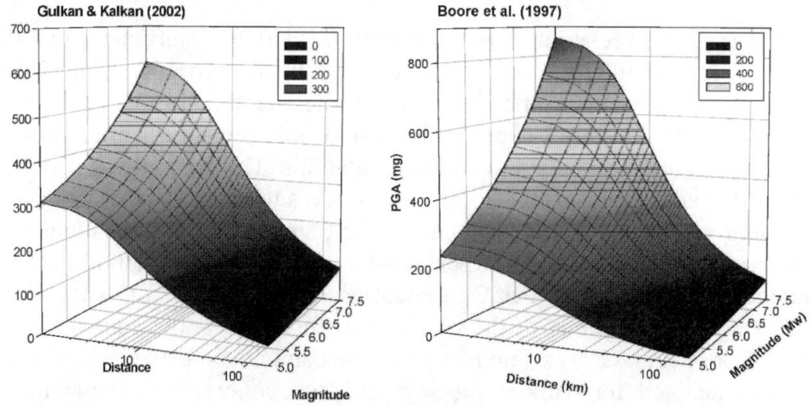

Figure 1. Comparison of our predictive model with that of Boore et al. (1997) on 3D plot for magnitude range of 5 to 7.5 and distance range of 0 to 150 km\

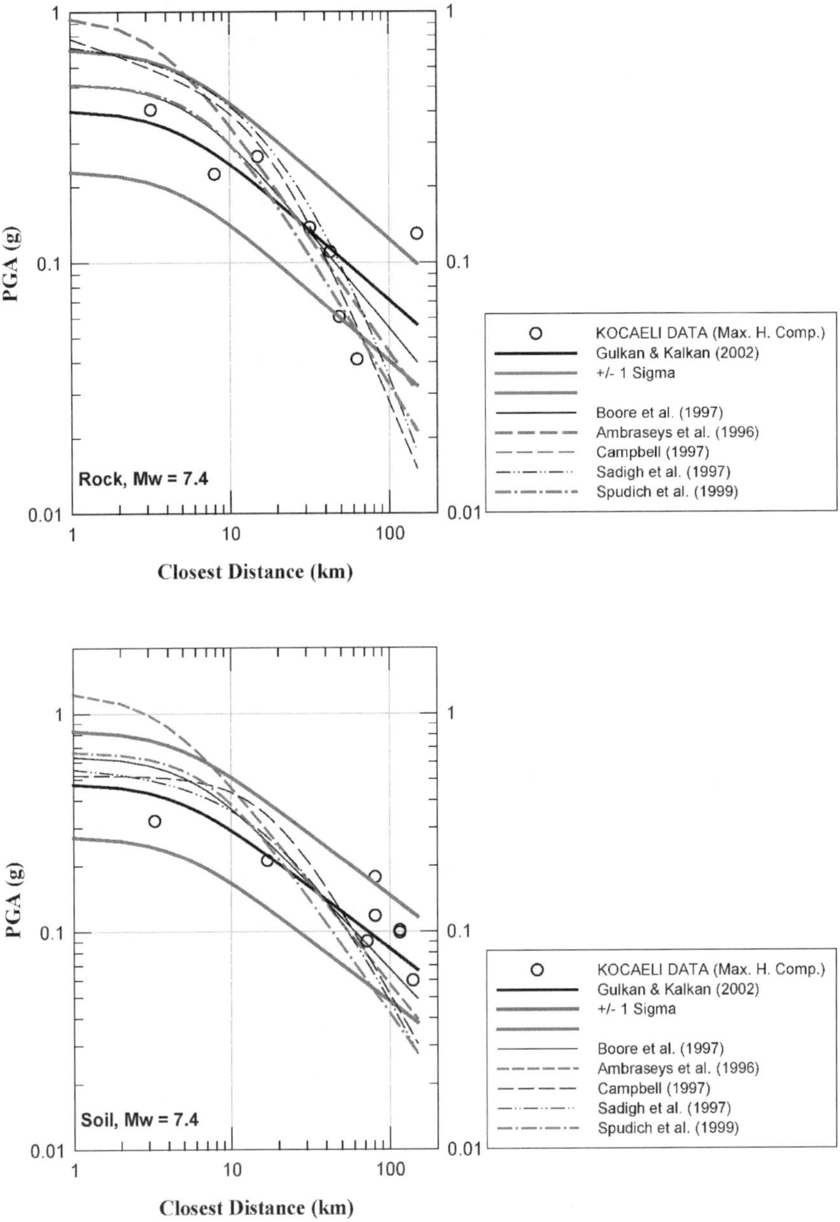

Figure 2. Curves of peak acceleration versus distance for a magnitude-7.4 earthquake at rock and soil sites

With S_a, the spectral acceleration at hand, it is a straightforward matter to convert to other spectral quantities. We believe that the use of S_v might lead to more useful results in highlighting the near-fault ground motion features on structural response. The pseudo-velocity spectrum has the ability to reflect the period of pulse-like near-fault waveform (T_p) that shifts the constant acceleration plateau of the spectrum, affecting both the elastic and inelastic structural behavior (Akkar and Gülkan, 2002). Pulse period also affects the structural behavior in terms of local and global near-fault ground motion demand (Alavi and Krawinkler, 2001). We did not attempt to classify the records with respect to their fault mechanisms. Limited to these observations, the discussions in the rest of the article will concentrate generally on these ranges.

Figure 3 displays the results of the prediction relationship for smoothed mean PSV in different distance and magnitude ranges. We have used the letter d instead of r_{cl} in these diagrams. These curves indicate that the velocity-sensitive region (intermediate periods where the response is related to ground velocity) seems to fall more quickly for increased magnitude values. This region is flatter and longer for $M_w = 6.5$ than the corresponding one for $M_w = 7.5$. The corner period that separates the acceleration-sensitive region (short periods where the response is related to peak ground acceleration) from the velocity-sensitive region ranges between 0.7-0.8 seconds. For comparison purposes we include the UBC curves that are discussed in the next section.

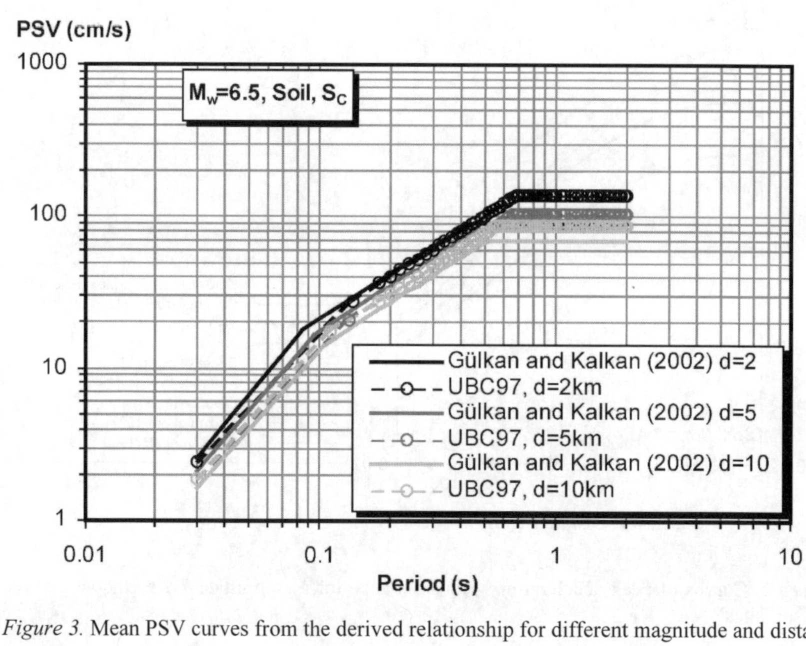

Figure 3. Mean PSV curves from the derived relationship for different magnitude and distance ranges.

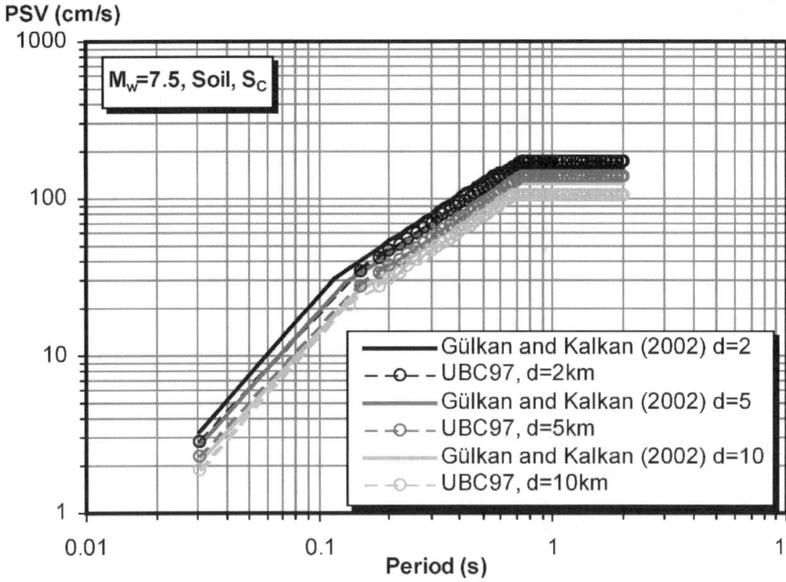

Figure 3. (Continued) Mean PSV curves from the derived relationship for different magnitude and distance ranges.

6. COMPARISONS WITH UNIFORM BUILDING CODE PROVISIONS

The UBC97 design spectrum is a function of two coefficients, namely C_a and C_v that control the acceleration- and velocity-sensitive spectral regions. The coefficient C_a represents the effective peak acceleration (EPA) of the ground whereas C_v is the spectral response of a T = 1 s system. It is related to the definition of effective peak velocity (EPV), as described by Kramer (1996). Evolutionary to other design spectra, this code incorporates near-source factors for the short and intermediate period spectral regions regarding the near-fault ground motion effects. These factors are designated as N_a and N_v, and are applicable only for most earthquake prone sites of the United States that are located within the near-field area of an active fault. The factor N_a modifies C_a to account for the near-fault effects in the short period range. Similarly, N_v amplifies C_v in order to modify the seismic hazard level of velocity sensitive region against near-fault earthquakes. The modification factors N_a and N_v attain values considering the fault activity and closest site distance to the fault. Specifically, UBC97 considers near-fault effects as relevant for $M_w \geq 6.5$ and distances less than 15 km. The N_a and N_v factors are assumed to change linearly between distances steps (e.g. between 2 to 5 km, and 5 to 10 km, etc.), and range from 1 to 1.5 and 1 to 2, respectively. Regardless of the fault seismicity, both N_a and N_v factors ignore the amplification of spectral ordinates for distances less than 2 km. We compare in Figure 4 the near-fault demand estimates of UBC97 with our findings. This comparison is relevant for the ongoing discussions that question the performance of UBC97 amplification factors in enveloping the demand of particular ground motions either for large or small magnitude events. We have fitted smooth curves to our spectral quantities via EPA concept as in the case of UBC97. Figure 4 displays the comparisons of our demand estimates with the corresponding UBC97 spectra in terms of spectral displacement, SD. The comparisons are exhibited for distances 2, 5, and 10 km as UBC97 defines specific amplification values at these distances, and assumes a linear amplification variation for the corresponding intermediate distances.

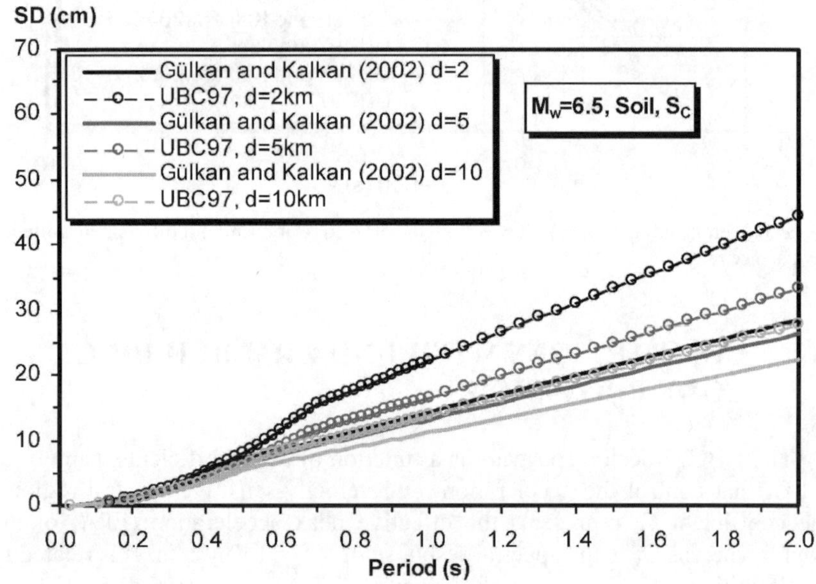

Figure 4. Comparison of displacement demand estimates with the corresponding UBC97 spectra.

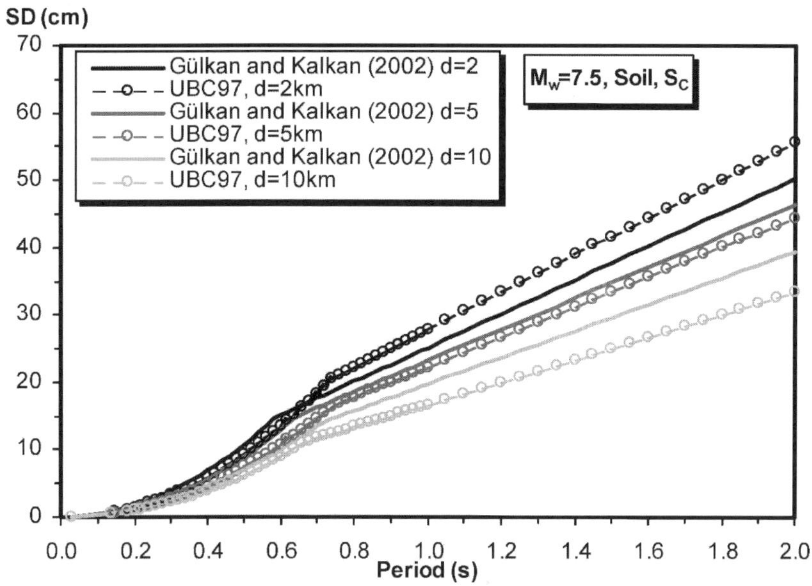

Figure 4 (Continued) Comparison of displacement demand estimates with the corresponding UBC97 spectra.

Our results fall below the UBC97 estimates for M_w = 6.5 at all distances ranges. The difference between these two approaches is less substantial for M_w = 7.5 and sites 2 km from the fault rupture beyond which range a reversal occurs. The discrepancy becomes less as the magnitude diminishes. For M_w = 6.5 and all distances greater than 2 km, UBC97 estimates yield far more conservative values. The two approaches overlap when the sites are at distances greater than 5 km from the fault rupture for M_w = 7.5. Our ground motion relationship appears to yield more conservative spectral values for magnitudes 7+ and distances greater than 5 km. We refrain from commenting on whether similar discrepancies would have resulted if we had used a different set of curves for comparison purposes.

In view of these results, we can calculate a set of distance-dependent amplification factors to provide a near-fault response spectrum. They are presented in the two-part Table 3. The first part of the table lists EPA and EPVA values at mean + 1 σ level at d = 10 km for different magnitude and site classes. Confined to the limits of our expression, we have accepted the d = 10 km EPA and EPV related acceleration (EPVA) values as the basis. The latter term is the spectral acceleration of a T = 1 s oscillator, and controls the intermediate periods of the spectrum (e.g., Kramer, 1996). The ratios of the other EPA and EPVA values at closer distances with respect to those at 10 km distance serve as the amplification factors (κ). For a given magnitude and site classification the ratio of two numbers calculated using Equation (1) becomes dependent only on distance. The entries for κ_{EPA} and κ_{EPVA} in the second part of the table have been derived from a different premise, but they may be considered to be analogous to N_A and N_V in the UBC97 lexicon.

Table 3. Effective peak acceleration (EPA), effective peak velocity related acceleration (EPVA) and computed amplification factors to construct the near-fault sensitive spectrum

M_w	EPA (g) (Mean + SD of 0.1-0.5 s)			EPVA (g) Mean + SD T = 1 s		
	Rock	Soil	Soft Soil	Rock	Soil	Soft Soil
6.5	0.30	0.35	0.43	0.40	0.45	0.52
7.0	0.34	0.40	0.49	0.56	0.62	0.72
7.5	0.38	0.44	0.54	0.71	0.79	0.91

	d^*									
	1	**2**	**3**	**4**	**5**	**6**	**7**	**8**	**9**	**10**
κ_{EPA}	1.66	1.57	1.47	1.37	1.29	1.22	1.15	1.10	1.05	1.00
κ_{EPVA}	1.29	1.28	1.25	1.21	1.18	1.14	1.10	1.07	1.03	1.00

Closest distance from the fault rupture (km)

7. DRIFT LIMITS

The displacement control in evaluating the structural performance against a seismic action is an important design objective. Recent structural performance assessment methods place priority on structural displacement capacity. The main goal of these DBD procedures is to achieve a satisfactory structural performance under the earthquake-induced deformation. They constitute an important topic in PBSD, and have persuaded engineers of their relevance especially in local displacement terms (i.e. inter-story drift) caused by the near-fault ground motions.

The drift spectrum concept was introduced by Iwan (1997). It utilizes the mechanical analog for a shear beam where velocity and displacement waves travel, causing shear deformations for which the corresponding structural engineering parameter is the inter-story drift. Chopra and Chintanapakdee (2001) have confirmed that by including a proper number of modes in direct integration for modal response, the base level drift can be estimated accurately. Recent research has shown that the drift spectrum can be obtained through spectral quantities more easily. Gülkan and Akkar (2002) and Heidebrecht and Rutenberg (2000) have derived drift formulations that depend on the fundamental mode contribution of spectral quantities. In this article, we focus on an elastic base level drift formulation by Gülkan and Akkar (2002):

$$D_r = 1.27 \frac{SD(\xi, T)}{h} \sin \frac{2\pi h}{Tc} \tag{3}$$

The term h is the story height and the constant 1.27 is the modal participation factor of the fundamental mode when a half sine displacement pattern is used to represent the deformed shape, equal to $4/\pi$. In Equation (3), T represents the fundamental period and c corresponds to the shear wave velocity traveling along the structure, generally in the range 120-180 m/s for most structural framing systems. We use c = 120 m/s and h = 3 m in the following.

We test the applicability of Equation (3) for calculation of drift in structural systems situated in the near-field in two ways: First, the spectral acceleration is estimated using Equation (1) and Table 2 for each of the 40 entries into Table 1 at the same magnitude, distance and approximate site classification. This is then converted into SD through the spectral relationship. Equation (3) is next evaluated for the drift estimate for that record. For comparison, a time history analysis is made for each record to calculate the exact drift in a framed system with the same fundamental period. The ratios of the drift estimates calculated this way are displayed in Figure 5. It is immediately noted that Equation (3) errs on the unsafe side in this comparison. There are two principal reasons for this. First, the derived spectral displacement is based on a prediction relationship whose data set is fundamentally foreign to the data set in Table 1. An issue is thereby raised in relation to the application of importing different datasets for estimation of drift. The other reasons are the omission of higher modes in the calculation of drift and site character.

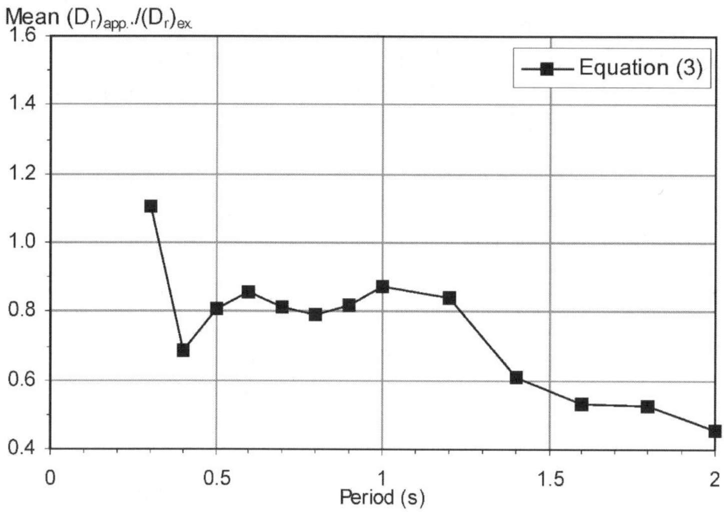

Figure 5. Estimated vs. Calculated Drift

A more rigorous test for the accuracy of Equation (3) is possible by utilizing the calculated spectral displacement for each record in Table 1, and comparing the approximate drift with its counterpart found from a time-history analysis. This way, the uncertainty caused by using the prediction relationship can be eliminated. We display the results in Figure 6.

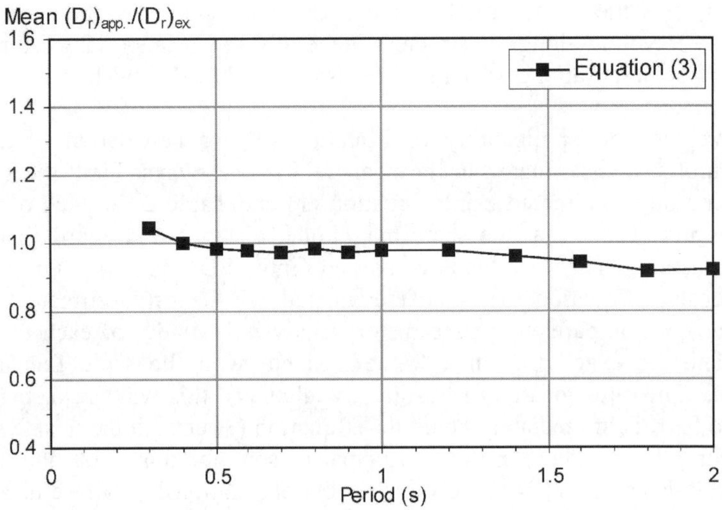

Figure 6. Estimated Drift Using Exact SD in Equation (3)

The information conveyed by Figure 6 is a strong confirmation of the prescience displayed by Equation (3) when the correct spectral displacement is used to estimate the drift. When the spectral displacement itself is an estimate, then the accuracy of Equation (3) is seriously compromised. Figure 6 suggests that higher mode effects are really not all that important in this context, and the pulse effect is subsumed in the spectral displacement.

8. INELASTIC DRIFT

Inelastic behavior of structures is likely in the case of near-fault ground motions that exhibit large deformation demands. This requires structural ductility as determined through the inelastic analysis. To this end, emphasis should be placed on the definition of inelastic drift limits for more realistic assessment of structural performance. Converting the elastic spectral displacement into inelastic range can be a one way of obtaining the inelastic base level drift from Equation (3). The inelastic spectral displacements are estimated by a constant ductility inelastic displacement equation (Miranda, 2000). This expression is given in Equation (4).

$$C_{\mu} = \frac{SD_{inelastic}}{SD_{elastic}} = \left[1 + (\frac{1}{\mu} - 1)\exp(-12T\mu^{-0.8})\right]^{-1} \tag{4}$$

In the above equation, C_{μ} designates the inelastic to elastic displacement modification factor that is dependent on period (T) and displacement ductility (μ). Equation (4) is valid for elasto-plastic hysteretic behavior.

In order to increase the versatility of the drift estimations, we modified Equation (3) by making use of code expressions such as $T=aH^b$. Among such expressions, we have utilized a recent proposal made by Chopra and Goel (2000). This equation reflects the building fundamental period for DBD approach. Equation (5) gives the upper bound regression results of the proposed regression formulas for reinforced concrete (RC) frames.

$$T_{RC} = 0.067H^{0.9} \tag{5}$$

In the above expression H is in meters. The expression for inelastic base level drift limit is given in Equation (6) where SD for near-fault conditions is calculated by the smoothed response spectrum by Gülkan and Kalkan (2002). The equation below can reflect distance and magnitude dependency of near-fault ground motions when appropriate spectral displacement is used.

$$[D_r(\xi, T, d, M_w)]_{inelastic} = C_{\mu} 1.27 \frac{SD(\xi, T, d, M_w)}{h} \sin\frac{\pi\sqrt[b]{ah}}{2\sqrt[b]{T}} \tag{6}$$

Constant ductility, inelastic drift limits for $\mu = 2$, 4 and 6 are displayed in Figure 7. The results are computed for RC frames by employing Equations (5) and (6) at 2 and 10 km. As in other frames the curves are drawn for periods between 0.3 to 2 s. The plots emphasize the frequency dependency of drift that is generally overlooked by the seismic design codes. The drift increases until the end of the constant acceleration spectral region due to the increase in seismic energy. This demand gradually decreases in the velocity sensitive region, indicating that the stabilized earthquake input energy can be balanced more easily. These observations derive from principles of structural dynamics. In the long period range, the displacement ductility does not cause large differences in base level drift for a given distance. These curves describe high drift demands for large magnitude and short distance cases. Within the general perspective of these curves, one may conclude that the drift levels of UBC97 and NEHRP may fall short for close distances and large magnitude events ($M_w > 7$). The drift limits of these provisions are on the safe side for larger distances and smaller magnitudes. The added value of serendipitous arrays established within close distances to active faults for engineering use is evident.

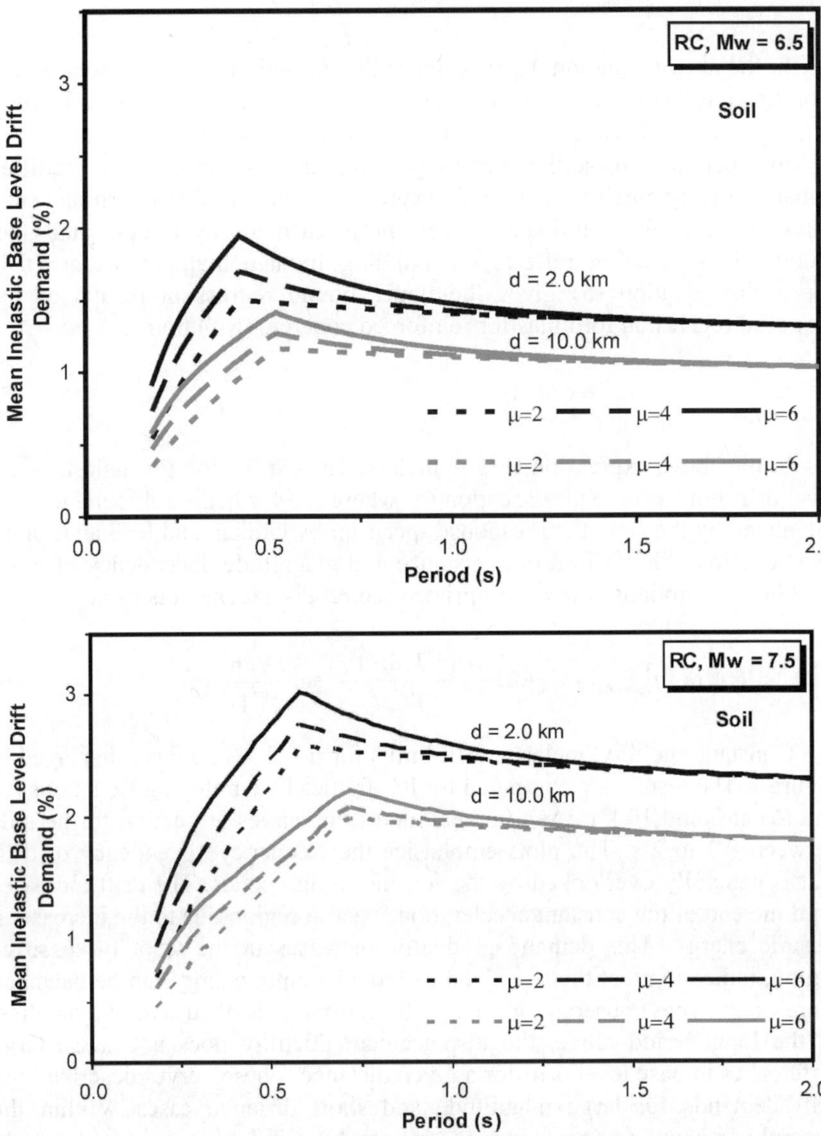

Figure 7. Distance and magnitude dependent inelastic base level drift limits for RC frame structures

9. CONCLUSIONS

This study has dealt with estimating elastic and inelastic drift in framed systems in near-fault distances. This has been done by a simplified expression that uses the spectral displacement as its prime input. As spectral displacement is calculated using only the acceleration record effects of different processing and filtering processes are minimized. When spectral displacement is estimated by making use of ground motion relations that do not adequately represent the tectonic framework then estimates of drift can be quite unsafe. When the exact spectral displacement is substituted in Equation (3) estimates for drift are very good, indicating the applicability of the expression for near-field conditions.

The results are expected to highlight some important aspects of near-fault ground motions in terms of engineering issues. The derived spectral quantities indicate that UBC97 can underestimate the short period near-fault drift demand. This observation is valid for large magnitude events (i.e. $M_w > 7$) and distances less than 5 km from the fault rupture. The drift limits computed by using the proposed spectral values display a frequency dependent character of local demands. According to our findings, for sites located in the very near vicinity of active faults, the code drift requirements underestimate the actual demand. A corollary may be appended here to suggest that loss estimates based on building stock properties, as reflected in the drift limits corresponding to various damage states, must not be judged independently of ground motion intensity measures applicable to where that building stock is situated.

ACKNOWLEDGMENTS

Appreciation is extended to the Graduate School of Science at Middle East Technical University for the grant to the second author that enabled him to work on this project.

REFERENCES

1. Akkar, S. and P. Gülkan (2001). "Near-field earthquakes and their implications on seismic design codes," *Earthquake Engineering and Research Center, Report 2001-01*, Department of Civil Engineering, Middle East Technical University, Ankara, Turkey.
2. Akkar, S. and P. Gülkan (2002). "Spectral velocity and displacement in performance based earthquake resistant design," *Proc. 12th European Conference on Earthquake Engineering-CD ROM*, Paper No. 137, London.
3. Akkar, S. and Gülkan, P. (2002). "A critical examination of near-field accelerograms from the Sea of Marmara region earthquakes," *Bulletin of the Seismological Society of America* 92(1), 428-447.

4. Akkar, S., D.M. Boore, and P. Gülkan (2004). "An evaluation of the strong ground motion recorded during the May 1, 2003 Bingöl, Turkey, earthquake," in press, *Journal of Earthquake Engineering*.

5. Alavi, B. and H. Krawinkler (2001). "Effects of near-fault ground motions on frame structures," *The John A. Blume Earthquake Engineering Center, Report No*. 138, Stanford University, California.

6. Ambraseys, N.N., K.A. Simpson and J. Bommer (1996). "Prediction of horizontal response spectra in Europe," *Earthquake Engineering and Structural Dynamics*, 25(4), 371-400.

7. Bolt, B.A. (1996). *From earthquake acceleration to seismic displacement, The Fifth Mallet-Milne Lecture*, Wiley.

8. Boore, D.M., W.B. Joyner and T.E. Fumal (1997). "Equations for estimating horizontal response spectra and peak acceleration from Western North American earthquakes: A Summary of Recent Work," *Seismological Research Letters*, 68(1), 128-153.

9. Campbell, K.W. (1997). "Empirical near source attenuation relationships for horizontal and vertical components of peak ground acceleration, peak ground velocity, and pseudo-absolute acceleration response spectra," *Seismological Research Letters*, 68(1), 154-179.

10. Chopra, A.K. and R.K. Goel (2000). "Building period formulas for estimating seismic displacements," *Earthquake Spectra*; 16(2): 533-536.

11. Chopra, A.K. and C. Chintanapakdee (2001). "Drift spectrum vs. modal analysis of structural response to near-fault ground motions," *Earthquake Spectra*; 17(2):221-234.

12. Douglas, J.A. (2001). "Comprehensive worldwide summary of strong motion attenuation relationships for peak ground acceleration and spectral ordinates (1969-2000)," *Engineering Seismology and Earthquake Engineering (ESEE), Report 01-1,* Civil Engineering Department, Imperial College of Science, Technology and Medicine, London.

13. Federal Emergency Management Agency, FEMA (1997). NEHRP recommended provisions for seismic regulations for new buildings and other structures. *Report No. FEMA-302*, Federal Emergency Management Agency, Washington, DC.

14. Gülkan, P. and S. Akkar (2002). "A simple replacement for the drift spectrum," *Engineering Structures*; 24(11):1477-1484.

15. Gülkan, P., and Kalkan, E. (2002). "Attenuation modeling of recent earthquakes in Turkey," *Journal of Seismology,* 6, 397-409.

16. Hall, J.F., T.H. Heaton, M.W. Halling and D.J. Wald (1995). "Near-source ground motion and its effects on flexible buildings," *Earthquake Spectra*; 11(4):569-605.

17. Heidebrecht, A.C. and A. Rutenberg (2000). "Applications of drift spectra in seismic design," *Proc. 12th World Conference on Earthquake Engineering*, Paper No. 290, Auckland, New Zealand.

18. International Conference of Building Officials, ICBO (1997). Uniform Building Code, Whittier, California.

19. Iwan, W.D. (1997). "Drift spectrums: measure of demand for earthquake ground motions," *Journal of Structural Engineering, ASCE*; 123(4): 397-404.

20. Kramer, S.L. (1996). *Geotechnical Earthquake Engineering.* Prentice-Hall Inc.: New Jersey.

21. Miranda, E. (2000). "Inelastic displacement ratios for displacement-based earthquake resistant design," *Proc. of the 12th World Conference on Earthquake Engineering,* Paper No. 1096, New Zealand.

22. Sadigh, K., C.Y. Chang, J.A. Egan, F. Makdisi and R.R. Youngs (1997). "Attenuation relationships for shallow crustal earthquakes based on California strong motion data," *Seismological Research Letters*, 68(1) 180-189.

23. Somerville, P.G., N.F. Smith, R.W. Graves and N.A. Abrahamson (1997). "Modification of empirical strong ground motion attenuation relations to include the amplitude and duration effects of rupture directivity," *Seismological Research Letters*; 68(1):199-222.

24. Spudich, P., W.B. Joyner, A.G. Lindh, D.M. Boore, B.M. Margaris and J.B. Fletcher (1999). "SEA99: A revised ground motion prediction relation for use in extensional tectonic regimes," *Bull. Seismological Soc. of Am.,* 89(5), 1156-1170.

IMPACT OF NEAR-FAULT PULSES ON ENGINEERING DESIGN

Helmut Krawinkler[1], Babak Alavi[2], and Farzin Zareian[1]
[1]*Stanford University, Stanford, California, USA*
[2]*Exponent Failure Analysis Associates, Wood Dale, Illinois, USA*

Abstract: This contribution attempts to shed light on the important issues that affect the response of frame structures to near-fault ground motions with forward directivity. There is evidence that ground shaking near a fault rupture is characterized by a small number of pulses with high input energy. Pulse response characteristics are utilized to describe behavior attributes of structures subjected to near-fault ground motions. The ultimate objective is to develop improved design procedures and guidelines that take advantage of the equivalent pulses. Relationships are developed that relate equivalent pulse properties to earthquake magnitude and distance from the fault rupture. The results suggest that using near-fault factors introduced in current seismic codes may not provide consistent protection against near-fault effects. The need exists for an explicit consideration of the near-fault effects in seismic hazard analysis.

Keywords: near-fault, near-field, pulse, frame structures, seismic demands, ground motions

1. INTRODUCTION

Ground motions close to a ruptured fault have become the subject of much concern in recent years. There is evidence indicating that ground shaking near a fault rupture may be characterized by a short-duration impulsive motion that exposes structures to high input energy at the beginning of the record. This pulse-type motion is particular to the "forward" direction, where the fault rupture propagates towards the site at a velocity close to the shear wave velocity, causing most of the seismic energy to arrive at the site within a short time. The radiation pattern of the shear dislocation of the fault causes the pulse to be mostly oriented perpendicular to the fault, i.e., the fault-normal component of the motion is more severe than the fault-parallel component (Somerville, 1998). The large differences between the fault normal and fault parallel components are evident in the spectra of various components of the 1994 Rinaldi Receiving Station record shown in Fig. 1. As

P. Gülkan and J.G. Anderson (eds.), Directions in Strong Motion Instrumentation, 83–106.

can be seen, rotating the components by 45° does not decrease the spectrum of the larger component by much. The near-fault phenomenon requires consideration in the performance evaluation of structures that are located in the near-fault region, which is usually assumed to extend about 10 to 15 km from the seismic source.

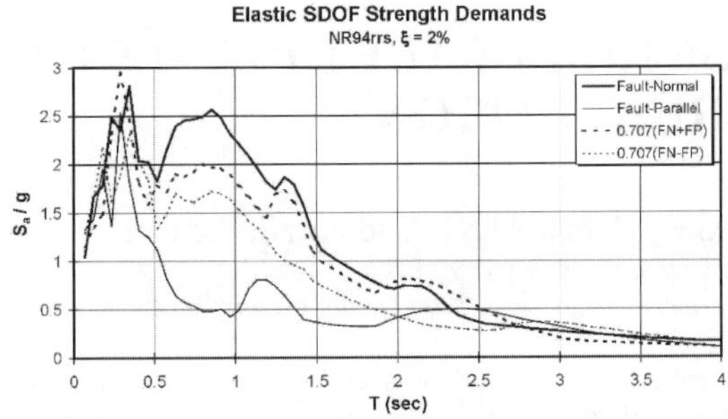

Figure 1. Acceleration spectra of components of a near-fault record (Alavi and Krawinkler, 2001, 2004a)

Near-fault effects have been known for several decades (e.g., Anderson and Bertero, 1978) but have become a focus of research only after the 1994 Northridge earthquake (e.g., Iwan and Chen, 1994, and Hall et al. 1995). The list of references at the end of this paper provides representative examples of research results published over the last ten years. A summary of the near-fault seismological phenomenon is presented in Somerville et al. (1997). General characteristics of near-fault ground motions are discussed extensively in the report by Stewart et al. (2001). In this section and the following one, comments are made as needed to emphasize the large intensity of these ground motions and to understand the dependence of seismic demands on the characteristics of these motions. The discussion is concerned only with the fault normal component of near-fault ground motions with forward directivity. The comments made here are based primarily on the work summarized in the references Alavi and Krawinkler (2000, 2001, 2004a, and 2004b).

The latter parts of the paper are concerned with the impact of near-fault effects on design and the importance of considering near-fault effects explicitly in seismic hazard analysis. The reader is referred to the growing literature on near-fault effects for more in-depth discussions on structural design issues. For instance, Hall et al. (1995) employed wave propagation theory to study the response of a continuous shear building to pulse-type ground motions. They warned about the damaging effects of near-fault ground motions and the inadequacy of current code provisions to address the problem effectively. Iwan (1997) utilized a similar elastic shear building to obtain the "drift spectrum" (maximum story drift plotted vs. structure period) as a measure of seismic demand for MDOF structures subjected to near-fault ground motions with pulse-type characteristics. He showed that even for elastic structures near-fault effects cannot be accounted for simply by multiplying the code base shear coefficient by a near-fault factor that is constant beyond a relatively short period (as in UBC'97). Rodriguez-Marek (2000) presents interesting work on near-fault pulse attenuation relationships and on the site soil dependence of near-fault ground motions.

The large intensity of the near-fault ground motions is illustrated in Fig. 2, which shows individual acceleration (elastic strength demand) spectra of fault normal components of six recorded motions with a relatively short pulse period (between 1 and 2 seconds). Also shown is the median spectrum as well as a NEHRP MCE spectrum for Los Angeles. The latter is obtained from the USGS web site for a 2/50 hazard level. For reasons discussed later, it may not be appropriate to get statistical measures for spectra of near-fault records, but nevertheless, a comparison of the median spectrum and the MCE spectrum provides an assessment of the intensity. The median near-fault spectrum is much larger than the MCE spectrum in the range of the pulse period of the records, which indicates very large "average" demands, even though the earthquakes causing these motions are only of magnitude 6.2 to 7.0.

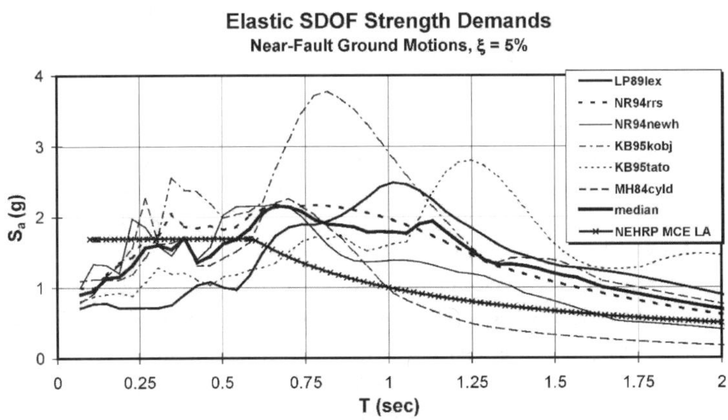

Figure 2. Acceleration spectra of several near-fault records, fault-normal component

A similar perspective of the intensity of near-fault ground motions is obtained from Fig. 3, which presents velocity response spectra of near-fault records and ordinary "design ground motions". The light solid thick line (denoted as 15-D*(mean)) represents the mean velocity spectrum of a set of ordinary ground motions whose individual spectra resemble the UBC'97 soil type S_D spectrum. The figure illustrates significant variations in the response of elastic SDOF systems to near-fault ground motions. It also indicates that near-fault ground motions impose seismic demands on structures that may be several times those imposed by the mean of design level "ordinary" ground motions. Clearly, ordinary ground motions also show considerable scatter about the mean, but a comparison of the design spectrum with the median of the near-fault spectra is telling.

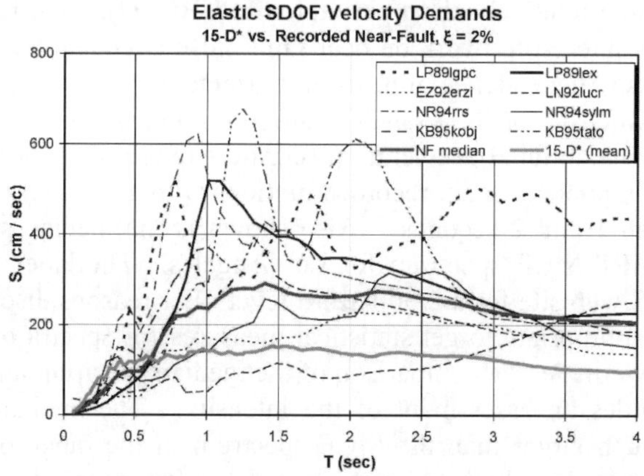

Figure 3. Velocity spectra of several near-fault records (Alavi & Krawinkler, 2001)

2. PULSE CHARACTERISTICS OF NEAR-FAULT GROUND MOTIONS

What sets near-fault ground motions with forward directivity apart from ordinary ones is not only their intensity but also their frequency content. The fault normal components exhibit pulse-like characteristics that can be defined by a pulse type, a pulse period, T_p, and an effective pulse velocity v_{eff}. Different investigators have matched different pulse types to recorded motions, all with many successes and some nonsuccesses. The pulse patterns are evident but not always consistent, and the consequence of selecting different pulse types is often smaller than the consequence of forcing pulse characteristics onto a near-fault ground motion. Alavi and Krawinkler (2001) have found that in many cases the simple square pulse shown in Fig. 4 is adequate to characterize near-fault ground motions and the dependence of seismic demands on such ground motions. Being able to represent a ground motion by only two parameters (T_p and $v_{eff} = a_{g,max}T_p/4$) has the advantage that comprehensive sensitivity studies can be performed for various T/T_p ratios and intensity levels.

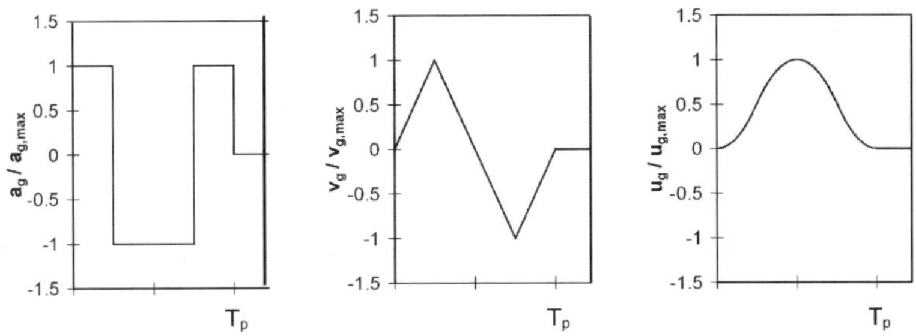

Figure 4. Acceleration, velocity, and displacement time histories of square pulse P2

2.1. Summary of Pulse Characteristics

With respect to pulse-like ground motion characteristics and their effect on elastic structural response of SDOF and MDOF systems, the following conclusions are drawn in the Alavi and Krawinkler references:

- A pulse representation is reasonable (but not necessarily accurate) for the fault-normal component of near-fault records with forward directivity for earthquakes of relatively large magnitude. In many cases the square pulse shown in Fig. 4 is appropriate, and in some cases a half-pulse (P1) or a double pulse (P3) is more appropriate. The pulse period T_p for a ground motion was obtained in these studies by inspection of the velocity and displacement time history traces and by matching the elastic velocity spectrum of the record with the spectra of pulses, as shown in Fig. 5. A more rigorous approach to the estimation of the pulse period is discussed in Section 2.2.

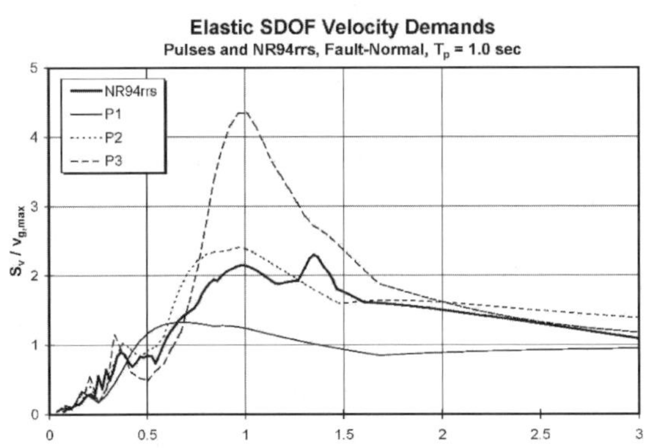

Figure 5. Matching of velocity spectra of pulses and ground motion (Alavi and Krawinkler, 2001)

- The pulse representation breaks down, and so does the dominance of the fault-normal component, as the magnitude approaches (from above) a value of 6.0 or smaller.

- For a set of 14 recorded and 6 generated near-fault records, for which a pulse period was identified by pulse matching, a regression of T_p against moment magnitude resulted in the following relationship:

$$\log_{10}T_p = -1.76 + 0.31\ M_w \quad (1)$$

There is significant scatter to the data on which this relationship is based, but there are good reasons to accept the concept that T_p is dependent on magnitude. Somerville showed that the period of the pulse contained in a near-fault ground motion is mostly affected by slip rise time, which is defined as the duration of slip at a given point on the fault (Somerville, 1998, Somerville et al., 2003). Clearly, the rise time is dependent on magnitude. If T_p is magnitude dependent, then records with vastly different T_p should not be statistically evaluated (i.e., by presenting median spectra) because they likely come from events with very different magnitude and have very different frequency characteristics. Considering the large scatter in T_p, there is some justification to present statistical data for records with a moderate variation in T_p, such as the median spectrum presented in Fig. 2.

- For the same set of records, a regression of v_{eff} against magnitude and distance resulted in the following relationship:

$$\log_{10}v_{eff} = -2.03 + 0.65\ M_w - 0.47\ \log_{10}R \quad (2)$$

This relationship indicates a strong increase in effective pulse velocity with magnitude. It was obtained from a relatively small data set, with none of the earthquakes of a magnitude greater than 7.0. It should be reasonable for the range of magnitudes between 6.0 and 7.0, but should not be extrapolated to larger magnitudes. It is anticipated that the effective velocity saturates with magnitude and may even decrease for large magnitudes for which T_p becomes long. More data points are needed to establish a relationship with confidence. A rigorous procedure for estimating the effective pulse velocity is discussed in Section 2.3.

- In almost all cases, the effective pulse velocity that minimized the error in pulse representation is very close to the PGV (Peak Ground Velocity) of the near-fault record. Thus, the PGV may be used as a surrogate for v_{eff} in the equation above.
- The pulse-like nature of the near-fault ground motions reflects frequency characteristics that will give rise to seismic demands that are strongly dependent on the ratio T/T_p. This is illustrated in Fig. 6 for elastic MDOF systems. The figure shows normalized story drift (interstory displacement over story height normalized by pulse max. displacement over structure height) profiles over the height for 20-story frame structures whose first mode period T takes on values corresponding to specific T/T_p values. The profiles change radically with T/T_p, and show a clear traveling wave

effect for $T/T_p > 1$. For large T/T_p values the maximum elastic story drifts occur in the upper portion of the structure. Since elastic story drifts are proportional to story shears, this implies that maximum story shears occur in these stories, which in turn means that these stories will yield much earlier than bottom stories.

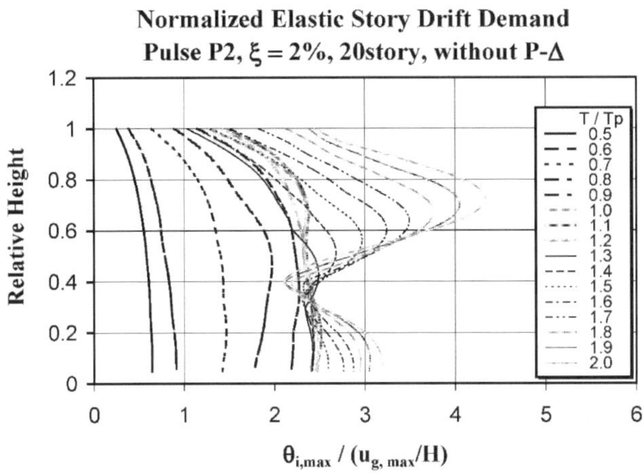

Figure 6. Elastic story drift demands for MDOF systems, various ratios of T/T_p, square pulse

- An SRSS combination of modal quantities does not represent accurately the traveling wave effects for MDOF structures with $T/T_p > 1.0$.
- For MDOF structures that respond inelastically to near-fault ground motions, the seismic demands depend strongly on the strength of the structure and on the ratio T/T_p. This is discussed in more detail in Section 4.

For the reasons enumerated here, and many more, there is a need to separate near-fault ground motions from ordinary ones, particularly as long as a single intensity measure, such as S_a, is employed to describe the seismic hazard. This separation is necessary to find out whether these ground motions do indeed affect seismic demands in a sufficiently different way to merit special consideration. This issue is discussed in Section 5.

2.2. Rigorous Approach for Estimation of Pulse Period

The process of identifying the pulse period by matching ground motion spectra with pulse spectra requires judgment and is based mostly on visual inspection. As an alternative, the following procedure is recommended as a rigorous approach for estimating T_p by finding the "best match" between the responses of an MDOF system with period T subjected to the near-fault record and a pulse, with a comprehensive sweep of T/T_p ratios. "Best" is defined here as the T_p that minimizes the "error" in predicting the elastic drift response of a representative structure to a near-fault ground motion from an equivalent pulse. A 20-story frame is used as a representative structure, and the "error" is defined as the sum of the square of the differences in story drifts obtained from the ground motion response and the pulse response, see Fig. 7. The process of obtaining the pulse period is as follows:

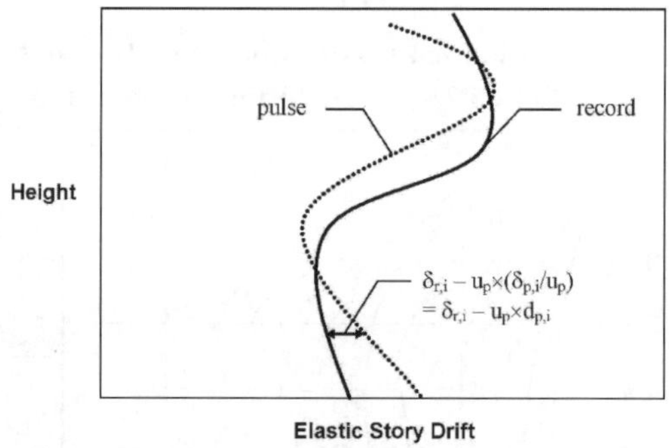

Figure 7. Drift matching for individual record and equivalent pulse

- For a pulse (defined by period T_p and effective pulse displacement $u_{p,eff}$) compute the elastic story drift in each story i, $\delta^{\alpha}_{p,i}$, for the 20-story structure with a first mode period $T_1 = \alpha T_p$, for $\alpha = 0.5$ to 2.0 with an increment in α of 0.1. The story drift ratios for the pulse responses ($\theta_{p,i} = \delta_{p,i}/h$) for the 16 α values are shown in Fig. 6.
- For near-fault ground motion:
  a. Select a safe upper and lower bound for T_p (e.g., 0.5 to 6.0 seconds)
  b. For a test value of T_p (starting with lower bound (e.g., 0.5 sec.)):
     - For the ground motion record, compute the elastic story drift in each story i, $\delta^{\alpha}_{r,i}$, for the 20-story structure with a first mode period $T_1 = \alpha T_p$, for $\alpha = 0.5$ to 2.0 with an increment in α of 0.1. This implies 16 T_1 cases for each test value of T_p.
     - For each T_1 (α) case:

       i. For each story i compute the difference in story drift between the ground motion response and the pulse response, i.e.,

$$diff = \delta^{\alpha}_{r,i} - \delta^{\alpha}_{p,i} = \delta^{\alpha}_{r,i} - u_{p,test}(\delta^{\alpha}_{p,i}/u_{p,test}) = \delta^{\alpha}_{r,i} - u_{p,test} d^{\alpha}_{p,i} \quad (3)$$
$$(d^{\alpha}_{p,i} = \delta^{\alpha}_{p,i}/u_{p,test})$$

Normalize this difference as follows:

$$e_i^\alpha = \frac{\left|\delta_{r,i}^\alpha - u_{p,test} d_{p,i}^\alpha\right|}{\delta_{r,i}^\alpha} \tag{4}$$

ii. The normalized difference is squared and summed up over the height (20 values), i.e.,

$$\varepsilon^\alpha = \sum_{i=1}^{20} (e_i^\alpha)^2 = \sum_{i=1}^{20}\left(1 - u_{p,test}\frac{d_{p,i}^\alpha}{\delta_{r,i}^\alpha}\right)^2 = 20 - 2u_{p,test}\sum_{i=1}^{20}\left(\frac{d_{p,i}^\alpha}{\delta_{r,i}^\alpha}\right) + u_{p,test}^2\sum_{i=1}^{20}\left(\frac{d_{p,i}^\alpha}{\delta_{r,i}^\alpha}\right)^2$$

- The so defined error is calculated for each T_1 (α) and is summed over all 16 α values, i.e.,

$$\varepsilon = \sum \varepsilon^\alpha = 320 - 2u_{p,test}\sum_{i=1}^{320}\left(\frac{d_{p,i}^\alpha}{\delta_{r,i}^\alpha}\right) + u_{p,test}^2\sum_{i=1}^{320}\left(\frac{d_{p,i}^\alpha}{\delta_{r,i}^\alpha}\right)^2 \tag{5}$$

c. The combined error is calculated for each test value of T_p and is plotted versus T_p. If this plot shows a clear valley (low point), it is postulated that the period associated with this valley is the pulse period for the near-fault record.

d. It further can be postulated that error minimization provides an estimate of the effective pulse displacement $u_{p,eff}$, i.e.,

$$\frac{\partial \varepsilon}{\partial u_{p,test}} = -2\sum_{i=1}^{320}\left(\frac{d_{p,i}^\alpha}{\delta_{r,i}^\alpha}\right) + 2u_{p,eff}\sum_{i=1}^{320}\left(\frac{d_{p,i}^\alpha}{\delta_{r,i}^\alpha}\right)^2 = 0 \tag{6}$$

$$u_{p,eff} = \sum_{i=1}^{320}\left(\frac{d_{p,i}^\alpha}{\delta_{r,i}^\alpha}\right) \Bigg/ \sum_{i=1}^{320}\left(\frac{d_{p,i}^\alpha}{\delta_{r,i}^\alpha}\right)^2 \tag{7}$$

In most cases this process resulted in an unambiguous identification of a pulse period, and in some (but not many) cases the process was less successful. Typical examples are shown in Fig. 8. The first graph shows a clear pulse period, whereas the second graph shows two error valleys, in which case the choice of T_p becomes a matter of judgment.

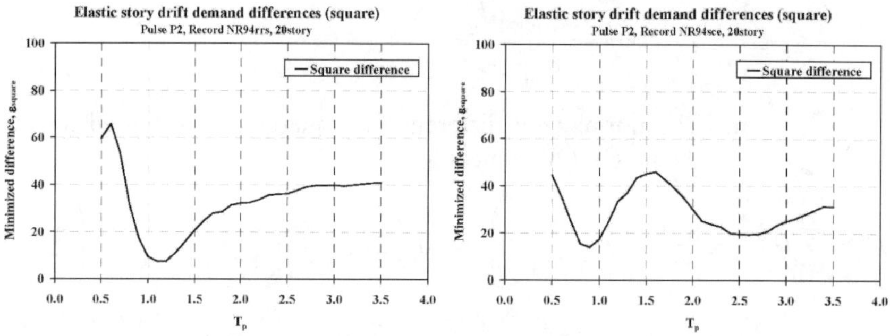

Figure 8. Examples of pulse period identification from error analysis

The process of estimating the effective pulse displacement (pulse intensity), as mentioned in d, above, is not the best choice because it utilizes only elastic structural systems. A better approach, which is based on the response of inelastic systems, is summarized next.

2.3. Rigorous Approach for Estimation of Pulse Velocity

To estimate the intensity of the equivalent pulse, a rigorous process can be employed (Alavi and Krawinkler, 2001, 2004a) whose objective is to minimize, for 20-story inelastic MDOF systems, the differences between the maximum story ductility demands obtained from the near-fault record and the equivalent pulse. In the following discussion the maximum acceleration, $a_{g,max}$, and velocity, $v_{g,max}$, of the equivalent pulse are referred to as the effective acceleration, a_{eff}, and velocity, v_{eff}. In summary, the procedure includes the following steps:

1. For the near-fault record, and using its previously determined T_p value, compute relationships between the MDOF base shear coefficient ($\gamma = V_y/W$) and the maximum story ductility, μ_{max}, using values of T/T_p = 0.375, 0.5, 0.75, 1.0, 1.5, 2.0, and 3.0. Figure 9 shows a representative example for a ground motion with T_p = 1.0 sec.
2. For the equivalent pulse compute relationships between the MDOF pulse base shear coefficient ($\eta = V_y/(m.a_{g,max})$) and the maximum story ductility, μ_{max}, using values of T/T_p = 0.375, 0.5, 0.75, 1.0, 1.5, 2.0, and 3.0. Figure 10 shows the result for the square pulse P2.
3. For each T/T_p value convert the pulse $\eta-\mu_{max}$ curve into a $\gamma-\mu_{max}$ curve [$\gamma = V_y/(m.g) = (a_{eff}/g)\eta$] and find best-fit values for a_{eff} by minimizing the relative differences between the pulse and ground motion $\gamma-\mu_{max}$ curves for $1 \leq \mu_{max} \leq 10$ (from nearly elastic to highly inelastic behavior). The differences between the two curves are minimized using the least-squares method.
4. Obtain final values for a_{eff} by averaging the best-fit values for the seven period ratios.

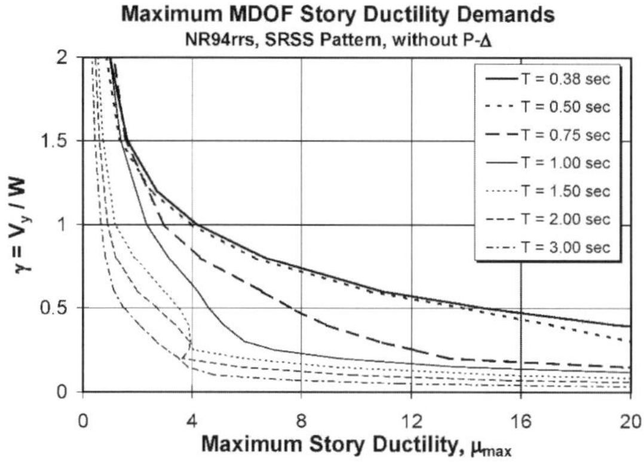

Figure 9. Base shear strength vs. maximum story ductility demands for structures with various T values, record NR94rrs (Alavi & Krawinkler, 2001, 2004a)

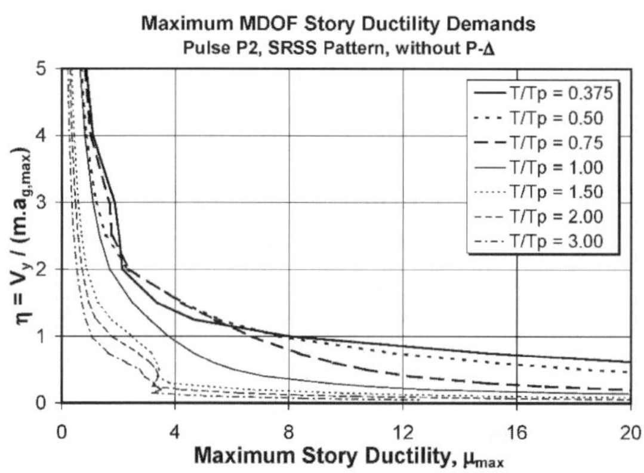

Figure 10. Base shear strength vs. maximum story ductility demands for structures with various T/T_p values, pulse P2 (Alavi & Krawinkler, 2001, 2004a)

Table 1 summarizes the results of this procedure for a set of recorded near-fault ground motions. The tables list the pulse type, pulse period, and maximum acceleration and velocity of the equivalent pulse. The peak velocity of the pulse is computed from the equation $v_{eff} = a_{eff}T_p/4$.

Table 1. Equivalent pulses for a set of near-fault ground motions (Alavi & Krawinkler, 2001, 2004a)

Designation	M_w	R (km)	Pulse Type	T_p (sec)	a_{eff} (g)	v_{eff} (cm/sec)	PGV (cm/sec)
LP89lgpc	7.0	3.5	P3	3.0	0.23	169	173
LP89lex	7.0	6.3	P2	1.0	0.70	172	179
EZ92erzi	6.7	2.0	P2	2.3	0.17	96	119
NR94rrs	6.7	7.5	P2	1.0	0.72	177	174
NR94sylm	6.7	6.4	P3	2.4	0.18	106	122
KB95kobj	6.9	0.6	P3	0.9	0.86	190	160
KB95tato	6.9	1.5	P3	2.0	0.40	196	174
IV79ar06	6.5	1.2	P2	3.4	0.13	108	110
IV79melo	6.5	0.0	P2	2.8	0.15	103	117
KB95kpi1	6.9	3.7	P2	1.8	0.25	110	100
MH84andd	6.2	4.5	P2	0.8	0.16	31	27
MH84cyld	6.2	0.1	P3	0.8	0.47	92	65
NR94newh	6.7	7.1	P2	1.3	0.37	118	119
NR94spva	6.7	8.9	P3	2.7	0.09	60	63

Comparing the so computed v_{eff} with the recorded peak ground velocity (PGV) listed in the last column of Table 1 indicates that in most of the cases the peak velocity of the equivalent pulse is very close to the PGV of the near-fault record. Only for the MH84cyld record does the difference exceed 20%. Thus, it appears to be feasible to use the PGV of the near-fault record to estimate the pulse intensity rather than following the elaborate procedure outlined previously.

Examples of story ductility distributions over the height for the NR94rrs record and its equivalent pulse are shown in Fig. 11 for cases of strong and weak structures. Although some differences exist, the equivalent pulse appears to reasonably capture the important response characteristics of the near-fault record, particularly the migration of ductility demands from the top to the bottom portion of the long-period structure ($T/T_p = 2.0$) if the strength of the structure is decreased.

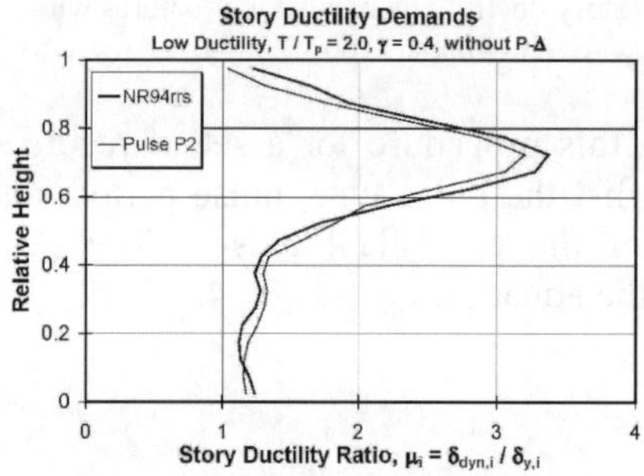

(a) High strength, low ductility

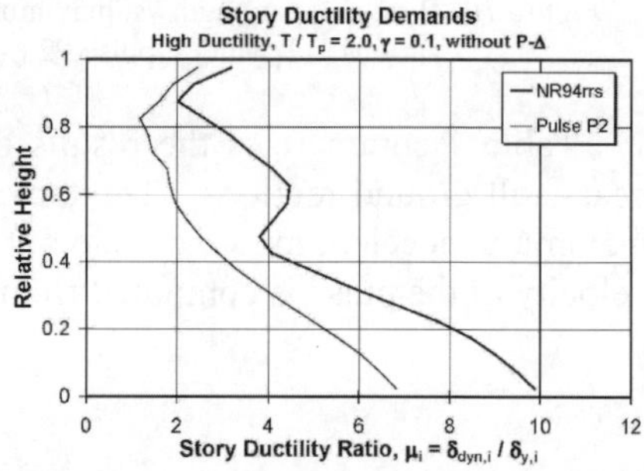

(b) Low strength, high ductility

Figure 11. Comparison of story ductility demands obtained from record NR94rrs and its equivalent pulse, $T/T_p = 2.0$ (Alavi & Krawinkler (2001, 2004a)

3. INELASTIC RESPONSE OF SDOF SYSTEMS TO NEAR-FAULT GROUND MOTIONS

In Section 2.1 regression equations are presented that relate pulse period to moment magnitude (Eq. 1) and pulse intensity, expressed by an effective velocity, to moment magnitude and distance (Eq. 2). The tendencies argued for in that section do exist, but the patterns are not clean and consistent, and the regression equations (1) and (2) exhibit large scatter. Considering that these patterns may have a dominant effect on seismic demands, they are pursued further in this section and in Section 4. But it is emphasized that more work, and particularly more evidence (recordings), are needed to gain confidence into quantitative information on near-fault ground motion patterns.

The ground motion example used here for illustration, the fault-normal component of the NR94rrs (Northridge 94 Rinaldi Receiving Station) record, has pulse characteristics that are rather clear. This is evident from Figures 12 to 14, which show, in sequence, R-μ relationships, displacement ratios, and strength demand spectra, for the NR94rrs record and for pulse P2. The correlation between record and pulse is excellent, using an equivalent pulse period of $T_p = 1.0$ seconds for the NR94rrs record.

The R-μ relationships for selected periods presented in Fig. 12 show that for short period SDOF systems (T on the order of $0.5T_p$ and smaller) μ is much larger than R, demonstrating very large ductility demands for small strength reduction factors. For relatively long period structures ($T \geq T_p$), the opposite is observed (i.e., $\mu < R$). The same observations are made from the δ_{in}/δ_{el} relationships presented in Fig. 13, since $\delta_{in}/\delta_{el} = \mu/R$. In fact, the δ_{in}/δ_{el} relationships look very similar to those for ordinary ground motions, but with one critical difference. The relationships of Fig. 13(b) are for normalized periods T/T_p, and indicate that displacement amplification (larger displacement of inelastic system than of elastic system) exists for T/T_p less than about 0.75. The implication is that displacement amplification exists for $T < 1.5$ sec. if $T_p = 2.0$ seconds. As Eq. (1) indicates, a T_p of 2.0 sec. is expected for an earthquake with $M_w = 6.7$. The conclusion is that seismic demands will depend <u>strongly</u> on the pulse period contained in the near-fault ground motion. Moreover, relatively short period structures ($T < 0.75T_p$) will experience larger displacements that equivalent elastic systems. The shapes of the δ_{in}/δ_{el} relationships also are very similar to those reported in Rahnama and Krawinkler (1993) for soft soil ground motions, in which displacement amplification is observed for $T/T_s < 0.75$, with T_s denoting the predominant soil period.

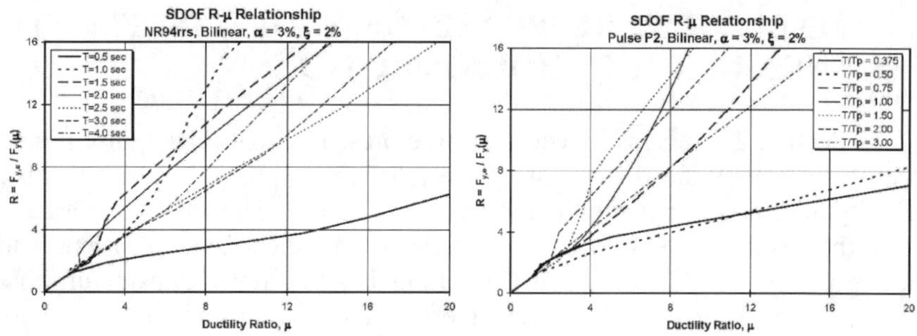

Figure 12. R-μ relationships for NR94rrs record and pulse P2 (Alavi & Krawinkler, 2001)
(a) NR94rrs record, various periods T, *(b)* Pulse P2, various period ratios T/T_p

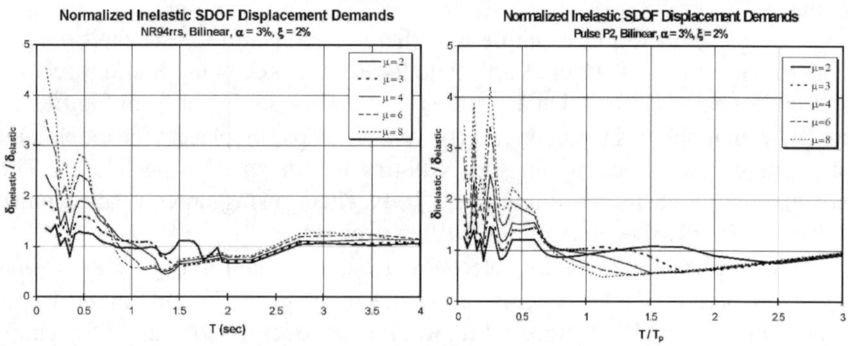

Figure 13. Displacement ratios δ_{in}/δ_{el} for various ductilities, NR94rrs record and pulse P2 (Alavi & Krawinkler, 2001)
(a) NR94rrs record, versus period T, *(b)* Pulse P2, versus period ratiosT/T_p

The large seismic demands for relatively short period structures are evident also in the constant ductility strength demand spectra presented in Figs. 14(a) and (b) for the NR94rrs record and pulse P2. Using $T_p = 1.0$ sec for the NR94rrs record, both figures show the same patterns. The inelastic spectra are presented for target ductility ratios μ = 2, 3, 4, 6, and 8. Similar to observations made in other studies (e.g., Rahnama and Krawinkler, 1993), the humps of the elastic spectra diminish and even disappear at large ductility ratios. At the same time, the peaks of the inelastic spectra shift to lower periods, which can be rationalized by the fact that the "effective period" of the structure elongates when the ductility increases and approaches the period associated with the large hump of the elastic spectra for structures with shorter elastic periods.

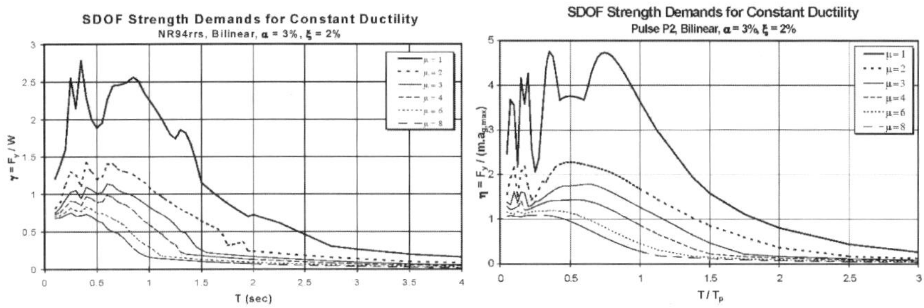

Figure 14. Strength demand spectra for NR94rrs record and pulse P2 (Alavi & Krawinkler, 2001). *(a)* NR94rrs record, versus period T, *(b)* Pulse P2, versus period ratio T/T_p

The strength demand spectra of Fig. 14(b) demonstrate the large dependence of seismic demands on the ratio of structure period T to pulse period T_p. If the equivalent pulse concept is indeed a realistic one, and if pulse periods increase with magnitude, then seismic demands for SDOF systems (and more so for MDOF systems) cannot be evaluated realistically by just scaling records to the widely used intensity measure $S_a(T)$. The implication is that pulse period T_p should become part of a vector that represents the ground motion intensity.

4. INELASTIC RESPONSE OF MDOF SYSTEMS TO NEAR-FAULT GROUND MOTIONS

The story ductility ratio, defined as the maximum story drift normalized by the story yield drift, i.e. $\mu_i = \delta_{max,i}/\delta_{y,i}$, can be used to quantify the response of MDOF structures to near-fault ground motions. Figure 15 shows distributions of story ductility demands over the height of MDOF frame structures subjected to the near-fault records whose velocity response spectra are shown in Figure 3. The demands are computed for MDOF systems with a fundamental period T = 2.0 sec. and base shear coefficients of $\gamma = 0.4$ and $\gamma = 0.15$, which correspond to a relatively strong and a relatively weak structure, respectively. The relatively long period of 2 sec. is chosen because the period of the pulse contained in most of the near-fault records is shorter than 2.0 sec. For comparison purposes, the mean story ductility demands obtained from the reference record set 15-D* are superimposed.

An important observation is that for most of the near-fault records, the maximum story ductility demand occurs in the upper portion of the structure when the structure is strong (large γ). However, a migration of ductility demands toward the base takes place when the structure becomes weaker (small γ). The non-uniformity of the story ductility demands observed for individual near-fault records holds true also in the mean, whereas the same story shear strengths on average result in a relatively uniform ductility distribution for ordinary ground motions (15-D*). Figure 15 also indicates the severity of near-fault ground motions in comparison with mean demands for the ordinary ground motions scaled to the UBC spectrum.

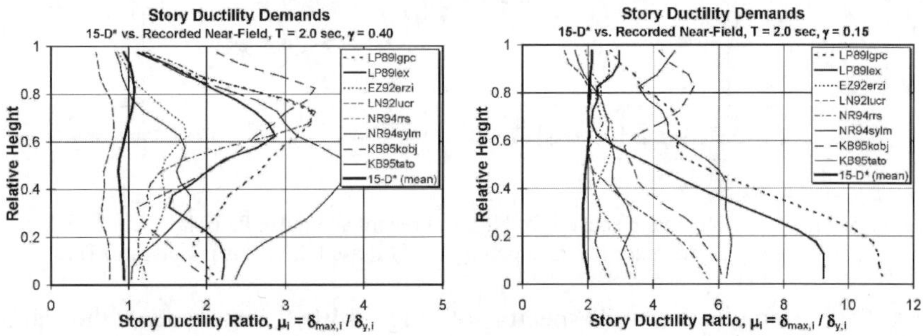

Figure 15. Story ductility demands for several near-fault and reference ground motions, T = 2.0 sec. (Alavi & Krawinkler (2001, 2004a) *(a)* γ = 0.40, *(b)* γ = 0.15

The strength and period dependence of story ductility distributions is illustrated in Figure 16 for the record NR94rrs. The distributions are shown for relatively short (T = 0.5 sec.) and relatively long (T = 2.0 sec.) period structures whose base shear strength coefficient, γ, is varied progressively. As with the previous observation from Fig. 15, for the structures with T = 2.0 sec. (a period that clearly exceeds the effective pulse period of 1.0 sec. for NR94rrs), maximum ductility demands occur in the upper stories when the structure is strong. With a reduction in the base shear strength, the ductility demands in the upper portion stabilize and grow no more. Further strength reductions result in a migration of the maximum demand toward the base. This phenomenon is not observed for the structures with T = 0.5 sec., a period that is shorter than the effective pulse period. For these structures the maximum ductility demands occur close to the base regardless of the base shear strength.

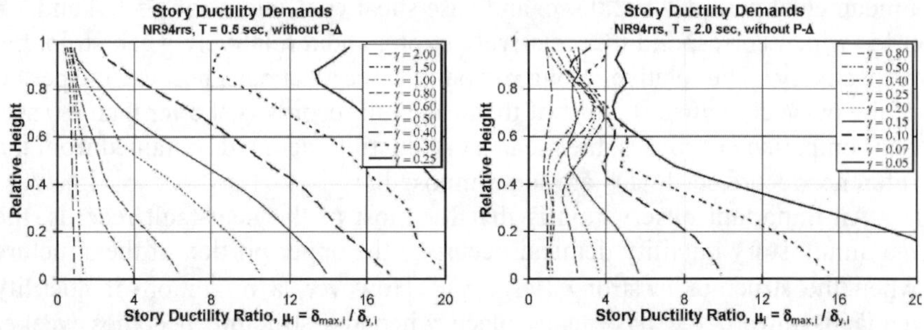

Figure 16. Dependence of distribution of story ductility demands on base shear strength for record NR94rrs (Alavi & Krawinkler (2001, 2004a). *(a)* T = 0.5 sec., *(b)* T = 2.0 sec.

5. DESIGN IMPLICATIONS OF NEAR-FAULT EFFECTS

Equations (1) and (2) provide an estimate (within limits of applicability) of the dependence of the equivalent near-fault pulse properties (T_p and $v_{eff} \approx$ PGV) on magnitude M_w and distance R. For a given scenario, this information can be utilized to develop MDOF design spectra for targeted ductility ratios. What is needed is to develop generic MDOF pulse strength demand spectra of the type shown in Fig. 14 for SDOF systems. This can be achieved by using the information in Fig. 10 and re-plotting the data in the η – T/T_p domain, as shown in Fig. 17.

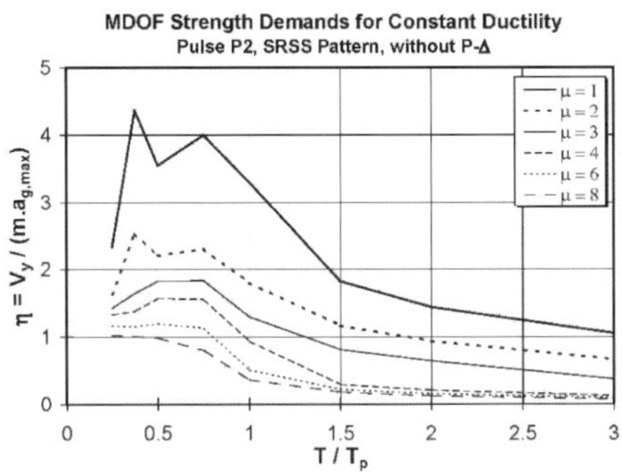

Figure 17. MDOF base shear strength demand spectra for target maximum story ductility ratios, pulse P2 (Alavi & Krawinkler (2001, 2004a)

These MDOF pulse strength demand spectra can be converted into design spectra for specific earthquake scenarios. For instance, using $M_w = 7.0$ and Eq. (1) gives $T_p = 2.6$ sec., and selecting distances of 10 and 3 km together with Eq. (2) provides $v_{eff} = 112$ and 198 cm/sec., which from $a_{eff} = 4v_{eff}/T_p$ results in $a_{eff} = 0.176g$ and 0.310g, respectively, for R = 10 and 3 km. Recalling that the base shear coefficient $\gamma = V_y/W = (a_{eff}/g)\eta$, the inelastic design spectra shown in Fig. 18 are obtained for the two earthquake scenarios. For the selected scenarios (defined by M_w and R) the so obtained γ–T curves represent base shear strength demand spectra for specified target ductilities, suitable for the types of frames represented by the generic frame used in this study.

To put these demands in perspective with present design practice, superimposed on both graphs are two design strength curves related to the UBC-97 (1997) soil Type S_D design spectra for Seismic Zone 4, one with and one without the code specified near-fault factor (denoted as "UBC/4"). The UBC near-fault factor is computed using Source Type A. The design strength curves are obtained by scaling down the code base shear by a factor of 4, which is arrived at by choosing a response modification factor of R = 8 (close to that for special moment frames) and assuming a period independent overstrength factor of 2. These values are purely for comparison, as the real strength of a structure will vary on a case-by-case basis.

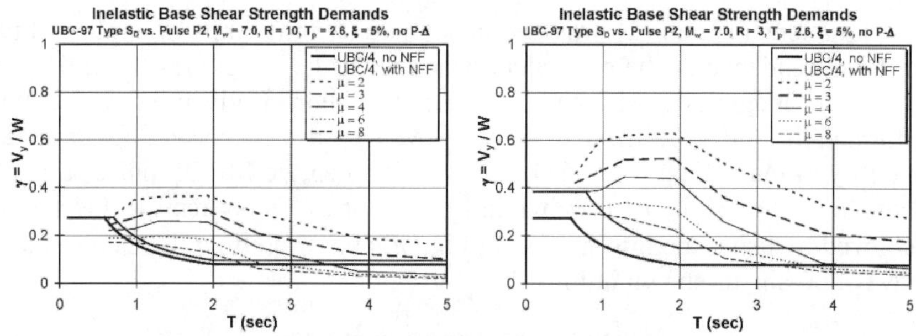

Figure 18. Magnitude and distance dependence of MDOF base shear strength demand spectra for constant ductilities, pulse P2; $M_W = 7.0$; *(a)* R = 10 km, *(b)* R = 3 km *(Alavi & Krawinkler (2001, 2004a)*

A comparison of the near-fault strength demand spectra with the UBC/4 curves reveals that for given magnitude and distance values, a structure designed according to present code provisions will experience quite different levels of inelasticity depending on the fundamental period. For instance, Figure 18(b) indicates that for $M_W = 7$, R = 3 km, and a ground motion of the type represented by pulse P2, structures designed according to the UBC'97 provisions (including the near-fault factor) may experience story ductility demands less than 4 when the period is longer than 3.4 sec. or shorter than about 0.8 sec, but may experience maximum ductility demands larger than 8 if the period is between 1.0 and 2.2 sec. This indicates a very inconsistent level of protection against near-fault effects for structures of different periods.

The problem cannot be addressed adequately by introducing period insensitive factors to conventional design spectra. A more rigorous approach appears to be necessary. The equivalent pulse concept can provide the basis for such an approach.

6. THE NEED FOR EXPLICIT CONSIDERATION OF NEAR-FAULT EFFECTS IN HAZARD ANALYSIS

As Fig. 19(a) illustrates, the shape of near-fault spectra is very sensitive to magnitude, because T_p is dependent on magnitude, and the seismic demand for a structure therefore will depend strongly on the positioning of the structure's period (T_1) with respect to the peak of the spectra. Site specific hazard analysis usually identifies a dominant (characteristic) event in terms of magnitude and distance. If this dominant event is a M = 6.5 event and T_1 coincides with the peak of the M = 6.5 spectrum, the structural demand will be very large, whereas it will be small if the dominant event is a M = 7.5

event. The reverse will be the case if T_1 coincides with the peak of the M = 7.5 spectrum. Thus, it would be grossly misleading to use both the M = 6.5 and 7.5 ground motions in the same record bin, even if both records are scaled to the same spectral acceleration at T_1. This is illustrated in Fig. 19(b), which clearly demonstrates that, because of period elongation, the inelastic demands for the M = 7.5 record would be vastly greater than the demands for the M = 6.5 record.

This points out the need to explicitly consider magnitude and distance in seismic hazard analysis if the site is prone to near-fault effects with forward directivity. The conclusion is that in such cases a single intensity measure, such as the spectral acceleration at the first mode period, is inadequate to define the seismic hazard, and a multi-parameter (vector) intensity measure has to be found. Spectral acceleration could be part of this vector, but so should be the pulse period T_p and the pulse intensity defined by v_{eff} or in the simplest way by PGV. The rigorous approaches for estimating these quantities, as described in Sections 2.2 and 2.3, should prove helpful for this purpose.

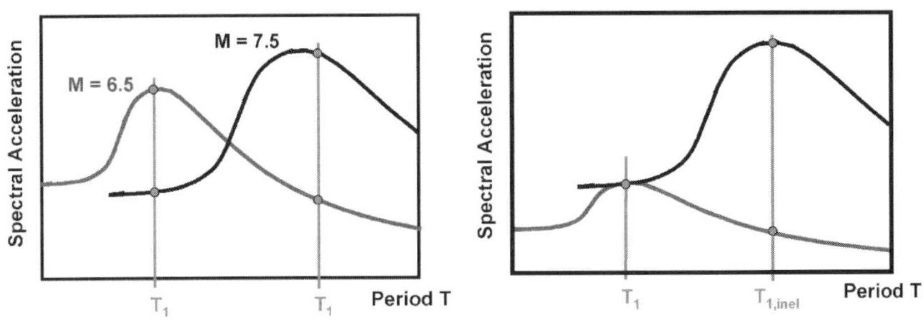

Figure 19. Sensitivity of seismic demands to magnitude dependent shape of near-fault spectra; *(a)* unscaled records, *(b)* records scaled to same S_a

The following discussion serves to demonstrate that the use of near-fault record will lead to seismic demand predictions that clearly differ from those for ordinary records. This demonstration is done on hand of IDAs (Incremental Dynamic Analyses, Vamvatsikos and Cornell, 2002) in which structures (the generic multi-story frames described in Medina and Krawinkler, 2003 are employed) are subjected to two sets of ground motions whose intensity, defined by $S_a(T_1)$, is increased in small increments.

One of the two sets of records used is the near-fault record set listed in Table 1, and the other is a set of ordinary (no near-fault characteristics) records employed in Medina and Krawinkler (2003). The differences in the two sets of records are evident from Fig. 20, in which individual spectra are scaled to a common S_a of 1.0g at the period of 1.2 sec. The median spectra of the two sets (bold, solid lines) have a clearly different frequency content, with the near-fault medium spectrum showing a much wider short period plateau (the T_p of most of these records is between 1 and 2.5 sec.) than the medium spectrum of the ordinary records. The near-fault records do not show the rapid decrease in S_a for periods exceeding 0.5 seconds.

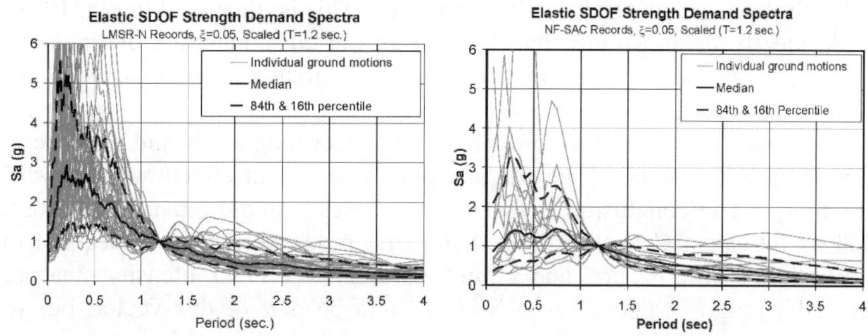

Figure 20. Spectral shapes of *(a)* ordinary records, *(b)* near-fault records

Figure 21 shows median (and 16th & 84th percentile) IDA curves for the maximum interstory drift ratio (IDR), plotted against the $S_a(T_1)$ of the ground motions, with T_1 being the first mode period of the frame structure. Results for three frame structures are shown, with each structure identified by a four digit code, the first two digits identifying the number of stories, and the last two the first mode period, i.e., 0612 stands for a 6-story structure with T_1 = 1.2 sec. The base shear strength of the structure is identified by the base shear coefficient $\gamma = V_y/W$. The individual data points shown identify the IDR for each individual unscaled near-fault ground motion.

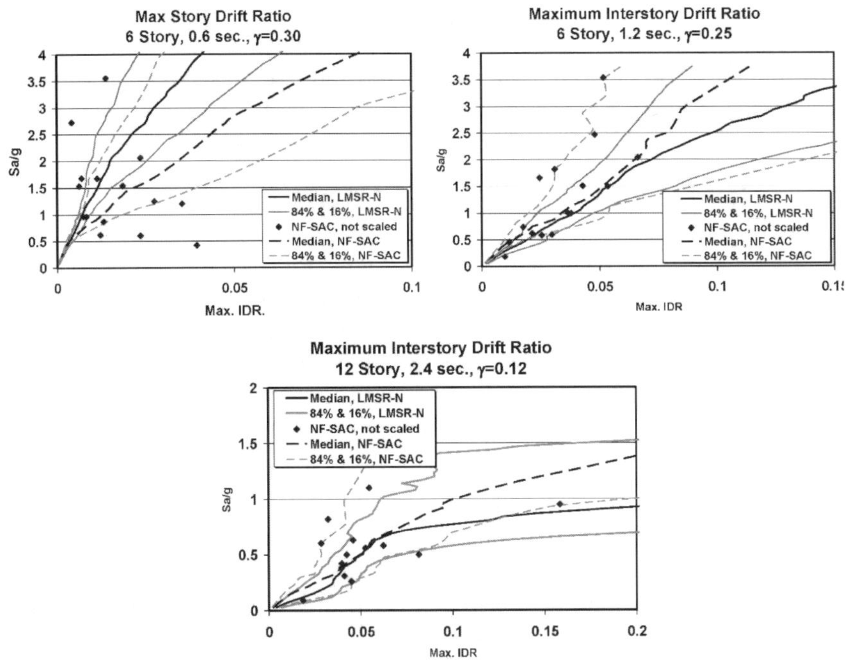

Figure 21. Median (and 16th & 84th percentile) IDA curves for ordinary and near-fault ground motions; *(a)* 0606 frame, $\gamma = 0.3$, *(b)* 0612 frame, $\gamma = 0.25$, *(c)* 1224 frame, $\gamma = 0.12$

The median IDRs show very distinct patterns. For the short period structure 0606, the IDR demands for the near-fault ground motions are much larger than for the ordinary ground motions. For the median period structure 0612 the IDRs are about the same for both records except for very large spectral accelerations where the demands for the ordinary records exceed those for the near-fault records. For the long period structure 1224 the IDR demands for the ordinary records become much larger than those for the near-fault records at spectral accelerations exceeding about 0.8g. Around this spectral acceleration the slope of the median IDR for the structure subjected to the ordinary record takes on a small value, which implies that unstable behavior due to P-delta effects is approached. On the other hand, the spectral acceleration of the near-fault ground motions can be increased by a significant amount without causing dynamic instability.

The reason for the disparate behavior is the different frequency content of the ordinary and near-fault ground motions. For relatively short period structures ($T_1 < T_p$) the near-fault ground motions exhibit a climbing spectrum, and period elongation will drive the response into the range of the pulse period, requiring high IDR demands. The reverse is true for long period structures ($T_1 > T_p$), where the elastic response is already on the descending branch of the spectrum and period elongation has a mitigating effect (more so than for ordinary records).

Again, these examples are intended to demonstrate the need for explicit consideration of near-fault effects in hazard analysis. It would be very misleading to treat near-fault and ordinary ground motions equally in seismic hazard analysis.

7. SUMMARY AND CONCLUSIONS

The response of SDOF and MDOF systems to near-fault ground motions and pulse-type inputs is evaluated. Equivalent pulses are introduced to represent near-fault ground motions. Preliminary relationships are proposed for the magnitude- and distance-dependence of pulse period and severity. A design methodology is suggested based on the response characteristics of the equivalent pulses. The main findings can be summarized as follows:

- The fault-normal component of near-fault ground motions with forward directivity is severe and shows pulse-like characteristics. If the two components of the motion are rotated by 45 degrees, one of the 45-degree components is almost as severe as the fault-normal component.
- The response of long-period MDOF structures ($T/T_p > 1.0$) to pulse-type inputs exhibits a traveling wave effect that involves higher modes of vibration, resulting in early yielding in the upper portion of strong structures.
- As the structure weakens, maximum ductility demands migrate to the bottom of the structure.
- Similarities exist between the response properties of near-fault ground motions and those of basic pulses. Within the period range $0.375 < T/T_p < 3.0$, most of the near-fault ground motions can be represented by equivalent pulses.
- Design can be based on the response characteristics of equivalent pulses, which are defined by the pulse type, period and intensity.
- Relationships between pulse properties and earthquake magnitude and distance, together with pulse strength demand spectra, can be utilized to develop base shear strength demand spectra for design.
- The near-source factors recommended by present codes are not believed to provide consistent protection against near-fault ground motions.
- The need exists for an explicit consideration of near-fault effects in seismic hazard analysis. If this is not done, bias will be introduced in seismic demand predictions.

It is important to note that conclusions drawn here should be interpreted and generalized in the context of the assumptions made. Much work remains to be done before a well-established design procedure can be developed for structures in near-fault regions. Most important, more data on near-fault ground motions needs to be obtained, either from actual ground motion recordings or from physically justifiable simulations. The authors hope that the ideas and concepts presented here will help in forming a foundation on which to base further research and development in this important field.

ACKNOWLEDGEMENTS

The research summarized here has been supported by grants from the California Strong Motion Instrumentation Program, the CUREe/Kajima research program, and by the National Science Foundation through Grant CMS-9812478. More recently, the work has been supported by the NSF sponsored Pacific Earthquake Engineering Research (PEER) Center.

REFERENCES

1. Abrahamson. N.A. and W.J. Silva (1997). Empirical response spectral attenuation relations for shallow crustal earthquakes. *Seismological Research Letters*, 68: 94-127.
2. Abrahamson, N.A. (2000). Effects of rupture directivity on probabilistic seismic hazard analysis. *Proceedings of the 6th International Conference on Seismic Zonation*, Palm Springs, Earthquake Engineering Research Institute.
3. Alavi, B. and H. Krawinkler (2000). Design considerations for near-fault ground motions. *Proceedings of the U.S. – Japan Workshop on the Effects of Near-Fault Earthquake Shaking*, San Francisco, March 20-21.
4. Alavi B, and Krawinkler H. (2001). Effects of near-fault ground motions on frame structures. *John A. Blume Earthquake Engineering Center Report No. 138*, Stanford University.
5. Alavi, B., and Krawinkler, H. (2004a). Behavior of moment-resisting frame structures subjected to near-fault ground motions, *International Journal for Earthquake Engineering and Structural Dynamics, **33**; 687-706*
6. Alavi, B., and Krawinkler, H., Strengthening of frame structures against near-fault effects, *International Journal for Earthquake Engineering and Structural Dynamics, **33**; 707-722.*
7. Anderson, J., and Bertero, V. (1978). Uncertainties in establishing design earthquakes," *ASCE Journal of Structural Engineering*, Vol. 113, No. 8, 1709-1724.
8. Anderson, J., Bertero, V., and Bertero, R. (1999). Performance improvement of long period building structures subjected to severe pulse-type ground motions, *Pacific Earthquake Engineering Research Center, University of California, Berkeley*, Report No. PEER-1999/09, Oct. 1999.
9. Archuleta, R. J. and S. H. Hartzell (1981). Effects of fault finiteness on near-source ground motion. Bull. Seismol. Soc. Am., 71: 939-957.
10. Hall, J., Heaton, T., Halling, M., Wald, D. (1995). Near-source ground motion and its effects on flexible buildings, *Earthquake Spectra*, Vol. 11, No. 4, 569-605.
11. Iwan, W., and Chen, X. (1994). Important near-field ground motion data from the Landers earthquake," *Proceedings, 10th European Conference on Earthquake Engineering*, Vienna.

12. Iwan, W. (1997). Drift spectrum: measure of demand for earthquake ground motions," *ASCE Journal of Structural Engineering*, Vol. 123, No. 4, 397-404.
13. MacRae, G., Morrow, D., and Roeder, C. (1998). Near-field ground motion effects on short structures, *PG&E-PEER Program, University of Washington,* Draft Report.
14. Mavroeidis, G. and A. Papageourgiou (2003). A mathematical representation of near fault ground motions. *Bull. Seismol. Soc. Am. 93,* 1099-1131.
15. Mavroeidis, G., Dong, G., and Papageourgiou, A. (2003). Near-fault ground motions, and the response of elastic and inelastic single-degree-freedom (SDOF) systems, Earthquake Engineering and Structural Dynamics, 33, 1023-1049.
16. Medina, R., and Krawinkler, H. (2003). Seismic demands for nondeteriorating frame structures and their dependence on ground motions," *John A. Blume Earthquake Engineering Center Report No. TR 144,* Department of Civil Engineering, Stanford U.
17. Rahnama, M., and Krawinkler, H. (1993). Effects of soft soils and hysteresis model on seismic demands, *John A. Blume Earthquake Engineering Center Report No. 108*, Department of Civil Engineering, Stanford University.
18. Rodriguez-Marek, A. (2000). Near fault seismic site response. Ph.D. Thesis, Civil Engineering, University of California, Berkeley, 451 pp.
19. Sasani M, Bertero V (2000). Importance of severe pulse-type ground motions in performance-based engineering: historical and critical review. *12th World Conference on Earthquake Engineering*, New Zealand Society for Earthquake Engineering, New Zealand.
20. Somerville, P.G., N.F. Smith, R.W. Graves, and N.A. Abrahamson (1997). Modification of empirical strong ground motion attenuation relations to include the amplitude and duration effects of rupture directivity, *Seismological Research Letters* 68, 199-222.
21. Somerville, P.G. (1998). Development of an improved representation of near fault ground motions. Proc. SMIP98 Seminar on Utilization of Strong Ground Motion Data, p. 1-20.
22. Somerville, P.G., K. Irikura, R. Graves, S. Sawada, D. Wald, N. Abrahamson, Y. Iwasaki, T. Kagawa, N. Smith and A. Kowada (1999). Characterizing earthquake slip models for the prediction of strong ground motion. *Seismological Research Letters*, 70, 59-80.
23. Somerville P. (2000). Magnitude scaling of near fault ground motions. *Proceedings of International Workshop on Annual Commemoration of Chi-Chi Earthquake*, National Center for Research on Earthquake Engineering, Taipei, Taiwan, September 2000.
24. Somerville, P.G., H. Krawinkler and B. Alavi (2000). Development of improved ground motion representation and design procedures for near-fault ground motions. *Final Report to CSMIP Data Utilization Program*, Contract No. 1097-601.
25. Somerville, P.G. (2003). Magnitude scaling of the near fault rupture directivity pulse. *Physics of the Earth and Planetary Interiors*, 137, 201-212.
26. Stewart J, Chiou S, Bray J, Graves R, Somerville P, Abrahamson N. (2001). Ground motion evaluation procedures for performance-based design. *PEER-2001/09*, Pacific Earthquake Engineering Research Center, University of California, Berkeley, September.
27. Takeo, M. (1998). Ground rotational motions recorded in near-source region of earthquakes. *Geophysical Research Letters* 25, 789-792.
28. UBC-97 (1997). *Uniform building code*. International Conference of Building Officials. Whittier, California,
29. Vamvatsikos D, Cornell C A. (2002). Incremental Dynamic Analysis. *Earthquake Engineerin.g & Structural Dynamics* **31:**491-514.

RAPID ASSESSMENT OF BUILDING RESPONSE USING GENERALIZED INTERSTORY DRIFT SPECTRA

Eduardo Miranda[1] and Sinan Akkar[2]
[1]*Dept. of Civil and Environmental Engineering, Stanford University, USA.*
[2] *Dept. of Civil Engineering, Middle East Technical University, Ankara, Turkey.*

Abstract: A new tool for rapid building response assessment is presented. By using a continuum model this new tool, named generalized interstory drift spectrum, provides estimates of maximum interstory drift demands in buildings responding to earthquakes. The continuous model consists of a combination of a flexural beam and a shear beam. By modifying a single parameter the model used in the generalized interstory drift spectrum can consider lateral deformations varying from those of a flexural beam to those of a shear beam. Therefore, it permits to account for a wide range of modes of deformation that represent more closely those of multistory buildings. Because of its computationally efficiency, and because it only requires a minimum of information about the building, the new spectrum provides a powerful tool for rapid assessment of many buildings. Examples of interstory drift demands from various ground motions recorded in the United States and Turkey are presented.

Key words: Interstory drift ratio, seismic analysis, approximate methods, multistory buildings, dynamic characteristics, continuum model, rapid assessment.

1. INTRODUCTION

It is well know that structural damage and many kinds of nonstructural damage in buildings are the result of lateral deformations. In particular, several studies have concluded that the structural response parameter that is best correlated with seismic damage is the peak interstory drift ratio, which is defined as the difference in lateral displacements in between two consecutive floors normalized by the interstory height.

Although response spectral ordinates provide a much better measure of ground motion intensity for estimating possible damage in buildings than peak ground acceleration or peak ground velocity, they have a number of

P. Gülkan and J.G. Anderson (eds.), Directions in Strong Motion Instrumentation, 107–121.

important disadvantages. Namely: (1) a single response spectrum ordinate cannot account for the contribution of higher modes; (2) even if displacements response spectrum ordinates are used, they only provide a measure of the overall lateral deformation in the building and do not take into account concentration in lateral deformations in certain stories that usually occur in buildings.

In 1997, Iwan introduced a simple and direct measure of drift demand in buildings subjected to earthquake ground motions called the drift spectrum (Iwan 1997). Similarly to the response spectrum, the drift spectrum is based on a relatively simple linear model. However, the drift spectrum differs from the response spectrum in that it is based on a continuous shear beam rather than a single-degree-of-freedom (SDOF) system. Despite its many advantages, the drift spectrum has received very little attention from engineers. A possible explanation is that the drift spectrum is based on the propagation of non-dispersive damped waves traveling up and down the shear beam, and unlike seismologists, most structural engineers are not familiar with wave propagation. More recently, Kim and Collins (2002) showed that the model and formulation used in the drift spectrum when using damped models could result in residual drifts for certain ground motions, which is inconsistent with the linear elastic behavior which is assumed in the model.

Although the shear beam model has been used in the past to represent buildings, in many cases it does not adequately reproduce the lateral deformations of buildings. For example, the presence of shear walls or bracing systems will lead to lateral deformation profiles that differ significantly from those of a shear beam. Furthermore, even in buildings in which shear walls or bracing are not present, as for example in moment-resisting frame buildings, lateral deformations may still differ from those of a shear beam. Lateral deformations of moment-resisting frame buildings would match those of a shear beam only if the beams are infinitely rigid in bending and columns are infinitely rigid axially.

The objective of this paper is to present a new tool for rapid assessment of building response. The new tool, referred to as *generalized interstory drift spectrum* (GIDS), is based on a continuous model that consists of a combination of a flexural beam and a shear beam, rather than only a shear beam. By modifying a single parameter this model can consider lateral deformations varying from those of a flexural beam all the way to those of shear beam. Hence, it permits to account for a wide range of modes of deformation that represent more closely those of multistory buildings. Furthermore, the proposed generalized interstory drift spectrum is based on conventional modal analysis techniques that are very familiar to structural engineers and assumes classical modal damping, thus avoiding the residual drift problems reported by Kim and Collins (2002). Mode shapes, modal participation factors and period ratios required to compute the response of the model are all computed with closed-form solutions and are a function of only one parameter. Hence, it provides a high computational efficiency while at the same time requiring only a minimum amount of information about the building whose seismic response is being assessed.

2. SIMPLIFIED BUILDING MODEL

The model used in the generalized interstory drift spectrum consists of a linear elastic continuum model. Continuum models have been proposed before for approximating the response of buildings to wind or seismic forces. For a review of previously-proposed models the reader is referred to Miranda and Taghavi (2004) and Miranda and Akkar (2004). The model used in the generalized interstory drift spectrum consists of a combination of a flexural cantilever beam and a shear cantilever beam deforming in bending and shear configurations, respectively (Figure 1). It is assumed that along the entire length of the model, both beams undergo identical lateral deformations. Furthermore, masses and lateral stiffnesses are assumed to remain constant along the height of the building.

As shown by Miranda and Akkar (2004), the response of uniform shear-flexural model shown in Figure 1 when subjected to an horizontal acceleration at the base $\ddot{u}_g(t)$ is given by the following partial differential equation:

$$\frac{\rho}{EI}\frac{\partial^2 u(x,t)}{\partial t^2}+\frac{c}{EI}\frac{\partial u(x,t)}{\partial t}+\frac{1}{H^4}\frac{\partial^4 u(x,t)}{\partial x^4}-\frac{\alpha^2}{H^4}\frac{\partial^2 u(x,t)}{\partial x^2}=-\frac{\rho}{EI}\frac{\partial^2 u_g(t)}{\partial t^2} \quad (1)$$

where ρ is the mass per unit length in the model, H is the total height of the building, $u(x,t)$ is the lateral displacement at non-dimensional height $x=z/H$ (varying between zero at the base of the building and one at roof level) at time t, c is the damping coefficient per unit length, EI is the flexural rigidity of the flexural beam and α is the lateral stiffness ratio defined as

$$\alpha = H\sqrt{\frac{GA}{EI}} \quad (2)$$

where GA is the shear rigidity of the shear beam. The lateral stiffness ratio, α, is a dimensionless parameter that controls the degree of participation of overall flexural and overall shear deformations in the continuous model and thus, it controls the lateral deflected shape of the model. A value of α equal to zero represents a pure flexural model (Euler-Bernoulli beam) and a value of α equal to ∞ corresponds to a pure shear model. Intermediate values of α correspond to multistory buildings that combine overall shear and flexural lateral deformations.

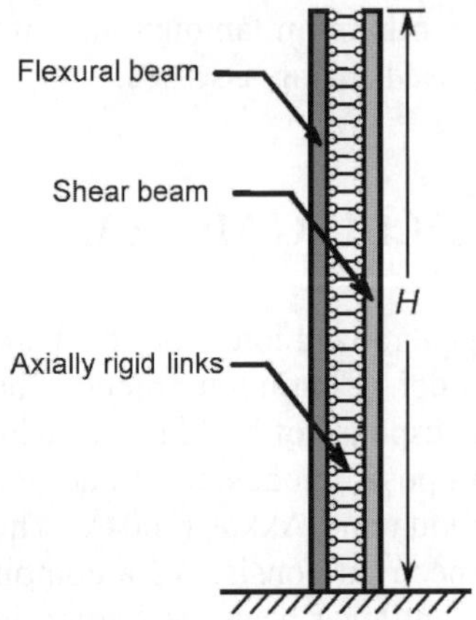

Figure 1. Continuum model used in the generalized interstory drift spectrum.

The mode shapes of the simplified model are given by (Miranda and Taghavi, 2004):

$$\phi_i(x) = \sin(\gamma_i x) - \gamma_i \beta_i^{-1} \sinh(x\beta_i) - \eta_i \cos(\gamma_i x) + \eta_i \cosh(\beta_i x) \tag{3}$$

where β_i and η_i are nondimensional parameters for the ith mode of vibration which are given by

$$\beta_i = \sqrt{\alpha^2 + \gamma_i^2} \tag{4}$$

$$\eta_i = \frac{\gamma_i^2 \sin(\gamma_i) + \gamma_i \beta_i \sinh(\beta_i)}{\gamma_i^2 \cos(\gamma_i) + \beta_i^2 \cosh(\beta_i)} \tag{5}$$

and γ_i is the eigenvalue of the ith mode of vibration corresponding to the ith root of the following characteristic equation:

$$2 + \left[2 + \frac{\alpha^4}{\gamma_i^2 \beta_i^2}\right] \cos(\gamma_i) \cosh(\beta_i) + \left[\frac{\alpha^2}{\gamma_i \beta_i}\right] \sin(\gamma_i) \sinh(\beta_i) = 0 \tag{6}$$

Periods of vibration corresponding to higher modes can be computed as a function of the fundamental period of vibration of the building T_1 by using period ratios computed as

$$\frac{T_i}{T_1} = \frac{\beta_1 \gamma_1}{\beta_i \gamma_i} \tag{7}$$

Since the masses are assumed to remain constant, the modal participation factors Γ_i can be computed with the following equation:

$$\Gamma_i = \frac{\int_0^1 \phi_i(x)dx}{\int_0^1 \phi_i^2(x)dx} \tag{8}$$

Integrals shown in equation (8) can be solved in closed-form solution. Readers interested in these closed-form solutions are referred to Miranda and Akkar (2004). As shown by these equations, mode shapes and modal participation factors depend only on a single parameter, the lateral stiffness ratio α.

While assuming the mass to remain constant along the height of buildings is reasonable in most cases, assuming that the lateral stiffness remains constant along the height of the building is perhaps only for a reasonable assumption for one to three-story buildings. However, Miranda and Taghavi (2004) have shown that the product of modal shapes and modal participation factors as well as period ratios are relatively robust and are not significantly affected by reductions in lateral stiffness. In the same study, it was also shown that reduction in masses along the height of the building also does not affect significantly the dynamic characteristics.

Taghavi and Miranda (2004) compared the response computed with the simplified model to that computed with detailed finite-element models of a ten-story steel moment resisting frame building and a twelve-story reinforce concrete building whose properties are available in the literature. Additionally, they compared the response computed with the model to that recorded in four instrumented buildings in California that have been subjected to earthquakes. In all cases, it was shown that the simplified model provided relatively good results. More recently, Reinoso and Miranda (2004) validated the model by comparing the response computed with the simplified continuous model to that recorded in five high rise buildings in California in various earthquakes.

3. GENERALIZED INTERSTORY DRIFT SPECTRUM

The contribution of the ith mode of vibration to the lateral displacement (relative to the ground) at non-dimensional height $x=z/H$ at time t is given by

$$u_i(x,t) = \Gamma_i \phi_i(x) D_i(t) \tag{9}$$

where Γ_i is the modal participation factor of the ith mode of vibration, $\phi_i(x)$ is the amplitude of ith mode at nondimensional height x, and $D_i(t)$ is the relative displacement response of a SDOF system, with period T_i and modal damping ratio ξ_i corresponding to those of the ith mode of vibration, subjected to ground acceleration $\ddot{u}_g(t)$. The $\Gamma_i \phi_i(x)$ product controls the spatial variation of the contribution of the *i*th mode to the total response while $D_i(t)$ controls its time variation.

In the generalized interstory drift spectrum, the interstory drift ratio at the *j*th story is approximated by the rotation in the model at the height corresponding to the middle of the story of interest as follows

$$IDR(j,t) \approx \theta(x,t) = \frac{1}{H}\sum_{i=1}^{\infty}\Gamma_i\phi_i'(x)D_i(t) \tag{10}$$

where x is the average height of the j+1 and j floors, and $\phi_i'(x)$ is the first derivative of the *i*th mode shape $\phi_i(x)$ with respect to non-dimensional height x. The derivative of the mode shapes with respect to non-dimensional height x is obtained by taking the derivative of Eq. (3) with respect to x as follows:

$$\phi_i'(x) = \gamma_i\cos(\gamma_i x) - \gamma_i\cosh(\beta_i x) + \eta_i\gamma_i\sin(\gamma_i x) + \eta_i\beta_i\sinh(\beta_i x) \tag{11}$$

Although Eq. (10) indicates that an infinite sum is required to obtain the rotation response history, in most cases only a relatively small number of modes is required to obtain good estimates of the peak rotation demand in the model. Hence, the interstory drift ratio at non-dimensional height x, can be approximated as

$$IDR(x,t) \approx \frac{1}{H}\sum_{i=1}^{m}\Gamma_i\phi_i'(x)D_i(t) \tag{12}$$

where m is the number of modes considered in the analysis. For buildings up to 20 stories m can be as small as 3 (Taghavi and Miranda, 2004), while for taller buildings the required number of modes to consider in the analysis m is typically 5 or 6 (Reinoso and Miranda, 2004). These validation studies indicate that although better results are obtained by specifying different damping ratios for each mode, relatively good approximations are obtained by setting the modal damping ratios ξ_i constant (i.e., using the same damping ratio for all m modes).

The ordinates of the *generalized interstory drift spectrum* (GIDS) are defined as the maximum peak interstory drift demand over the height of the building and are computed as

$$IDR_{\max} \equiv \max_{\forall t,x}\left|\theta(x,t)\right| \tag{13}$$

The generalized interstory drift spectrum is a plot of the fundamental period of the building in the abscissas versus $IDR_{\max}$ in the ordinates. Similarly to the response spectrum, the GIDS provides seismic demands for a family of systems with different periods of vibration. However, instead of having ordinates of maximum relative displacement, maximum relative velocity or maximum acceleration of a single degree of freedom system, the GIDS provides a measure of peak interstory drift demands, which is a demand parameter that is better correlated damage. In particular, the GIDS provides a rapid estimation of peak interstory drift demand in buildings with different periods of vibration.

If the same damping ratio is used for the *m* contributing modes, then the model is fully defined by using only four parameters: (1) the fundamental period of vibration of the building, T_1; (2) a modal damping ratio that represents the damping ratio in the building, ξ; (3) the lateral stiffness ratio α; and (4) the building height, *H*. Since the derivative of the modes, modal participation factors and period ratios can be computed in closed-form solution, the GIDS is computationally very efficient, requiring just a few seconds in most personal computers. If empirical relations between building height and fundamental period are used, the number of parameters is then reduced from four to three.

4. EFFECT OF STRUCTURAL PARAMETERS ON GIDS

As indicated in the previous section, the proposed method mainly depends on four parameters (*i.e.*, fundamental building period T_1, modal damping ratio ξ, lateral stiffness ratio α, and building height *H*) for the computation of maximum interstory demands on structural systems. The influence of these parameters on GIDS is illustrated by using some ground motions recorded during the 1999 Kocaeli and Düzce, Turkey, earthquakes.

One of the important advantages of the GIDS over response spectra is the consideration of higher modes. Figure 2 presents the generalized interstory drift spectra for the EW components of the motions recorded at he Sakarya and Bolu stations during the 1999 Kocaeli and Duzce earthquakes, respectively. Spectra on the left are computed for flexural beam models (α=0) while spectra on the right hand side display the maximum interstory drift demands for α=30 which represents a structural system deforming primarily as a shear model but with some small participation of flexural deformation. Both figures present the undamped GIDS computed by considering only a finite number of modes *m* = 1, 2, 4 and 6. The total building height required in Eq. (12) was computed as a function of the fundamental period of vibration by using the relationship for steel moment-resisting frames in the 1997 UBC code (ICBO, 1997), namely, $T_1 = 0.0853H^{0.75}$, where *H* is in meters. Figure 3 shows GIDS ordinates computed using *m* = 1, 2 and 4 normalized with maximum interstory drift demands computed using the contribution of six modes. These normalized GIDS provide a measure of the participation of higher modes in interstory drift demands. It can be seen that in both cases, the use of only fundamental mode tends to underestimate interstory drift demands, particularly for long period buildings. For the EW Sakarya component, underestimations are about 20 percent of the drift estimates computed with six modes for fundamental periods greater than 0.3s. Using only the fundamental mode leads to large underestimations of interstory drift demands for the EW Bolu component when fundamental periods are greater than 1.5s. The tendency in underestimation for this record increases rapidly and can reach about 75 percent of the drifts for fundamental period of 5.0s. Consideration of only the contribution of first two modes leads to relatively good estimates of drift demands for the EW Sakarya record. However, for the Bolu EW component using only two modes of vibrations produces important underestimations for fundamental periods are greater than 2.5s. For both records, the drift demands computed with four modes are practically the same as those computed with six or more modes. These records demonstrate the number of modes that need to be considered depends on the frequency content of the ground motion. In general, for most records using four to six modes of vibration lead to very good estimates of peak maximum interstory drift demands for a wide range of α and fundamental periods.

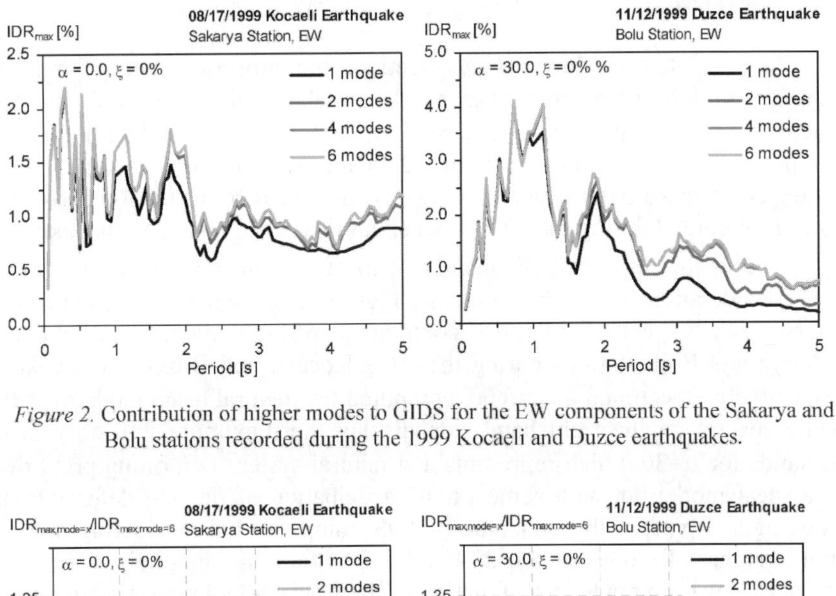

Figure 2. Contribution of higher modes to GIDS for the EW components of the Sakarya and Bolu stations recorded during the 1999 Kocaeli and Duzce earthquakes.

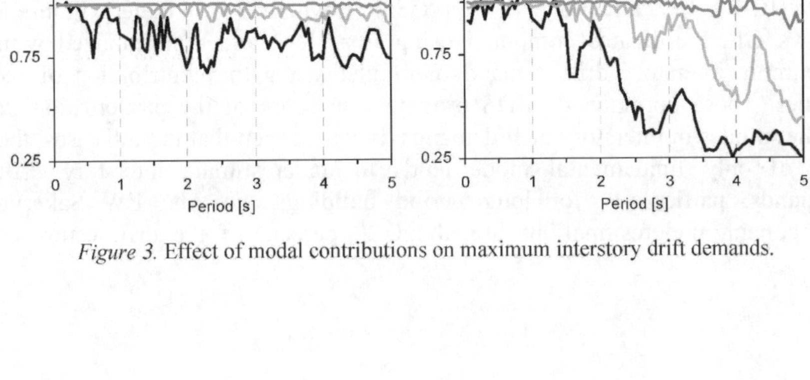

Figure 3. Effect of modal contributions on maximum interstory drift demands.

Figure 4 shows GIDS ordinates computed with α = 0, 5, and 30 normalized by interstory drift demands computed using α = 650. A value of α=650 represents buildings deforming laterally like shear beams, so ordinates less than one in this figure indicate interstory drift demands smaller than those computed for shear beams. The ground motions used in Figure 4 are the same ones presented in Figures 2 and 3, and the empirical relationship $T_1=0.0853H^{0.75}$ is used to compute the total height required for a given fundamental mode. It can be seen that the estimation of interstory drift demands using a shear beam model can lead to overestimations for a wide range of period of vibrations. For the EW Sakarya record, the interstory drift demands computed with α=5 yield interstory drift ratios that are approximately 15 percent smaller than those computed with the shear beam model for A wide range of periods of vibration. Furthermore, for this particular record, interstory drift demands computed with flexural beams (α=0) tend to have similar demands to those experienced by shear beam models with differences fluctuating between 15 to 25 percent. The influence of the lateral stiffness ratio on interstory drift demands is more significant for the EW Bolu record. For example, interstory drift demands computed with α=0 can be up to 40 percent smaller than those computed with the shear beam model.

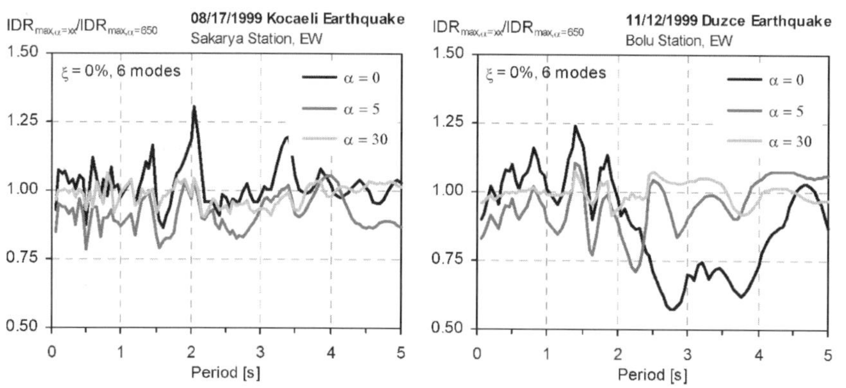

Figure 4. Effect of α on peak interstory drift demands.

Figure 5 shows the influence of damping ratio on maximum interstory drift ratios. The spectra on the left are computed with α=5 which approximately corresponds to buildings with dual lateral-resisting systems while the spectra on the right correspond to models with practically pure shear behavior (α=650). It is shown that, analogously to response spectra ordinates, an increase in damping ratio results in a decrease of interstory drift demands. Figure 6 presents GIDS computed for flexural beams (α=0) using different levels of damping normalized by interstory drift demands of structures deforming approximately like shear beams (α = 650). As previously mentioned, interstory drift demands at different levels of damping vary significantly depending on the ground motion, fundamental period of vibration and lateral stiffness ratio. The normalized GIDS presented on the left correspond to the EW Sakarya record where differences in interstory drift demands between flexural and shear beam models are typically within 25 percent. However, the normalized GIDS on the right that is for the EW Bolu record which exhibits larger differences in interstory drift demands between flexural and shear beam models, particularly for fundamental periods of vibration greater than 2.0s. The difference in interstory drifts demands between flexural and shear beam models increases with decreasing damping ratio.

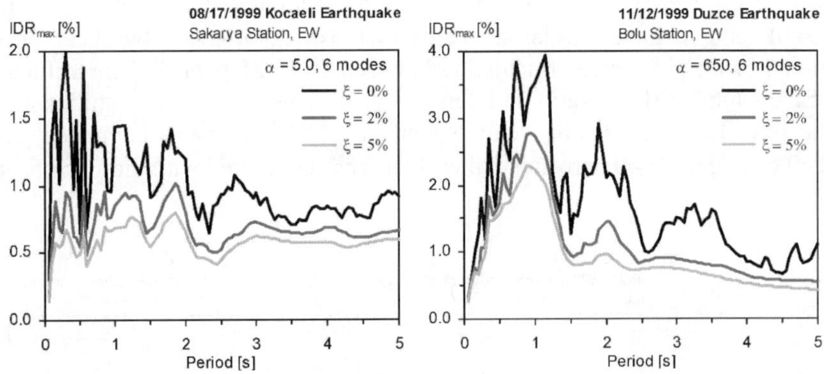

Figure 5. GIDS for different values of damping computed with the EW components recorded at the Sakarya and Bolu stations during the 1999 Kocaeli and Duzce earthquakes, respectively.

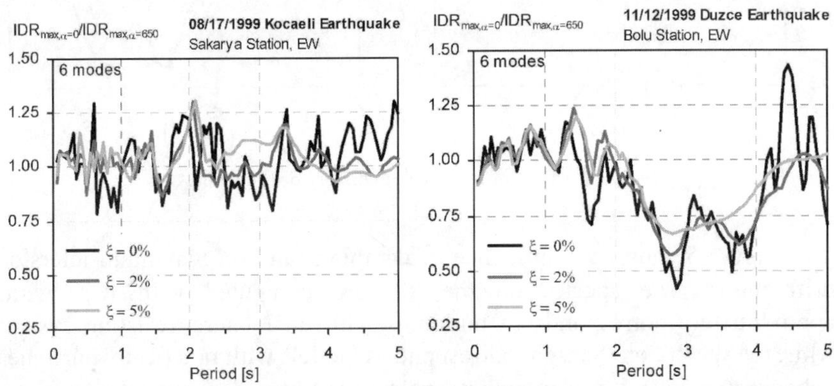

Figure 6. Effect of structural damping on maximum interstory drift demands.

Figure 7 shows generalized interstory drift spectra for buildings with specific lateral-resisting systems. The plots show interstory drift demands computed for the moment-resisting reinforced concrete frames (MRRCFs) subjected to the EW and NS components of the ground motions recorded in the Duzce station during the 1999 Kocaeli and Duzce earthquakes. MRRCF behavior is simulated by using α=12.5, which is a typical value for moment-resisting frames (Miranda and Taghavi, 2004), and considering the mean fundamental period empirical relationship $T_1 = 0.018H^{0.9}$ proposed for such buildings (Chopra and Goel, 2000). The spectra on the left shows interstory drift demands for the NS component of the ground motions recorded in the city of Duzce. The spectra on the right are drawn for the EW component of the same ground motions and display the interstory drift demands multiplied by the building heights corresponding to each fundamental period of vibration. Interstory drift demands computed for the NS components follow a similar trend except at two specific fundamental periods where the record from the 11/12/99 Duzce earthquake yield almost twice larger interstory drifts demands with respect to the 08/17/99 event. On the other hand, comparisons between the spectra computed for the EW components reveal that MRRCFs with fundamental periods greater than 2.0s experience substantially higher interstory drift demands when they are subjected to a ground motion with features similar to the 11/12/99 Duzce event. It should be noted that for this later case both peak ground acceleration and velocity values for the EW component of the 11/12/99 event are significantly larger than those of the 17/08/99 event. This does not occur for the NS components that have comparable peak ground values for both of these events.

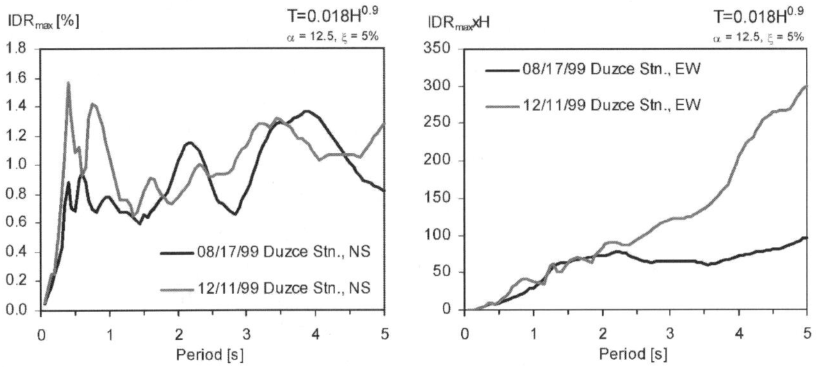

Figure 7. Mean maximum interstory drift estimations for RCMRFs subjected to the ground motions recorded in the Duzce station during the 1999 Kocaeli and Duzce earthquakes.

5. APPLICATION TO RAPID DAMAGE ASSESSMENT

Many studies have indicated that interstory drift ratios are correlated very well with structural damage in buildings. Because of its many computational advantages, the generalized interstory drift spectrum described in the previous sections, can be used in rapid damage assessment at urban areas with wide range of structural systems for a major earthquake event. This can be accomplished by analyzing continuous models corresponding to different structural systems at instrumented locations. A typical example is presented for the Los Angeles Metropolitan area in Figures 8 and 9. These three dimensional contour maps were prepared by using the ground motions recorded at 91 strong motion stations that were triggered during the 1994 Northridge earthquake in the city of Los Angeles and San Fernando Valley. Ground motions were downloaded from the website of COSMOS Virtual Data Center (http://db.cosmos-eq.org) and all of them are either free-field records or ground motions recorded at the ground floor of one-story buildings. Figure 9 shows interstory drift demands for buildings with fundamental periods T_1=1.0s corresponding to mid-rise MRRCFs in this region. The value of α was taken as 12.5 and the mean fundamental period empirical relationship by Chopra and Goel (2000) was used to compute the building height as a function of the fundamental period of vibration. Mean interstory drift ratios of both horizontal components are computed at each recording station using 5 percent damping ratio.

The maps presented in Figure 9 indicate that for the given distribution of strong motion stations, the MRRCFs with fundamental periods of approximately 1.0s located in the northern portion of the San Fernando Valley were subjected to large interstory drift demands and therefore more susceptible to serious structural/nonstructural damage compared to those of stiffer or more flexible buildings in the same region. Another important observation from this map is the consistency of computed interstory drift demands with the reported rupture direction.

It should be noted that even though the elaboration of the map shown in figure 9 involves the computation of the seismic response of both directions of building models located at more than 90 recording stations, they can be computed in personal computers within a few minutes after an earthquake provided that ground motions are set to a central location using telemetry. This map can provide a valuable tool in rapid damage assessment as well as for planning purposes using various scenarios.

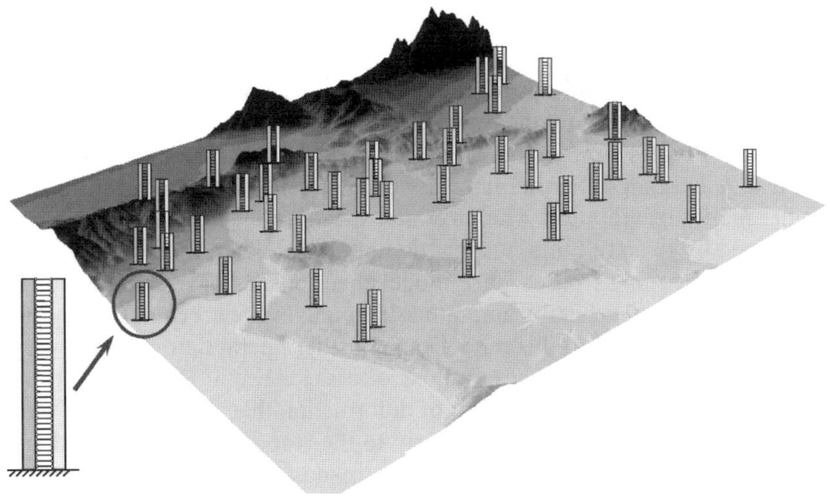

Figure 8. Maximum IDR_{max} contour maps at particular fundamental periods computed from the 91 recording stations deployed around the Los Angeles metropolitan area and triggered during the 1994 Northridge earthquake.

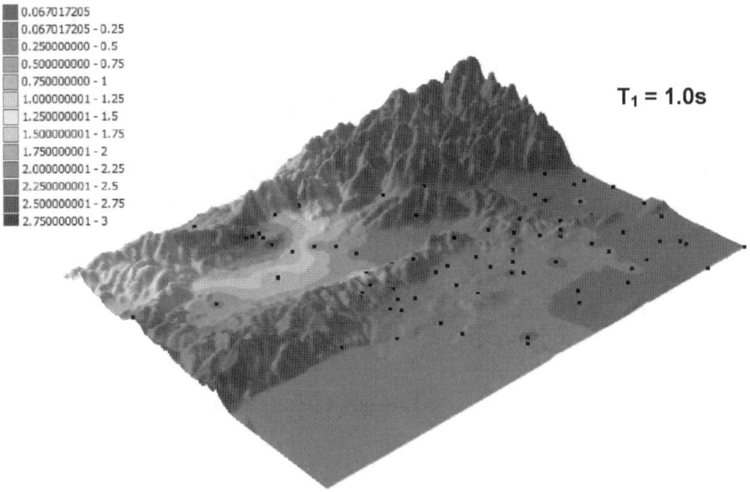

Figure 9. Maximum IDR_{max} contour map for moment-resisting buildings with fundamental periods of vibration of 1.0s computed using 91 recording stations deployed in the Los Angeles metropolitan area and triggered during the 1994 Northridge earthquake.

6. SUMMARY AND CONCLUSIONS

A new measure of seismic demand of earthquake ground motions has been presented. The new measure, referred to as generalized interstory drift spectrum provides a direct indication of maximum interstory drift demands in multistory buildings. Furthermore, its ordinates incorporate the influence of higher modes of vibration. The proposed spectrum extends the drift spectrum developed by Iwan to buildings that do not deform laterally like shear beams. The new spectrum is based on a model that consists of a combination of a flexural beam and a shear beam, rather than only a shear beam. By modifying the lateral stiffness ratio this model can consider lateral deformations varying from those of a flexural beam to those of a shear beam. Hence, it permits to account for a wide range of modes of deformation that represent more closely those of multistory buildings. Similarly to the response spectrum method the generalized interstory drift spectrum provides demands for a family of buildings, but in this case the ordinates are better correlated with structural and nonstructural damage.

The results presented in this study indicate that the maximum interstory drift demands are influenced by various structural parameters. The use of simplified models such as shear beam or considering only the fundamental mode behavior may lead to erroneous interstory drift estimates depending on the ground motion features, lateral stiffness and structural damping. The proposed method surmounts these shortcomings efficiently as it can capture an accurate maximum interstory drift estimate by defining only the three prominent parameters of a structural system: (1) the fundamental period of vibration of the building, T_1; (2) a modal damping ratio that represents the damping ratio in the building, ξ; (3) the lateral stiffness ratio α. Using the empirical relations between building height and fundamental period together with these fundamental parameters, the interstory drift demand estimates can be computed accurately for a wide variety of structural systems.

Computation of the generalized drift spectrum takes only fractions of a minute in most personal computers, hence allows for the rapid assessment of hundreds of buildings, within few minutes after an earthquake. A typical case study from the 1994 Northridge earthquake is presented to demonstrate how the proposed procedure can be used as an efficient tool to serve for such a purpose.

ACKNOWLEDGEMENTS

This work was initiated during a postdoctoral stay of the second author at Stanford University. Financial support for this postdoctoral stay was partially provided by the Scientific Research and Technical Council of Turkey. This financial support is greatly appreciated. Ground motion records used in this study are downloaded from the Earthquake Research Department of the General Directorate of Disaster Affairs, Republic of Turkey and COSMOS web site. The authors acknowledge the efforts of ERD and contributing agencies to COSMOS consortium in installing, maintaining and operating the seismic instruments as well as the dissemination of data. Special thanks are also given to Dr. Masayuki Kohiyama and Prof. Eduardo Reinoso for providing assistance in the use of GIS software.

REFERENCES

1. Chopra, A.K., and Goel, R.K. (2000). "Building period formulas for estimating seismic displacements," *Earthquake Spectra*, 16(2), 533-536.
2. Gülkan P and Akkar S. (2002). "A simple replacement for the drift spectrum," *Engineering Structures*, 24(11), 1477-1484.
3. International Conference of Building Officials (ICBO). (1997). "Uniform Building Code," International Conference of Building Officials, Whittier, California.
4. Iwan, W.D. (1997). "The drift spectrum: a measure of demand for earthquake ground motions," *J. Struct. Engrg.*, ASCE, 123(4), 397-404.
5. Kim, J., and Collins, K.R. (2002). "Closer look at the drift demand spectrum," *J. Struct. Engrg.*, ASCE, 128(7), 942-945.
6. Miranda, E., and Akkar, S. (2004). "Generalized Interstory Drift Spectrum." Submitted for publication to *J. of Structural Engineering,* ASCE.
7. Miranda, E., and Taghavi, S. (2004). "Approximate floor acceleration demands in multistory buildings I: formulation." To appear in *J. of Structural Engineering,* ASCE, 106(2).
8. Reinoso, E. and Miranda, E. (2004), "Estimation of floor acceleration demands in high-rise buildings during earthquakes," Accepted for publication *J. of Tall Buildings*, Wiley.
9. Taghavi, S. and Miranda, E. (2004). "Approximate floor acceleration demands in multistory buildings II: applications." To appear in *J. of Structural Engineering,* ASCE, 106(2).

INFLUENCE OF GROUND MOTION INTENSITY ON THE PERFORMANCE OF LOW- AND MID-RISE ORDINARY CONCRETE BUILDINGS

Sinan Akkar[1] Haluk Sucuoğlu[1] and Ahmet Yakut[1]
[1] *Dept. of Civil Engineering, Middle East Technical University, 06531 Ankara, Turkey.*

Abstract: Fragility functions are determined for low- and mid-rise ordinary concrete buildings. A hybrid approach is employed where building capacities are obtained from field data and their dynamic responses are calculated by response history analysis. Lateral stiffness, strength and deformation capacities of the sample buildings are determined by pushover analyses. Uncertainties in lateral stiffness, strength and damage limit states are expressed by using statistical distributions. The seismic deformation demands of the subject buildings are calculated under 82 ground motions. Peak ground velocity is selected as the measure of seismic intensity since it has been observed that maximum inelastic displacements are better correlated with PGV. The results have revealed that the type of investigated buildings is highly vulnerable to strong ground shaking expected during the future earthquakes. Moreover, their fragility increases with the number of stories

Key words: Fragility; Peak ground velocity; Nonlinear response; Ordinary concrete buildings

1. INTRODUCTION

Fragility functions are the essential tools for seismic loss estimation in built environments. They represent the probability of exceeding a damage limit state for a given structure type, subjected to a seismic excitation (Shinozuka et al., 1999). The damage limit states may be defined in terms of global structural capacity parameters such as global drift ratio (maximum roof drift normalized by the structural height), interstory drift ratio (maximum lateral displacement between two consecutive stories normalized by the story height), story shear force, etc. The ground motion intensity measures on the other hand can be spectral quantities, peak ground motion values, modified

P. Gülkan and J.G. Anderson (eds.), Directions in Strong Motion Instrumentation, 123–138.

Mercalli scale, etc. In this respect, fragility curves consider uncertainty associated with structural capacity, damage limit state definition and record-to-record variability of ground motion intensity.

A particular structure type is considered in this study, namely 2 to 5 story reinforced concrete buildings which generally do not comply with modern seismic resistant design and construction practices. These buildings constitute the majority of the vulnerable building stock in Turkey, which is revealed by the recent strong earthquakes in the last decade. The investigated type of buildings is further categorized into sub-groups with respect to their number of stories. Field observations after recent damageable earthquakes have clearly indicated a significantly increasing damage trend with the number of stories. Although this may not be expected in buildings conforming to seismic design regulations, it is very likely otherwise. Figure 1 shows the damage distribution in 6478, 2 to 5 story buildings, in Düzce after the 12 November 1999 M_w 7.1 Düzce earthquake (Sucuoğlu and Yılmaz, 2001).

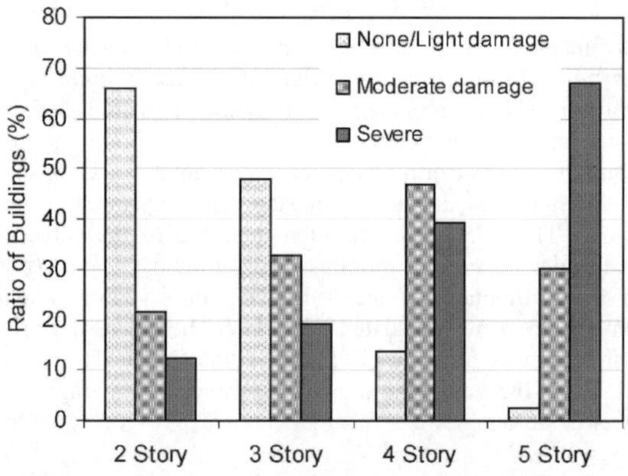

Figure 1. Damage distribution in the city of Düzce after the 1999 Marmara and Düzce earthquakes.

The objective of this study is to determine the fragility functions for 2 to 5 story substandard concrete buildings in Turkey. A realistic description of the fragilities for such buildings is important because almost 75 percent of approximately one million buildings in İstanbul are in this category, and İstanbul is under significant seismic risk (Parsons et al., 2000). The uncertainties in structural characteristics are taken into account by considering the variability in vibration periods, lateral strengths and damage limit states that are obtained from field data. Using a suite of 82 strong ground motions, spanning a broad range of intensities on the other hand, incorporates randomness of seismic excitations. The fragility functions are determined separately for 2 to 5 story concrete buildings as the probabilities of exceeding the specified displacement-based damage limit states under ground excitations with seismic intensities expressed in PGV terms.

2. FIELD DATA

Lateral load-deformation characteristics of typical reinforced concrete buildings in Turkey have been determined by investigating 32 sample buildings. These buildings were selected to represent a typical subset of a comprehensive database consisting of nearly 500 buildings that was compiled in the city of Düzce after the 1999 earthquakes, based on post-earthquake damage assessments (Yakut et al., 2003). All buildings were reinforced concrete frame structures with masonry infill walls, the most common and predominant construction type in Turkey. They were further classified into 4 height groups based on the number of stories. The average heights for 2, 3, 4 and 5 stories were 6.0, 8.9, 11.7 and 14.4 meters, respectively.

Three-dimensional models of each sample building were prepared and nonlinear static analyses were conducted to determine the base shear versus roof displacement relationship (capacity curve). Then modal properties were computed consistently, which conform to the initial linear part of the bi-linear representation of capacity curve. Flexural elements for beams, beam-column elements for columns, strut elements for infill walls and rigid diaphragms for floors were employed for modeling the structural components of the buildings. Nonlinear behavior of the components was modeled per guidelines given in FEMA-356 (ASCE, 2000), and general recommendations in the literature. The capacity curve of each building was approximated with a bilinear curve using the guidelines given in FEMA-356. Several parameters determined for each building are marked on the typical bi-linear capacity curve shown in Figure 2. A typical capacity curve in this study represents the variation of base shear coefficient against the global drift ratio. The ultimate drift ratio denoted by θ_u corresponds to the point at which the structure reaches the state of a mechanism. The yield drift ratio θ_y represents the global drift capacity when the yield base shear capacity (V_y) of the building is exceeded. The relative yield base shear capacity with respect to the total weight of the building W is defined as the yield base shear coefficient η.

Detailed information on the general dynamic characteristics of the investigated building stock is presented in Figure 3. The fundamental periods of the buildings in two principal directions range approximately from 0.15 s to 0.90 s. Apparent scatter for each story group reflects the natural characteristics of the representative building stock employed. The boxes shown in Figure 3.b represent effective period ranges selected for each story group (mean plus/minus one standard deviation) where the effective period ranges for 2, 3, 4 and 5 story buildings are 0.15-0.30, 0.25-0.45, 0.30-0.55 and 0.40-0.65 seconds, respectively. The effective building periods compared with the equations proposed by Chopra and Goel (2000) in Figure 3.a reveal good correlation for the upper limit and the mean values whereas their lower limit does not bound the data. A lower limit calculated by using mean minus standard deviation of the data constitutes a better representation.

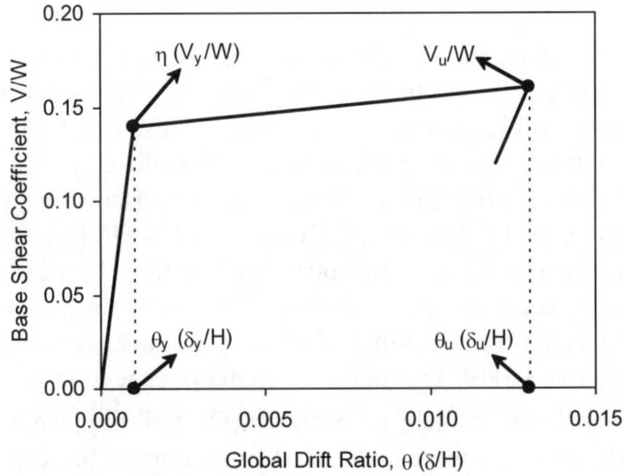

Figure 2. A typical bi-linear capacity curve.

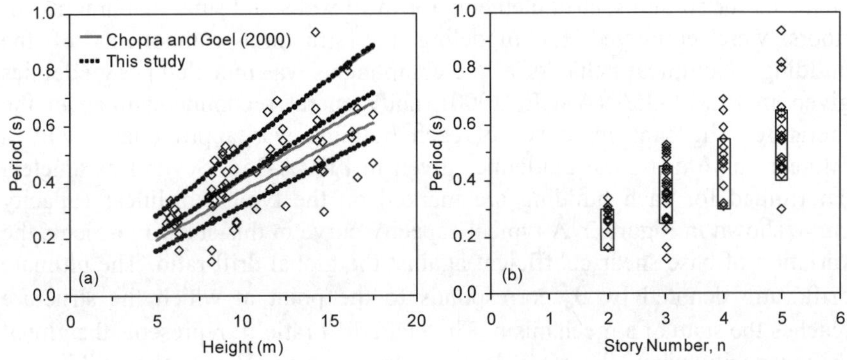

Figure 3. Variation of fundamental period with **(a)** building height, and **(b)** the number of stories.

The variation of yield base shear coefficient with effective period was obtained for the buildings in the database as shown in Figure 4. The spectral variation of the code yield base shear coefficient that was promulgated in 1975 and updated in 1998 is also displayed in Figure 4, and compared with the field data. There is relevant information to support that all of the buildings in the database were dated from the post-1975 period, accordingly their earthquake designs were expected to conform to the provisions of the 1975 edition of the Turkish Seismic Code (1975). It can be observed from Figure 4 that there is inherent over strength in the period range shorter than 0.4 seconds compared to the 1975 Code, however most sample buildings in the 0.4s-1.0s range fail to satisfy the minimum base shear capacity requirement of the Code.

In addition to the yield base shear coefficient η, two displacement parameters indicated on the capacity curve in Figure 2, namely the yield drift ratio θ_y and the ultimate drift ratio θ_u are of essential importance in the displacement-based performance evaluation of the buildings. The statistical properties of these three parameters are given in Table 1 separately for each number of stories. There is a clear trend that η and θ_u decrease with increase in the number of stories.

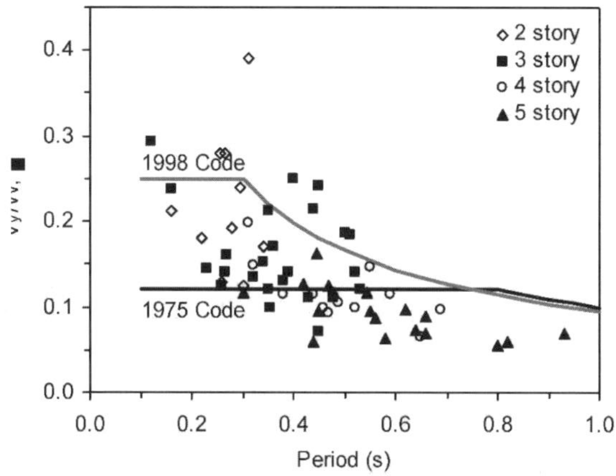

Figure 4. Comparison of the base shear capacities of sample buildings with code requirements.

When the representative probability density functions of θ_y and θ_u are examined, larger dispersion is observed for θ_u (Figure 5). The structural characteristics of the concrete building stock in Turkey lead to global structural yielding much sooner compared to the buildings in other countries such as US and Japan that in turn yields small values of yield drift ratio with small dispersions as evidenced in Figure 5. The peculiarities of Turkish buildings, which are reflected well in the data set, are taken into consideration in assigning performance limit states in terms of global drift ratios. Three performance limits, immediate occupancy, life safety and collapse prevention that are specified in several other international specifications are adopted here. The performance limits expressed with median drift ratios are presented in Table 2, and employed in the analytical procedure described later while developing the fragility curves. It is assumed that light, moderate and severe damage is experienced when the immediate occupancy, life safety and collapse prevention drift limit states are exceeded, respectively.

Table 1. Statistics of the capacity curve parameters

Parameter	Story Number	Mean	Median	Standard Deviation
η	2	0.21990	0.19300	0.08082
	3	0.16233	0.14100	0.05478
	4	0.11667	0.10600	0.03438
	5	0.09339	0.09000	0.02927
θ_y	2	0.001175	0.00100	0.00054
	3	0.001117	0.00110	0.00035
	4	0.001291	0.00120	0.00040
	5	0.001084	0.00110	0.00026
θ_u	2	0.01460	0.01600	0.00334
	3	0.01410	0.01500	0.00463
	4	0.01390	0.01500	0.00380
	5	0.01130	0.01200	0.00430

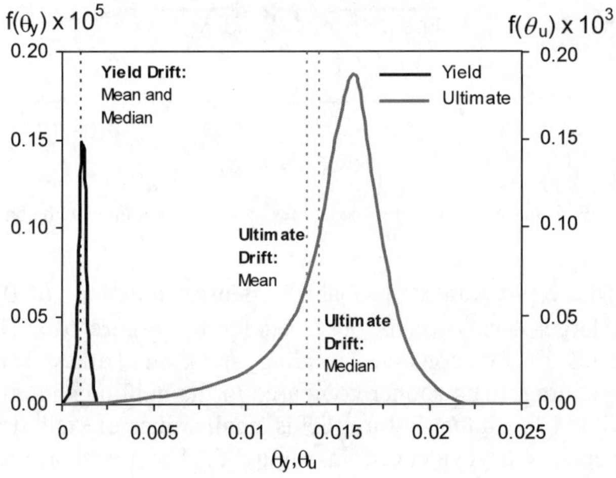

Figure 5. Representative sketch of the probability density functions for drift ratios.

Table 2. Drift ratio limits for performance objectives

Performance Objective	Limit
Immediate occupancy	$\theta \leq \theta_y$
Life safety	$\theta \leq 0.5\ (\theta_y + \theta_u)$
Collapse Prevention	$\theta \leq \theta_u$

A number of attempts have been made recently to recommend idealized capacity curves for the common building types in Turkey. A study conducted by Japan International Cooperation Agency (JICA, 2002) and the İstanbul Metropolitan Municipality (2002) focused on estimating losses from future earthquakes that are likely to impact İstanbul. In this study, the capacity curves were idealized by elasto-plastic curves obtained from simplified analyses. A further endeavor by Boğaziçi University (BU) dealt with earthquake risk assessment for İstanbul area (Boğaziçi, 2002). The capacity curves were represented by elastic-perfectly plastic behavior similar to JICA study.

In order to investigate the differences between the capacity curves of the buildings employed herein and those proposed in other studies, the mean capacity curves obtained for each height category are compared in Figures 6.a and 6.b including the HAZUS (National, 1997) recommendations for moderate code. The significant difference between HAZUS and Turkey is quite expected due to the differences in construction practices as well as code enforcement and compliance efforts. Another source for the differences between this study, and JICA and BU studies are the simplifications implemented in modeling. None of these explanations however justifies the over estimation of capacities in the JICA study, which is regarded as the basic document for loss estimation in İstanbul based on an expected M7.5 earthquake in the Western Marmara Sea. The curves in this study were obtained from the analyses that were based on rigorous 3-D modeling whereas JICA and BU studies are the results of simpler analyses using certain approximations and assumptions.

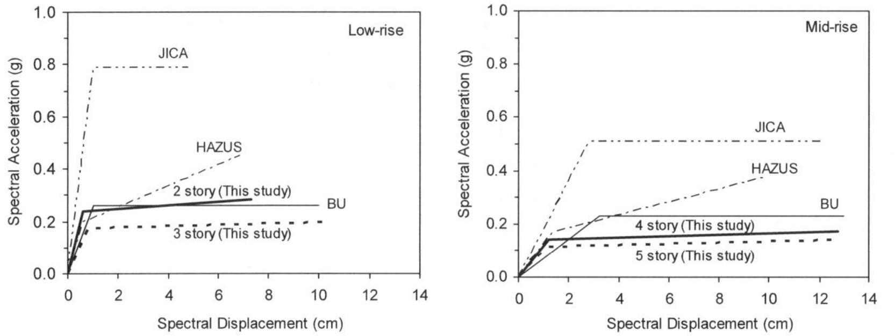

Figure 6. Comparison of capacity curves for low-rise (2-3 story) and mid-rise (4-7 story) buildings.

3. GROUND MOTION DATA

A set of 82 strong ground motion records are used to compute the empirical building fragility curves based on the building information given earlier. The strong ground motion data consists of dense-to-firm soil earthquake records with surface wave magnitudes (M_s) ranging from 5.2 to 7.6. The soil profiles correspond to NEHRP C and D sites with average shear waves of 360 m/s < v_s < 750 m/s and 180 m/s < v_s < 360 m/s, respectively (ASCE, 2000). These soil conditions represent local site geology of the building data set described in the previous section. The magnitude range is pertinent to ground motions of moderate to large earthquake scale. The site-to-source distance (d) is bounded by 20 km since longer distances do not create a serious concern for seismic loss estimation in large urban areas. Pulse dominant waveforms due to forward directivity effects in the proximity of causative sources are eliminated as such records have distinct effects on the general structural behavior. Figures 7 and 8 present the variation of peak ground acceleration (PGA) and peak ground velocity (PGV) of the ground motion data set with respect to magnitude and distance. Both figures show a clear trend that the increase in magnitude and decrease in distance are both associated with an increase in the PGA and PGV values.

Among various ground motion intensity measures, PGA and PGV are more popular than the others mainly because of their simplicity. These two intensity parameters are comparatively evaluated herein for obtaining the most consistent representation of structural damage variation in the empirical fragility curves. The inelastic (Δ_{ie}) spectral displacement scatter diagrams are drawn in Figure 9 as functions of PGA and PGV for comparative evaluations. The spectral displacement values are presented for elasto-plastic systems with constant strength reduction factor R that is defined as the ratio of elastic strength to yield strength of a single-degree-of-freedom (SDOF) system under a given earthquake ground motion. The left and right columns show the scatter diagrams as a function of PGA and PGV, respectively, whereas the rows correspond to the vibration periods of 0.2s, 0.5s and 1.0s simulating relatively short, medium and long structural periods. The inelastic deformation scatters in Figure 9 suggest the superiority of PGV over PGA regardless of the structural period. The weaker correlation of PGA with the increasing deformation demand is noteworthy. Thus, PGV appears to be more suitable ground motion intensity parameter for describing deformation demands in structures that deform beyond the elastic range. Though not shown in the context of the paper, the relation between the elastic structural deformation demand and ground motion intensity is described fairly well by PGA for short and medium period structural systems. As the structural period shifts to longer values (i.e. 1.0s), PGV also correlates well with the elastic structural deformation demand.

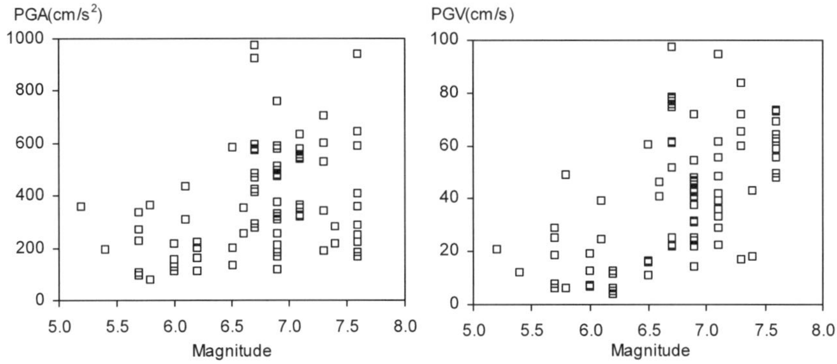

Figure7. Variation of PGA and PGV with respect to magnitude

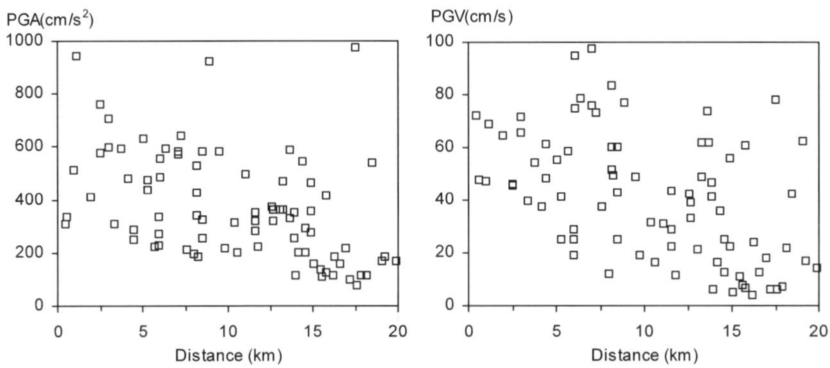

Figure 8. Variation of PGA and PGV with respect to distance

The observations highlighted in the preceding paragraph confirm the conclusions derived by Wald et al. (1999). In their study, Wald et al. indicated that low levels of structural damage identified by the modified Mercalli scale less than VII correlate well with PGA. As structural damage increases (i.e. modified Mercalli scale greater than VII), PGA values level off and PGV is more indicative for defining the correlation between structural damage and ground motion intensity. The merits of PGV for seismic hazard and loss estimation studies have been realized in various countries. Seismic contour maps based on PGV are available in earthquake prone countries like Canada and Australia and research efforts on the computation of more realistic, PGV based design spectrum are in progress (Lam et al., 2000). Under the guidance of these observations, PGV is selected as the ground motion intensity measure in the derivation of empirical fragility curves. A detailed description of the analytical method for obtaining the fragility functions is presented in the next section.

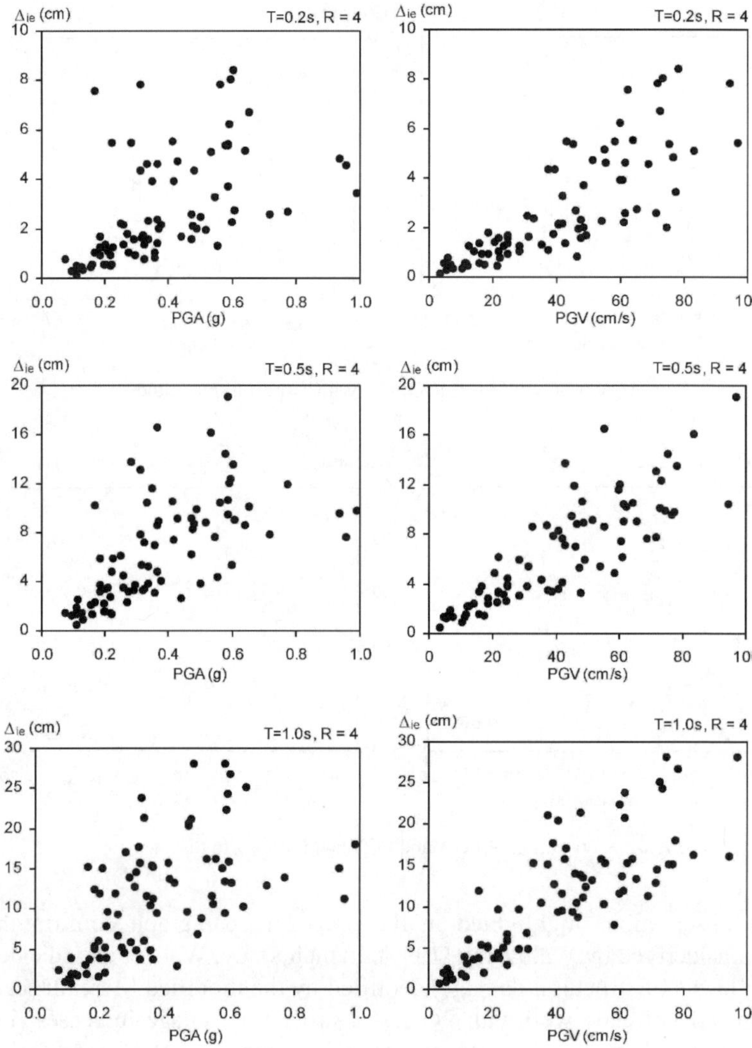

Figure 9. Scatter diagrams for inelastic spectral displacements as functions of PGA and PGV.

4. ANALYTICAL METHOD

Generically, fragility curves are conditional cumulative distribution functions that define the exceeding probability of a damage state for a given ground motion intensity level. The probability distribution function is the standard lognormal distribution in most cases, and the curves represent median fragility values. The lognormal distribution fit is assured by certain optimization algorithms and goodness-of-fit tests (Shinozuka et al., 2000; Kircher et al, 1997).

Table 3 lists the median damage threshold levels used for defining the performance levels of building groups with different number of stories. The details of the pertinent statistics and the computation of these thresholds were given in the previous sections.

Table 3. Median thresholds for performance levels and the associated damage limit states (in terms of global drift)

Story Number	Immediate Occupancy (Light Damage)	Life Safety (Moderate Damage)	Collapse Prevention (Severe Damage)
2	0.0011	0.0085	0.016
3	0.0011	0.0080	0.015
4	0.0012	0.0081	0.015
5	0.0011	0.0065	0.012

The median life safety and collapse prevention drift thresholds decrease with the increase in the number of stories. On the other hand, the threshold for immediate occupancy that can be considered as a transition boundary between global elastic to inelastic behavior practically attains the same regardless of the number of stories. It is noteworthy that all of the above drift-based performance levels are smaller than the ones proposed in the building rehabilitation standards of U.S. or Japan (ASCE, 2000; JICA, 2002; National, 1997). This must be understood as a serious concern for addressing the country specific vulnerabilities, and accordingly loss estimation policies. The damage levels presented in Table 3 are consistent with other individual studies that investigate the drift capacity of reinforced concrete frame buildings in Turkey (Sucuoğlu et al., 2004).

The relationship between the vibration period, base shear capacity and the number of stories is established by using the field data presented previously. For each building group identified with a different number of stories, the relevant mean and standard deviation statistics are computed for the associated base shear capacity and natural vibration period distribution. These statistics are then employed to find the effective base shear and period ranges that represent the group of buildings investigated. The corresponding ranges cover the intervals of mean plus/minus one standard deviation for the parameters of interest to account for their central dispersion. Figure 10 illustrates these intervals with respect to the number of stories when they are superimposed on the actual trend of the building data. Each block in Figure 10 represents the overall distribution of period and base shear for a building group with a particular number of stories. The overlapping boundaries of the blocks indicate the extent of interaction between the building groups due to their inherent dynamic features. It can be observed that the variation in base shear capacity decreases whereas the dispersion in period increases with the increasing number of stories.

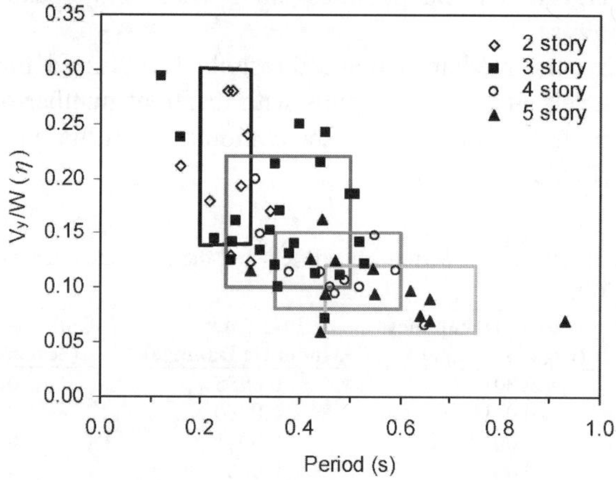

Figure 10. Effective period and base shear capacity ranges for different building groups.

The set of 82 records comprising the ground motion data is used to compute the inelastic displacement response history of SDOF systems representing the base shear capacities and natural vibration periods of building groups presented above. The variations in building periods and base shear coefficients are described more accurately by dividing the rectangular blocks of Figure 10 into finer meshes. For the sake of uniformity, the period range of each rectangular block is divided into equally spaced intervals of 0.05 seconds. Similarly, the corresponding base shear ranges are divided into four equal intervals. A total of 9020 SDOF inelastic response history analyses were carried out for the associated period and base shear combinations. The capacity curves computed from the building data are represented by a bilinear hysteretic model with 3 percent strain hardening, which corresponds to the median post yielding stiffness ratio of the building data set. Then the computed maximum inelastic SDOF displacements are converted to top story displacements by using the modal participation factors. This process assumes a fundamental mode dominant structural behavior, which can be considered as a reasonable assumption for the existing low- to mid-rise buildings in the data set. Finally, the global drift values are calculated by normalizing the top story displacements with the average building heights defined above for the building inventory.

The maximum global drift values computed by nonlinear response history analyses express the seismic performance of the investigated reinforced concrete frames. Using the defined damage threshold levels in Table 3, the exceeding probabilities of that particular fragility curve are computed from the PGV versus maximum global drift scatters specific to each building group (i.e. 2, 3, 4 and 5 story buildings). The scatter diagrams are clustered for different PGV intervals and the ratio of global drifts greater than a given damage threshold level to the total number of global drift values in each cluster is computed to estimate the exceeding probabilities of the fragility curves. Exponential functions are fit over the jaggedly varying exceeding probability points to achieve smooth fragility curves for that specific damage state and building group. A representative sketch for the above procedure is shown in Figure 11.

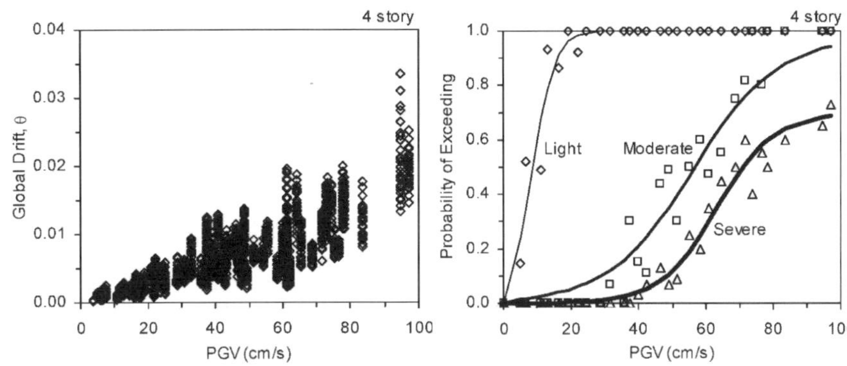

Figure 11. The scatter diagram and the corresponding fragility curves for four story buildings

5. RESULTS AND DISCUSSION

The fragility curves produced by the presented methodology are shown in Figures 12.a-d for 2, 3, 4 and 5 story reinforced concrete buildings, respectively. The three curves in each figure represent the probability of exceeding the immediate occupancy (light damage), life safety (moderate damage) and collapse prevention (severe damage) limit states respectively, from left to right. Comparisons of Figures 12.a-d also indicate that the number of stories has a significant effect on the probability of exceeding the moderate, and especially the severe damage limit states.

The results presented in Figure 12 in terms of the median fragilities can be compared and verified with the damage distribution observed in Düzce, which was shown in Figure 1. It is appropriate to exclude undamaged and lightly damaged buildings since they are not separated in Figure 1. The geometric mean of PGV values recorded along the two horizontal components in Düzce is 70.8 cm/s. The fragilities associated with this PGV value can be read from the related curves in Figures 12. A comparison of the calculated median probabilities of exceeding moderate and severe damage limit states, and the corresponding observed damage ratios are given in Table 4 for 2 to 5 story buildings. Median fragilities exceed the observed damages by 10 and 66 percent in 2 story buildings and by 25 and 85 percent in 3 story buildings for moderate and severe damage limit states, respectively. The fragilities slightly underestimate the observed damages for 4 story buildings where the differences are within 15 percent. For 5 story buildings the error bound for the fragility estimations is approximately 10 percent with respect to the observations in the field. One reason for the discrepancies between the observed and estimated damage probabilities could be the subjective building damage assessment in the field. The uncertainties inherent in the median fragilities and randomness of ground motion intensities also play a role for the differences between the observed and estimated damage probabilities. Nevertheless, these preliminary fragility results can be a useful guide for a rapid performance assessment of big building stocks after a large earthquake.

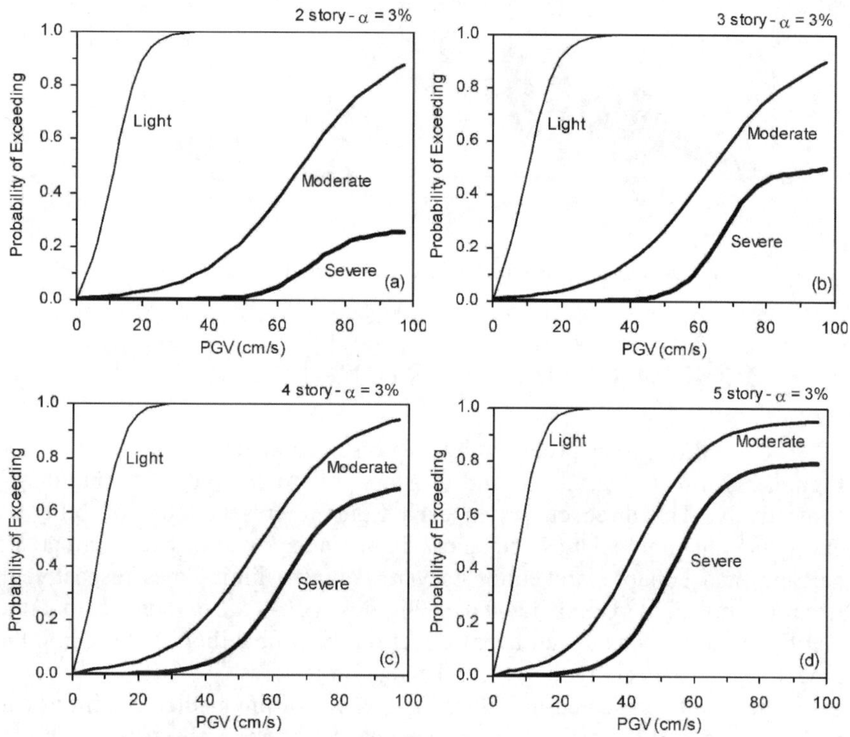

Figure 12. Median fragility curves for 2, 3, 4 and 5 story buildings.

Table 4. Comparison of predicted and observed damage distributions

Story Number	Damage Limit State	Prediction (Median)	Observed (Damage Ratio)
2	>M	56.7	34.1
	>S	13.5	12.3
3	>M	64.9	51.9
	>S	35.1	19
4	>M	75.7	86.4
	>S	35.7	39.4
5	>M	89.2	97.4
	>S	72.1	67.1

6. CONCLUSIONS

There are a number of simple conclusions that can be derived from this study. First, fragility curves have to take into consideration the regional characteristics of the investigated group of structures. Field data is the most valuable information source in identifying such characteristics. Furthermore, fragility curves must be expressed explicitly for the structure sub-groups that are classified with respect to the dominant structural parameters influencing their seismic performance. Here, this parameter was the number of stories. When these two objectives are achieved, fragility curves can serve for more accurate estimation of losses from ground shaking.

Second, when fragility curves are based on maximum inelastic dynamic response displacements calculated under the ground excitations, PGV is a reliable measure of strong motion intensity, and the post yield stiffness has a significant influence on maximum displacement response. Therefore a preference for PGV is suggested over PGA.

ACKNOWLEDGEMENTS

The research work presented in this study is supported in part by the Scientific and Research Council of Turkey (TUBITAK) under grant: YMAU-ICTAG-1574 and by NATO Scientific Affairs Division under grant: NATO SfP977231.

REFERENCES

1. American Society of Civil Engineers (ASCE), 2000. *Prestandard and Commentary for the Seismic Rehabilitation of Buildings, Report No. FEMA-356*, Washington, D.C.
2. Boğaziçi University, 2002. *Earthquake Risk Assessment for İstanbul Metropolitan Area, Final Report,* Kandilli Observatory and Earthquake Research Center, İstanbul.

3. Chopra, A. K. and Goel, R. K., 2000, Building period formulas for estimating seismic displacements, *Earthquake Spectra*, **16**, 533-536.
4. Japan International Cooperation Agency (JICA) and İstanbul Metropolitan Municipality, 2002. *The Study on A Disaster Prevention / Mitigation Basic Plan in İstanbul including Seismic Microzonation in the Republic of Turkey, Final Report*, İstanbul.
5. Kircher, C.A., Nassar, A.A., Kustu, O., and Holmes, W.T., 1997. Development of building damage functions for earthquake loss estimation, *Earthquake Spectra*, **12**, 663-682.
6. Lam, N.T.K., Wilson J.L., Chandler, A.M., and Hutchinson G.L., 2000. Response spectral relationships for rock sites derived from the component attenuation model, *Earthquake Engineering and Structural Dynamics,* **29**, 1457-1489.
7. National Institute of Building Sciences, 1997. *Earthquake Loss Estimation Methodology HAZUS97 User's Manual*, prepared for Emergency Management Agent Washington, D.C.
8. Parsons, T., Toda, S., Stein, R.S., Barka, A., and Dieterich, J.H., 2000. Heightened odds of large earthquakes near İstanbul: An interaction-based probability calculation, *Science,* **288**, 661-665.
9. Shinozuka, M., Feng, M.Q., Lee, J., and Naganuma, T., 2000. Statistical analysis of fragility curves, *Journal of Structural Engineering, ASCE*, **126**, 1224-1231.
10. Shinozuka, M., Grigoriu, M., Ingraffea, A.R., Billington, S.L., Feenstra, P., Soong, T.T., Reinhorn, A.M., and Maragakis, E., 1999. *Research Progress and Accomplishment 1999-2000: Selected Papers, MCEER Publication*, SUNY at Buffalo, NY.
11. Sucuoğlu, H., Gür, T., and Günay, M.S., 2004. Performance-based seismic rehabilitation of damaged buildings, *Journal of Structural Engineering, ASCE*, in press.
12. Sucuoğlu, H. and Yılmaz, T., 2001. Düzce, Turkey: A city hit by two major earthquakes in 1999 within three months, *Seismological Research Letters*, **72**(6), 679-689.
13. Turkish Ministry of Public Works and Settlement, 1975, revised in 1998, Specifications for Buildings Constructed in Disaster Areas, Ankara.
14. Wald, D.J., Quitariano, V., and Heaton, T.H., 1999. Relationships between peak ground acceleration, peak ground velocity and modified Mercalli intensity in California, *Earthquake Spectra,* **15**, pp. 557-564.
15. Yakut, A., Yılmaz N., and Bayılı, S., 2003. Analytical assessment of the seismic capacity of RC frame buildings, *International Conference in Earthquake Engineering to Mark 40 Years from Catastrophic 1963 Skopje Earthquake,* Skopje-Ohrid, Paper No. 50.

Part 3

ARRAYS AND OBSERVATIONS

INTEGRATED SURFACE AND BOREHOLE STRONG-MOTION, SOIL-RESPONSE ARRAYS IN SAN FRANCISCO, CALIFORNIA

Empirical Measurements of Low-Strain Site Coefficients at Site Class E and D Soil Sites

R.D. Borcherdt, G. Glassmoyer, C. Dietel and R.E. Westerlund

U.S. Geological Survey (MS 977), 345 Middlefield Road, Menlo Park, CA 94025

Abstract: An integrated set of four borehole arrays and ten surface installations is installed in the city of San Francisco, California to measure the response of soft-soil deposits to strong earthquake ground motions. The borehole arrays extend through thick layers of soft water-saturated soils of Holocene age and older more consolidated soils of Pleistocene age into bedrock at depths up to 90 m. The surface installations are configured in pairs to provide simultaneous comparative surface measurements of soft soils and nearby rock. The rock locations also permit comparative measurements of rock as observed at the surface and in nearby boreholes. The arrays are designed to address a wide variety of scientific and engineering issues, and especially the issue of anelastic and nonlinear soil response at high strain levels as might be recorded during a large regional earthquake. Recordings of ground motions from the largest regional earthquakes which have occurred since the installation of the arrays show marked evidence of amplification as measured on the borehole and surface arrays. Implications of the results for low-strain site coefficients in present U.S. building codes are discussed.

Key words: site coefficients, building codes, site response, San Francisco array, borehole array, soft soil, Yountville earthquake

1. INTRODUCTION

Structural damage and consequent loss of life from earthquakes is often concentrated in areas underlain by soft soils. In the city of San Francisco, damage from both the great California earthquake of 1906 and the Loma

P. Gülkan and J.G. Anderson (eds.), Directions in Strong Motion Instrumentation, 139–155.

Prieta earthquake of 1989 was concentrated in areas underlain by soft soil deposits. Many of these areas are along the margins of San Francisco Bay and are densely urbanized. However, no *in-situ* borehole measurements of the amplification characteristics of these deposits in the SF Bay region have yet been obtained at damaging levels of motion. Such measurements are of special importance for quantifying the response of the built environment in a major metropolitan area for improving earthquake resistant design.

Several important questions exist concerning the *in-situ* response of soft soil deposits at damaging levels of shaking. A question of particular interest concerns the role of non-linearity in modifying the *in-situ* amplification characteristics of soft soil deposits as a function of shaking amplitude. This is a major issue yet to be resolved with *in-situ* evidence. Other important issues concern empirical estimates of site coefficients as a function of maximum strain and material constitutive properties, changes in constitutive properties with duration, ground motion characteristics necessary for the onset of liquefaction, and the influence of basin geometry.

This paper describes a set of integrated borehole arrays and surface installations in the city of San Francisco designed to address these issues. It provides a brief summary of the design, location, geotechnical information, and instrument configuration for the arrays. It summarizes results from two of the largest events recorded to date on the arrays. These recordings are used to estimate site coefficients as presently specified for U.S. building codes. Implications of these low-strain estimates in regard to present low-strain site coefficients for site class E and D soils are discussed.

2. ARRAY DESIGN

The integrated strong-motion soil response arrays in San Francisco operated by the U.S. Geological Survey are designed to provide comprehensive strong-motion data not yet obtained on the *in-situ* response of "soft soils". These data are to be collected using four borehole arrays at locations underlain by thick sections of soft water saturated clays, sands, and silts, and at ten surface locations selected as soil-rock pairs to provide ground-response measurements for comparison. Complementary surface and structural response installations at several nearby sites permit a wide variety of additional scientific and engineering issues to be addressed.

Locations of the borehole arrays and surface installations are shown on maps of the city (Figures 1a and 1b). The map shows the location of the sites in relation to the geographic distribution of site classes as defined in present versions of U.S. building codes. The arrays also were designed to augment data recorded on dense arrays of strong-motion instrumentation in nearby buildings as installed by the Strong Motion Instrumentation Programs of the California Geological Survey and the United States Geological Survey (Figure 1b).

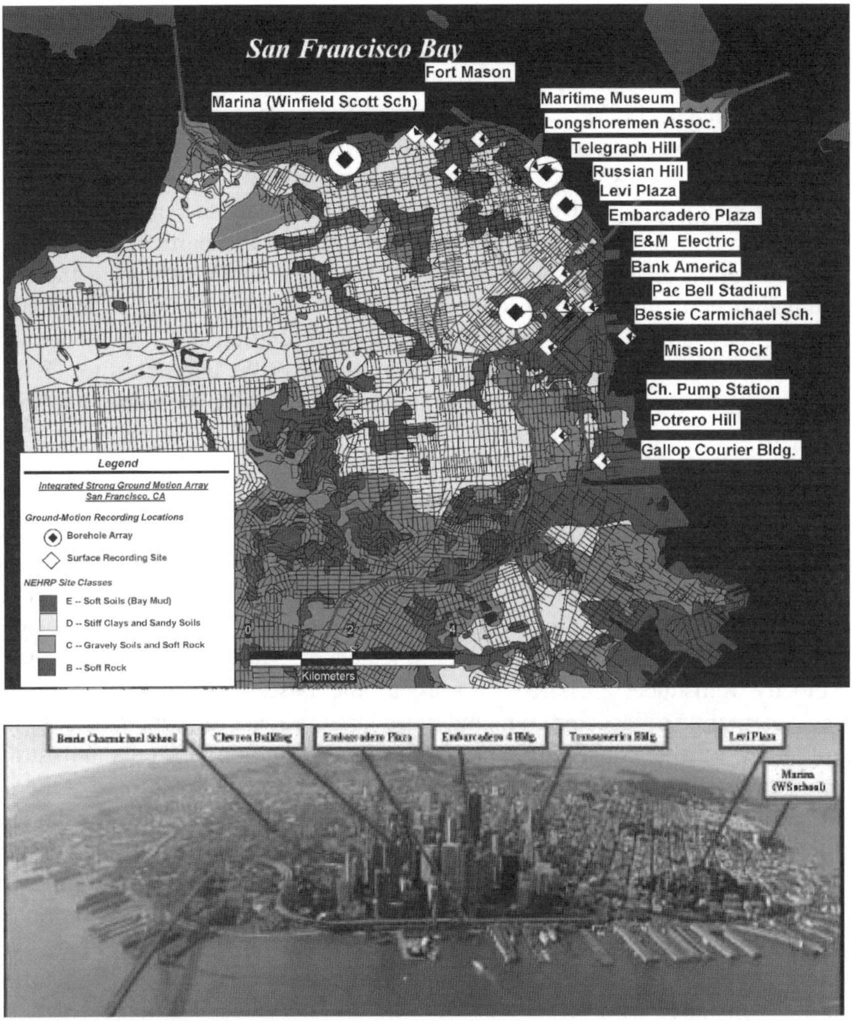

Figure 1. Maps showing location of borehole arrays and surface installations in the Integrated San Francisco Array superimposed on a map showing streets and the site classes adopted in US building code provisions and location of borehole arrays and nearby structural response arrays superimposed on aerial view of San Francisco.

The borehole arrays are located in the vicinity of Embarcadero Plaza (EMB) near lower market street, in Levi Strauss Plaza near the base of Telegraph Hill (LP), at Winfield Scott School in the Marina district (WSS), and at Bessie Charmichael School (BCS) in an area south of Market Street in the city of San Francisco. Each of the borehole array locations is underlain by a thin compacted layer of man-made fill overlying deposits of soft clay, silty clay, and silts termed the Younger Bay Mud , which in turn is underlain by stiff older more consolidated soils, termed the Older Bay Sediments, overlying rock of the Franciscan Formation. Detailed geotechnical and seismic velocity logs for each of the borehole locations are described by Borcherdt, *et al.* (1999). In general, the shear wave velocities range from greater than 150 m/s for the Younger Bay Mud to 350 m/s for the Older Bay Sediments to over 800 m/s for the underlying rock. Seismic, geologic, and sensor configurations are shown for each borehole array site (Figure 2).

Measurements in these areas of San Francisco are particularly relevant for evaluation of performance of major man-made facilities during the next damaging earthquake. Major concentrations of damage occurred in each of these areas during both the 1906 and 1989 earthquakes. Evidence for liquefaction during the earthquakes exists at three of the borehole sites (EMB, WSS, and BCS). Important facilities in the areas include high-rise buildings, apartment buildings, rapid transit facilities, bridges, and underground utilities.

The arrays are designed to provide an integrated set of data pertinent to addressing a wide variety of scientific and engineering issues. The borehole arrays are designed to provide *in-situ* data not previously available on the non-linear response of "soft soils" as a function of strain amplitude with and without liquefaction-induced failure. Three of the borehole arrays include four three-component forced-balance accelerometers (FBA), two co-located wide dynamic-range pore-pressure transducers, and at least one vertical velocity transducer installed in rock at the base. The co-located *in-situ* measurements of acceleration and pore pressure are designed to provide estimates of the constitutive properties as a function of time and strain amplitude that cannot be inferred from either measurement alone (Borcherdt, *et al.* 1988 and 1989). A separate borehole array of velocity transducers co-located with three FBAs is installed at one of the sites (WSS). These collated velocity and acceleration sensors provide high resolution recordings of signals ranging in amplitude from seismic background noise levels to damaging levels of shaking up to 2g. These signals ranging in amplitude over more than 150 dB are useful for studies of non-linearity. The WSS array, initially installed with velocity transducers, has provided an extensive data set interpreted by Liu, *et al.* (1992).

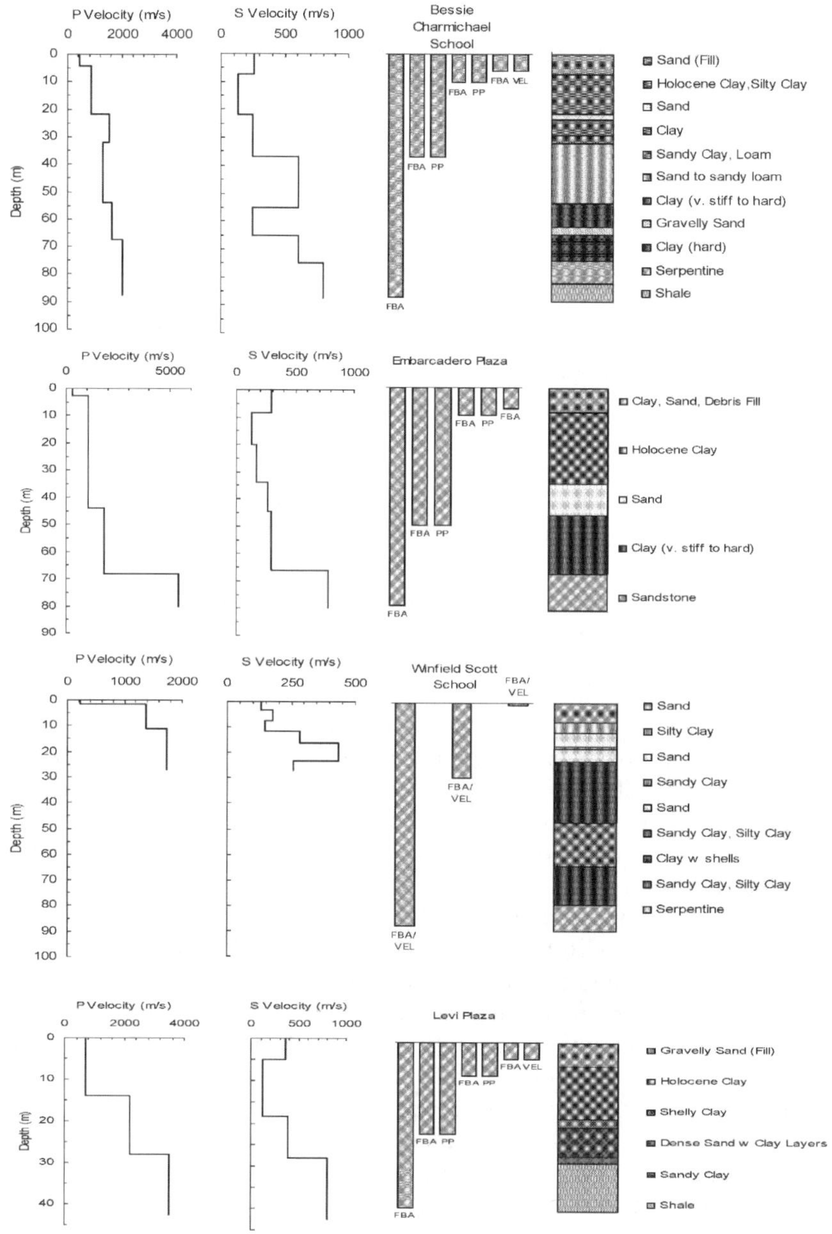

Figure 2. Seismic log (P and S), geologic log, and sensor borehole configuration for the Bessie Charmichael School (BCS), Embarcadero Plaza (EMB), Winfield Scot School (WSS) and Levi Plaza (LP).

The recorders at each borehole array are set to trigger simultaneously in a master-slave configuration using the signal from a borehole velocity transducer in rock. This configuration permits simultaneous recording of small events on all components of the array.

The surface array is designed to provide comparative measurements of response at pairs of sites on rock and soil. These instruments provide additional response measurements of soft soil sites in San Francisco. The surface rock locations also permit comparative studies of motions as observed at the surface and in the boreholes at the soil-rock interface. These measurements are useful for quantifying the response of nearby surface rock sites.

The sensors deployed for direct measurement of acceleration are three component borehole force-balance accelerometers with full scale at $\pm$ 1 or $\pm$ 2 g , and frequency response from DC to 100 Hz, models FBA 13 and FBA 23 DH produced by Kinemetrics, Inc. The velocity transducers with a natural frequency of 2 Hz have been developed for borehole deployment (Liu, *et al.*, 1991). The pore-pressure transducers are wide dynamic range (~140 dB) transducers with digital output proportional to pressure applied to a quartz crystal resonator, model 8DP depth sensor produced by Parascientific, Inc.

Signals from each of the sensors at each location are recorded on site using broadband digital recorders (General Earthquake Observation Systems, GEOS, Borcherdt, *et al.*, 1985) with the capability for event derived parameters transmitted and recorded in near real time via GOES satellite (Mueller *et al.* 1993). Signals from the various sensors are recorded at various gain levels depending on depth and type of sensor. Gains up to 84dB selectable in 6dB steps are available to permit signals over a dynamic range near 180 dB at frequencies near 1 Hz to be resolved from a variety of sensor types. In general, gains for similar sensors in a single borehole array vary between 6 and 18 dB.

Events are recorded in event triggered mode at 200 samples per second per channel using 16-bit linear analog to digital conversion. Absolute time and recording parameters are recorded simultaneously. Event and instrument parameters can be transmitted at a selectable time interval usually chosen to be less than 10 minutes via GOES satellite. Parameters such as trigger time, peak amplitude, time of peak amplitude and duration together with a variety of instrument status parameters are transmitted. The satellite transmissions can permit the retrieval of critical event and instrument data in near real time (Mueller, *et al.* 1993). Sensor and recording system calibrations are recorded before and after each storage media change.

3. EARTHQUAKE RECORDINGS

Twenty five earthquakes with magnitudes in the range M 2.9 - 5.1 have been recorded on the arrays since completion of installation of the borehole arrays in 1992. The two largest of these events are the Yountville (M 5.1) and the Gilroy (M 4.9) earthquake, which occurred north and southeast of the arrays, respectively at distances of 60 to 70 km and 110 to 120 km.

An example of the borehole recordings is shown for the Yountville earthquake as recorded on the Embarcadero Plaza borehole array (Figure 3). The three-component recordings of acceleration show that the motions recorded at the surface and at a depths of 10 meters are significantly larger in amplitude and richer in lower frequencies than those recorded at larger depths in the stiffer and more consolidated soil deposits at 49m and in rock at 79 m. Similar observations are apparent from plots of the recordings obtained on the other borehole arrays for these events (not shown).

The north-south component of velocity plotted at the same scale for the Yountville earthquake as recorded from sensors on the surface are shown in Figure 4. As an additional example, similar plots of acceleration as recorded for the Gilroy earthquake are shown in Figure 5. The equiscaled plots show that the motions at the surface of the soft soil sites are significantly larger than those recorded at the surface of sites underlain by firm to hard rock.

Amplitude spectral ratios for the radial component of motion are shown for eleven sites that recorded the Yountville earthquake (Figure 6). The spectral ratios show that the amplitude response is strongly frequency dependent especially for sites underlain by soft soils. Several of the softer sites indicate dominant periods of amplification in the period band 0.5 to 2 seconds. The well defined peaks in some of the spectral ratios suggest that resonance plays a major role in determining the final surface motion.

Amplitude spectral ratios for the recordings obtained at stations underlain by rock (PBS, P50, RHL, M9A C9A, and E8A) also show significant variations. Those computed for the PBS and P50 sites show evidence of dominant periods which are suspected of being associated with structural response and topographic amplification, respectively. The RHL site on Russian Hill is similar in response to that of FM3. The base motion of the Marina (WSS) site (M9) is consistently larger than that recorded at the other borehole locations in rock at their base. The ratio computed for the recordings at the base of the borehole arrays at the Embarcadero Plaza site and the Charmichael school site provide an estimate of the response of the near-surface rock layers beneath the FM3 site.

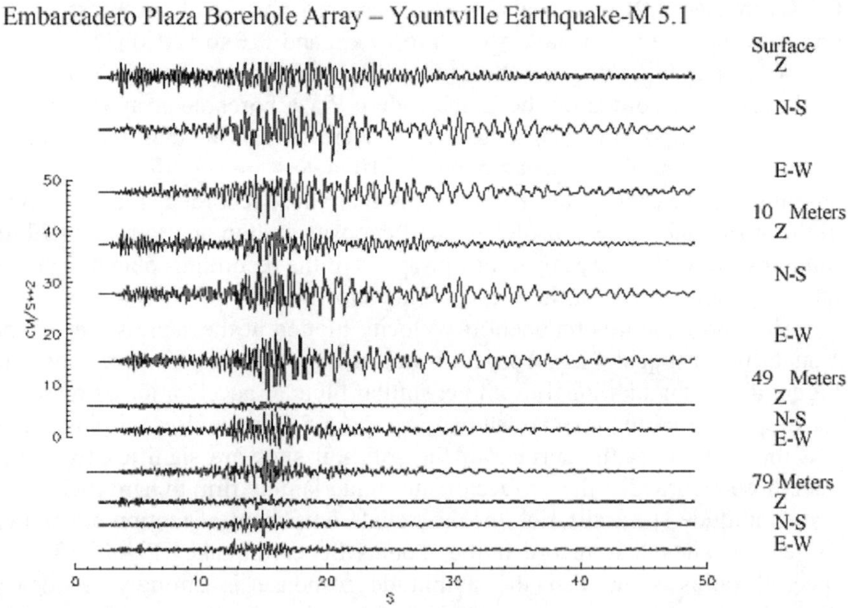

Figure 3. Vertical (Z), North-South (N-S), and East-West (E-W) recordings of the Yountville earthquake (M 5.1) on the Embarcadero Plaza Borehole Array at the surface and depths of 10 m, 49.4 m, and 79.2 m.

The Fourier amplitude spectral ratios for the radial component of motion recorded at the Embarcadero site (not shown) computed with respect to the corresponding component of motion recorded in rock at the base of the borehole array show a well defined fundamental period and successively higher modes decreasing in amplitude. The well defined modes provide strong evidence of resonance motion in the surface layer with the amount of damping in the soil increasing with increasing frequency. This type of response is explained well using a simple one-dimensional model of vertically propagating homogenous S waves and a viscoelastic soil layer over a viscoelastic half space (Borcherdt, 2004).

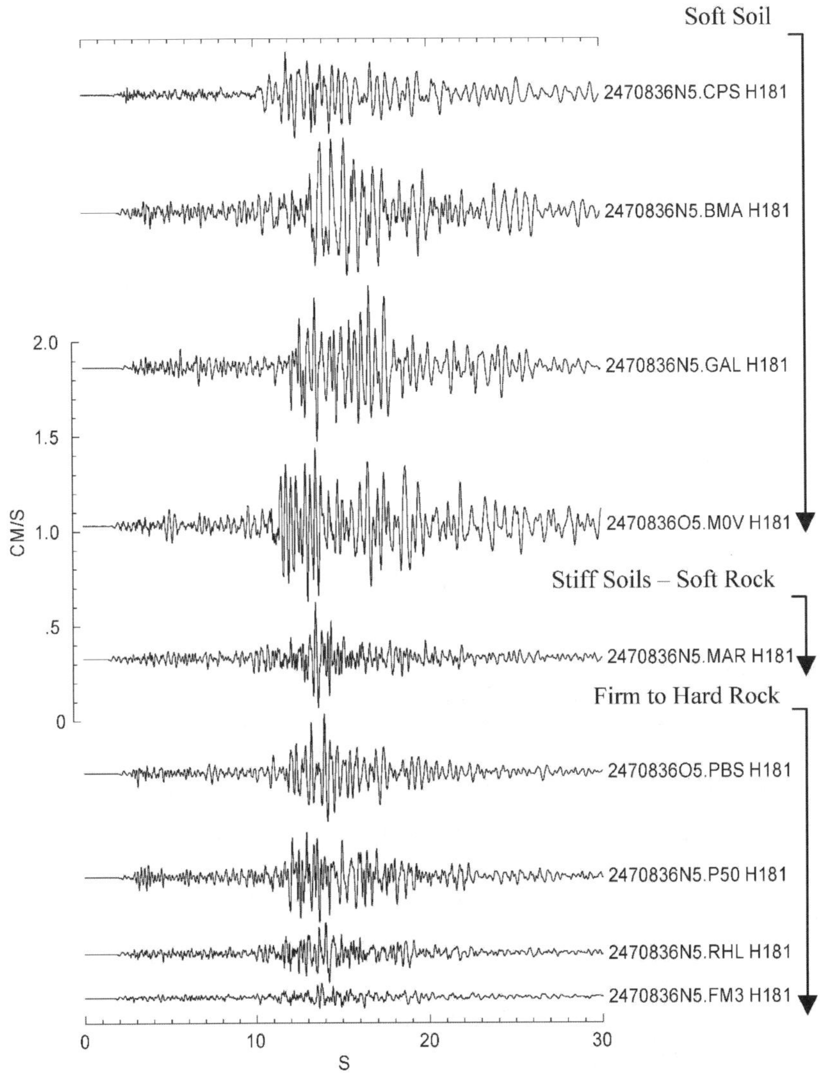

recorded at the surface at nine sites in the San Francisco Integrated Array. The equiscaled recordings show that the velocities recorded at the surface of sites underlain by soft soil are significantly greater than those recorded at the surface of firm to hard rock sites.

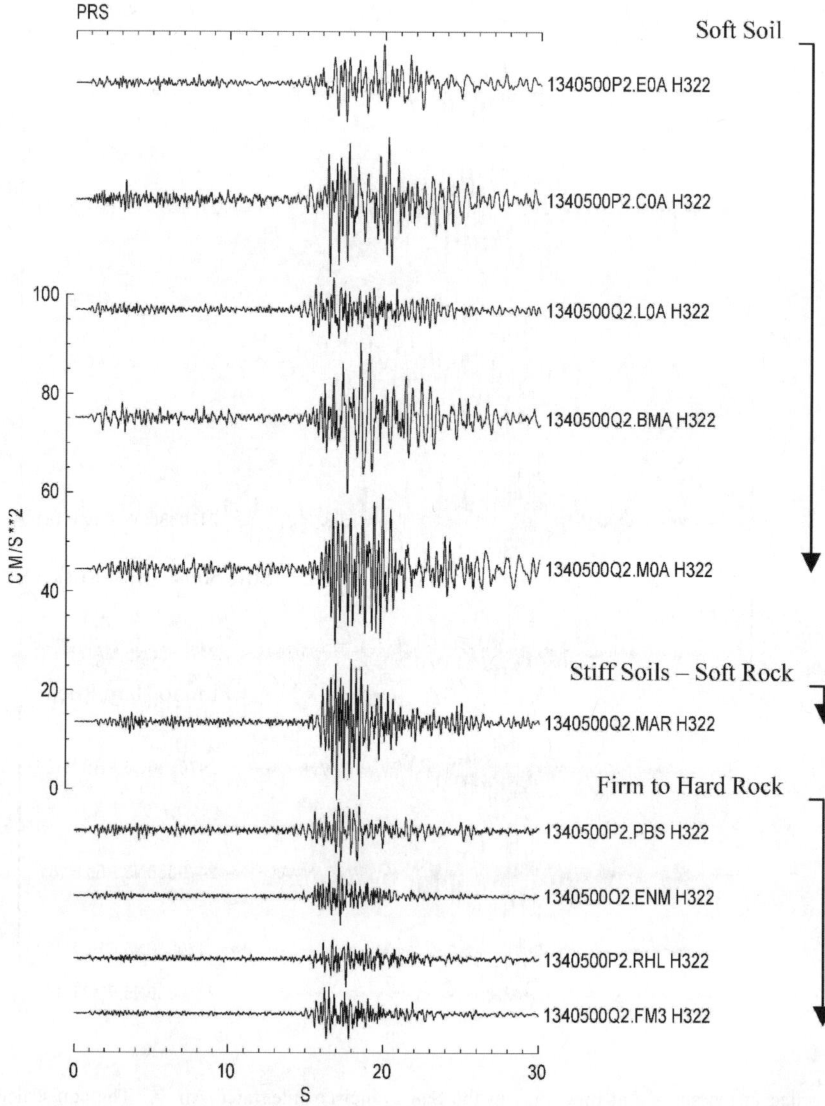

Figure 5. North-South component of acceleration for the Gilroy earthquake (M 4.9) as recorded at the surface at ten sites in the San Francisco Integrated Array. The recordings show that the accelerations recorded at the surface of sites underlain by soft soil are significantly greater than those recorded at the surface of firm to hard rock sites.

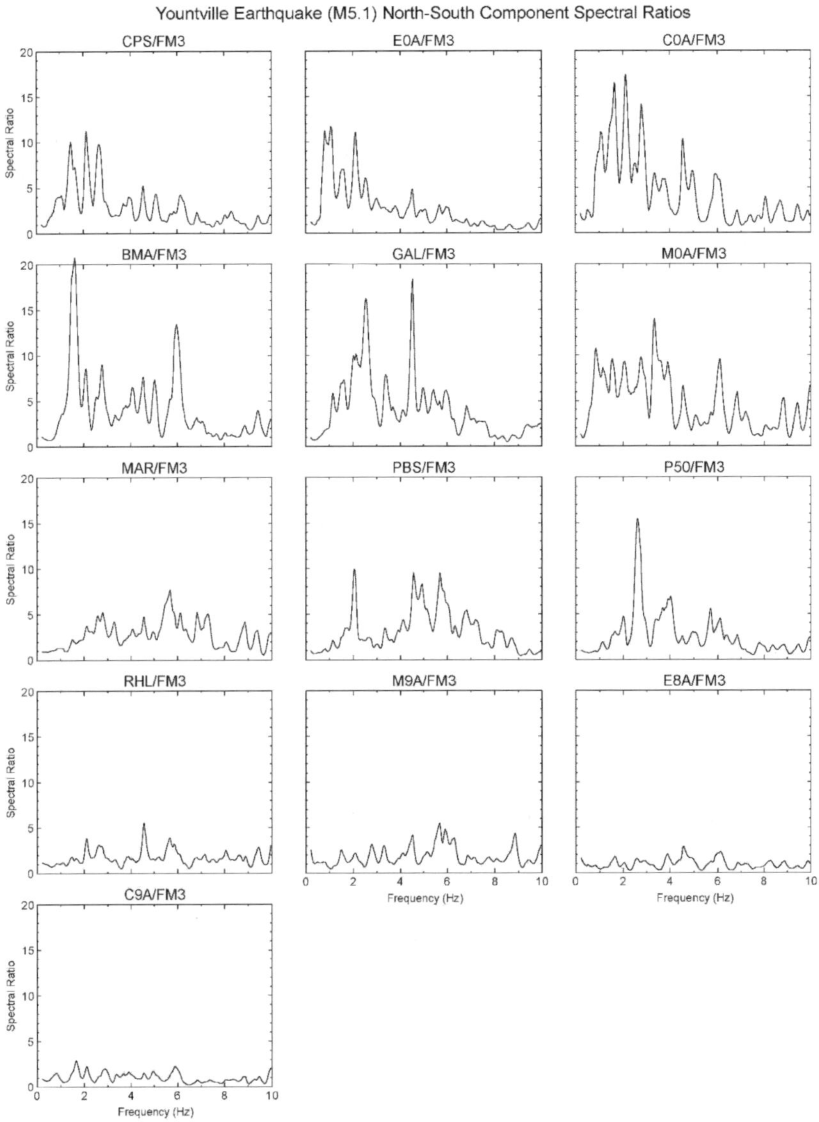

Figure 6. Amplitude spectral ratios for the north-south (radial) component of acceleration as recorded from the Yountville earthquake at 10 sites in the San Francisco Array and computed with respect to the hard-rock, surface recording at the FM3 (Fort Mason) site.

4. EMPIRICAL MEASUREMENTS OF SITE COEFFICIENTS

To derive empirical measurements of site coefficients, Fa and Fv, as presently specified in U.S. building codes, Fourier amplitude spectral ratios were derived for each of the recordings with respect to the corresponding component of motion recorded at the Fort Mason site (FM3). Averages of these amplitude spectral ratios over appropriate period bands provide estimates of site coefficients with respect to a common surface rock location consistent with the procedure used to empirically estimate current site coefficients for base input motions less that 0.1g. Estimates of the short-period amplification factor Fa and the mid-period factor Fv for the sites underlain by Younger Bay Mud and a surface rock site are tabulated (Table 1). The values inferred from the recordings obtained at the surface are plotted in Figures 7a and 7b. The figure show the empirical regression curves and corresponding 95 percent confidence limits on the ordinate to the true population regression curve derived from the recordings of the Loma Prieta earthquake (Borcherdt, 1992, 1994).

Establishing the null hypothesis that the empirical estimates are equal to the value of the mean predicted by the Loma Prieta sample regression line shows that the hypothesis can be rejected at the 95 percent confidence limit for the estimates of Fa at sites, COA, LOA, BMA, GAL, and MOA. A similar null hypothesis for Fv at sites EOA, COA, BMA, and MOA may be rejected, suggesting that these values differ significantly from those predicted empirically using the recordings available from the Loma Prieta earthquake. Perhaps of most significance are the larger estimates of Fv at the borehole array sites at Bessie Charmichael (COA), Embarcadero (EOA), and the San Francisco Marina (denoted by MOA and initially by WSS). Each of these borehole sites is underlain by a thick section of Younger Bay Mud, which in turn is underlain by a thick section of Older Bay Sediments (see Figure 2).
The sites used to derive the empirical regression curves for the Loma Prieta earthquake are predominantly sites underlain by thinner sections of soft soil and Older Bay sediments than the borehole sites in San Francisco (Borcherdt 1994). The site at the Oakland Harbor Wharf that recorded the Loma Prieta earthquake is underlain by thick sections of soft soil and Older Bay Sediments. Estimates of Fa and Fv for this site from the Loma Prieta earthquake are comparable to those obtained here from the Yountville and Gilroy earthquakes, namely 2.93 and 6.12 (Borcherdt, 1994). These results suggest that the sample of recordings obtained from the Loma Prieta earthquake provides estimates the average response most appropriate for sites on thinner sections of Younger Bay Mud and Older Bay Sediments than those as at the borehole locations in San Francisco.

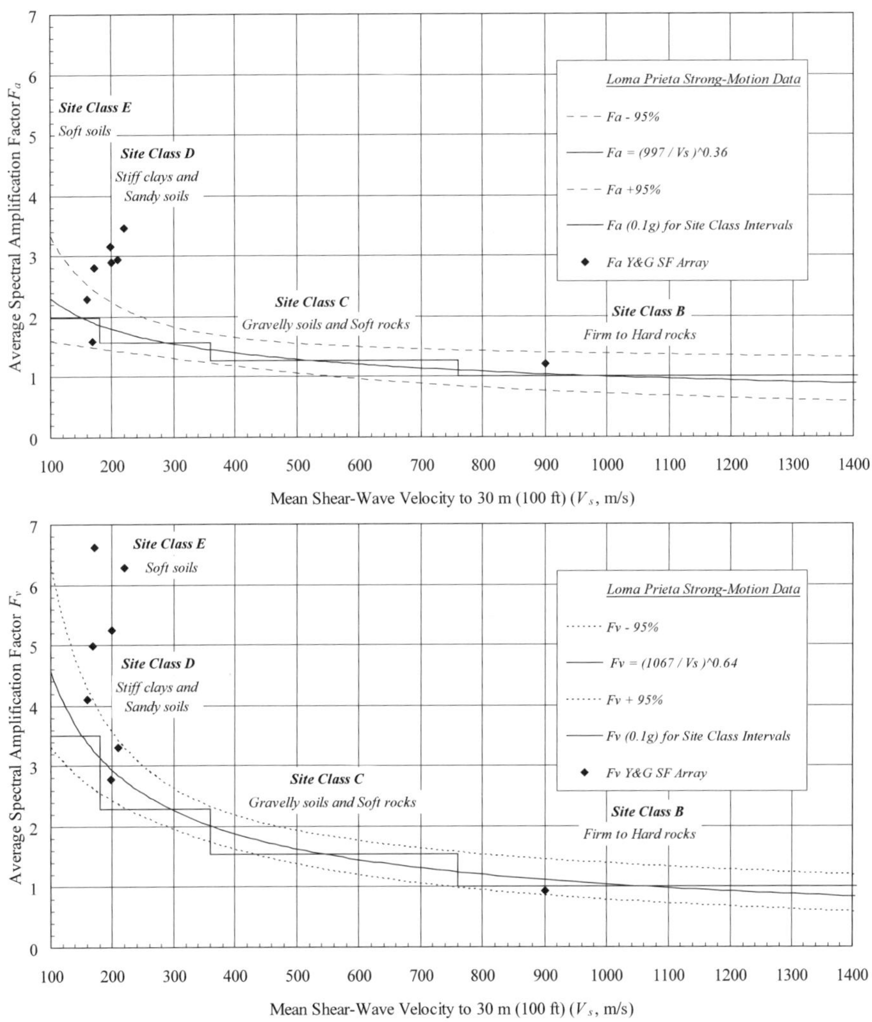

Figure 7. Empirical estimates of the short-period (Fa) and the mid-period (Fv) site coefficients from recordings of the Yountville and Gilroy earthquakes on the San Francisco array. The estimates are superimposed on empirical regression curves derived from strong motion data recorded from the Loma Prieta earthquake (from Borcherdt, 1994).

Table I. Average spectral ratios for the San Francisco array computed with respect to the surface recordings at site FM3

Station	ID	Location Long	Location Lat	Elev. m	Vs30 m/s	Ratio to FM3 (Yountville) Vertical Fa	Vertical Fv	Radial Fa	Radial Fv	Trans Fa	Trans Fv	Avg Horiz Fa	Avg Horiz Fv	Ratio to FM3 (Gilroy) Vertical Fa	Vertical Fv	Radial Fa	Radial Fv	Trans Fa	Trans Fv	Avg Horiz Fa	Avg Horiz Fv	Avg Evt Ratio Fa	Avg Evt Ratio Fv
Ch Pump Sta	CPS	-122.399	37.771		160	5.4	2.1	2.5	4.9	2.1	3.3	2.3	4.1										
Emb. Plaza	E0A	-122.395	37.796	2	168	2.7	4.5	2.1	6.3	1.7	5.0	1.9	5.6	3.1	4.2	1.2	5.0	1.4	3.6	1.3	4.3	1.6	5.0
Emb. Plaza	E1A	-122.395	37.796			2.5	3.7	2.9	5.3	2.3	4.5	2.6	4.9	2.7	4.0	1.6	4.7	1.5	3.2	1.5	3.9	2.1	4.4
Emb. Plaza	E5A	-122.395	37.796			1.1	1.1	1.8	1.9	1.4	1.3	1.6	1.6										
Emb. Plaza	E8A	-122.395	37.796			0.7	1.0	1.0	0.8	0.7	0.7	0.9	0.8	0.7	0.8			0.6	0.6	0.6	0.6	0.7	0.7
BC School	C0A	-122.405	37.777	4	172	4.0	5.1	3.9	8.6	2.7	8.5	3.3	8.5	3.9	4.8	2.2	4.5	2.5	4.9	2.3	4.7	2.8	6.6
BC School	C1A	-122.405	37.777											3.2	4.2	1.7	4.0	1.3	3.8	1.5	3.9		
BC School	C4A	-122.405	37.777			1.9	1.8	1.8	3.6	1.6	3.1	1.7	3.4	0.9	0.9	0.5	0.8	0.6	0.7	0.5	0.7	1.1	2.0
BC School	C9A	-122.405	37.777			0.9	0.9	1.0	1.3	1.0	1.0	1.0	1.1	1.1	1.0	0.7	0.7	0.6	0.6	0.6	0.6	0.8	0.9
Mean (ratio surface site class E)					170	4.1	3.9	2.8	6.6	2.1	5.6	2.5	6.1	3.5	4.5	1.7	4.8	1.9	4.3	1.8	4.5	2.1	5.3
Levi Plaza	L0A	-122.399	37.802	2	198									4.9	2.8	3.1	3.3	3.2	2.3	3.2	2.8		
Levi Plaza	L1A	-122.399	37.802											3.2	1.8	1.6	2.4	2.1	1.6	1.8	2.0		
Levi Plaza	L2A	-122.399	37.802											1.3	1.2	0.0	0.0	1.0	1.1	0.5	0.6		
Longshore Bldg.	BMA	-122.413	37.807	4	200	2.8	1.9	3.6	6.3	2.7	4.8	3.2	5.5	3.4	2.9	2.7	5.0	2.6	4.9	2.6	5.0	2.9	5.2
Gallop Bldg.	GAL	-122.387	37.752	6	210	10.0	2.3	4.3	5.5	1.6	1.2	2.9	3.3										
Marina	M0A	-122.442	37.804	4	220	3.3	4.9	4.4	7.0	3.7	6.3	4.0	6.7	3.5	7.2	2.8	6.7	3.0	5.1	2.9	5.9	3.5	6.3
Marina	M3A	-122.442	37.804			2.1	3.8	2.5	4.5	2.2	3.8	2.4	4.1	2.3	6.2	1.3	4.1	1.6	3.0	1.5	3.5	1.9	3.8
Marina	M9A	-122.442	37.804			1.7	1.2	2.0	1.3	1.7	1.2	1.8	1.2	1.6	1.7	0.9	1.2	1.0	0.9	1.0	1.0	1.4	1.1
Mean (ratio surface site class D)						5.4	3.0	4.1	6.3	2.7	4.1	3.4	5.2	3.9	4.3	2.9	5.0	2.9	4.1	2.9	4.5	3.1	4.9
Mar Museum	MAR	-122.423	37.807	8	350	2.1	1.2	3.0	1.9	2.2	1.6	2.6	1.7										
PB Stadium	PBS	-122.390	37.778		700	2.0	0.9	2.9	2.2	2.2	2.2	2.5	2.2	1.7	1.3	1.2	1.4	1.2	1.4	1.2	1.4	1.9	1.8
Mission Rock	P50	-122.382	37.773	2	900	1.8	0.8	3.4	2.7	2.5	2.4	3.0	2.5										
Russian Hill	RHL	-122.419	37.801	88	900	1.1	0.9	1.8	1.5	0.9	0.3	1.3	0.9	0.9	0.8	1.1	0.9	1.1	1.0	1.1	1.0	1.2	0.9
Mean (ratio borehole rock to surface rock FM3)					207	1.1	1.0	1.3	1.1	1.1	1.0	1.2	1.0	1.2	1.2	0.5	0.6	0.8	0.8	0.7	0.7	0.9	0.9

Comparison of the estimates of Fa and Fv as derived herein with those presently in the codes as plotted in Figure 7 shows that several of the estimates are significantly greater than Fa and Fv code values for both site class E and D sites. The estimates suggest that a site class boundary at 200, 210 or even possibly 220 m/s would be more consistent with these measurements than that of 180 m/s as presently specified in the code (see Borcherdt, 1994 for further discussion of this issue).

In light of present code provisions, these preliminary results suggest that site class E and D sites underlain by materials similar to those at the borehole array sites should be given additional consideration. Specifically, the results suggest that the new site-specific procedure (Proposal 3-6 R; http://www.bssconline.org/2003NEHRPproposals.html) is appropriate and likely provide a more accurate estimate of the response of the site than that based on ascribing site coefficients based on site class and input ground motion level.

5. CONCLUSIONS

The borehole arrays, designed to provide *in-situ* data not previously available on the non-linear response of "soft soils" as a function of strain amplitude with and without liquefaction-induced failure, have yielded several low-strain recordings of small to moderate earthquakes since their installation. These recordings provide useful measurements of low-strain site response and constitutive parameters for consideration in developing a better understanding of the response of soft soil deposits in the city of San Francisco. Critically needed recordings of the moderate- to high-strain response of the deposits must await the occurrence of a significantly larger earthquake on one of the nearby faults.

Twenty-five earthquakes in the magnitude range M 2.9 to 5.1 have been recorded on the San Francisco Array since its installation. Two of these events have been well recorded above urban background noise levels on the borehole arrays and at several of the surface installations. These quantitative *in-situ* measurements of the response of the near-surface deposits at the various sites show significant variations associated with variations in geotechnical characteristics of the sites. They permit empirical measurements of the short- and mid- period site coefficients Fa and Fv as presently adopted for use in US building codes. significantly greater statistically at the 95% confidence level than the corresponding values adopted for site class E and D sites in the current 2003 edition of the NEHRP provisions. The borehole sites in the San Francisco array are underlain by unusually thick sections of soft soil overlying firm to hard rock. No empirical measurements of the response of these types of sites during the Loma Prieta earthquake were obtained in San Francisco. The preliminary results obtained herein suggest that site class E and D sites underlain by materials similar to those at the borehole array sites should be given additional consideration for purposes of seismic design and retrofit consideration. Specifically, the results suggest that the new site-specific procedure (Proposal 3-6 R; http://www.bssconline.org/2003NEHRPproposals.html) would be appropriate for these sites. The results suggest that this new site specific procedure being adopted in the 2003 NEHRP (National Earthquake Hazard Reduction Program) can be used to provide better estimates of appropriate site coefficients for these types of sites than those that could be derived on the basis of site classification and input ground motion level alone.

ACKNOWLEDGEMENTS

The efforts of the late H. P. Liu and R. E. Warrick in the installation of arrays are especially appreciated. Conscientious review comments by Janise Rodgers and Carlo Di Alessandro were helpful.

REFERENCES

1. Borcherdt, R.D, 1970, Effects of local geology on ground motion near San Francisco Bay, *Bull. Seismol. Soc. Am.,* **60**, 29-61.
2. Borcherdt, R.D, 1994, Estimates of site-dependent response spectra for design (methodology and justification), *Earthquake Spectra*, **10**, 617-653.
3. Borcherdt, R.D. and G. Glassmoyer, 1992, On the characteristics of local geology and their influence on ground motions generated by the Loma Prieta earthquake in the San Francisco Bay region, California. *Bull. Seismol. Soc. Am.,* **82**, 603-641.
4. Borcherdt, R.D. and G. Glassmoyer, 1994, Influences of local geology on strong and weak ground motions in the San Francisco Bay region, California and their implications for site-specific code provisions, *in* The Loma Prieta earthquake of October 17, 1989 -- strong ground motion, R. D. Borcherdt, *ed., U.S. Geol. Surv. Prof. Paper 1551-A,* 77-108
5. Borcherdt, R.D., and J. F. Gibbs, J.F., 1976, Effects of local geological conditions in the San Francisco Bay region on ground motions and the intensities of the 1906 earthquake, *Bull. Seismol. Soc. Am*, **66**, 467-500.Borcherdt, R.D, 1988, Volumetric strain and particle displacements for body and surface waves in a general viscoelastic half-space: *Geophys. J. Royal Astr. Soc.,* **93**, 215-228.
6. Borcherdt, R. D., Liu, H. P., Westerlund, R. E., Dietel, C., Gibbs, J. F., and Warrick, R. E., 1999, Integrated surface and borehole strong-motion soil response arrays in San Francisco, California, *Procs. NATO Advanced Res. Workshop, Strong-motion Inst. for Civil Eng. structures*, June 2 -5, Istanbul, Turkey.
7. Borcherdt, R. D., 2004, A theoretical model for site coefficients in building code provisions, *Procs. 13th World Conference on Earthquake Engineering,* Vancouver, B.C., Canada, August 1-6, *in press*.
8. Borcherdt, R.D., Johnston, M.J.S., and G. Glassmoyer, 1989, On the use of volumetric strain meters to infer additional characteristics of short-period seismic radiation. *Bull. Seismol. Soc. Am*, **79,** 1006-1023.

9. Borcherdt, R.D., Fletcher, J.B., Jensen, E.G., Maxwell, G.L., Van Schaack, J.R., Warrick, R.E., Cranswick, E., Johnston, M.J.S., and McClearn, R., 1985, A general earthquake observation system (GEOS), *Bull. Seismol. Soc. Am.*, **75**, 1783-1825.
10. Borcherdt, R.D., and Wennerberg, L., 1985, General P, type-I S, and type-II S waves in anelastic solids: Inhomogeneous wave fields in low-loss solids, *Bull. Seismol. Soc. Am.*, **75**, 1729-1763.
11. Fumal, T.E, 1978, Correlations between seismic wave velocities and physical properties of geologic materials in the San Francisco Bay region, California, *U.S. Geol. Surv. Open-File Rept. 78-1067.*
12. Fumal, T.E., 1991, Shear-wave velocity estimates and site geology for strong-motion recordings sites of the Loma Prieta earthquake of October 17, 1989, *U.S. Geol. Surv. Open-File Rept. 91-311.*
13. Gibbs, J.F., Fumal, T.E., Borcherdt, R.D. Warrick, R.E., Liu, H.-P., and R.E., Westerlund, 1975, Seismic velocites and geologic logs from boreholes at three downhole arrays in San Francisco, California, *U.S. Geol. Surv. Open-File Rept. 94- 706.*
14. Helley, E.J., and K.R. Lajoie, 1979, Flatland deposits of the San Francisco Bay region, California---their geology and engineering properties, and their importance to comprehensive planning, *U.S. Geol. Surv. Prof. Paper 943.*
15. Kayan , R.E., H.-P., Liu, Fumal, T.E., Westerlund, R.E., Warrick, R.E., Gibbs, J.F., and H.J. Lee., 1990, Engineering and seismic properties of the soil column at Winfield Scott School, San Francisco, California, *U. S. Geol., Surv., Open-File Rept. 90-253*, G1-G18.
16. Liu, H.-P., Warrick, R.E., Westerlund, and J.B. Fletcher, 1991, A three-component borehole seismometer for earthquiake seismology. *Bull. Seismol. Soc. Am.* **81**, *2458-2485.*
17. Liu, H.-P., Warrick, R.E., Westerlund, R.E., Sembera, E.D., and L. Wennerberg, 1992, Observation of local site effects at a downhole -and surface station in the Marina district of San Francisco, *Bull. Seismol. Soc. Am.,* **82**, 1563 - 1591. Mueller, R.J., Lee, M., Johnston, M.J.S, Borcherdt, R.D., Glassmoyer, G., and Silverman, S., 1993, Near-real time monitoring of seismic events and status of portable digital recorders using satellite telemetry, *Bull. Seismol. Soc. Am.*, **85**, 640-645.

STRUCTURAL MONITORING ARRAYS – PAST, PRESENT AND FUTURE

M. Çelebi
Research Civil Engineer, U.S. Geological Survey (MS977), 345 Middefield Road, Menlo Park, Ca. 94025

Abstract : This paper presents a summary of the seismic monitoring issues as practiced in the past, as well as current applications and new developments to meet the needs of the engineering and user community. A number of examples exhibit the most recent applications that can be used for verification of design and construction practices, real-time applications for the functionality of built environment and assessment of damage conditions of structures.

Key Words : Seismic monitoring, structural response, global positioning system, real-time, acceleration, displacement, drift ratio, performance

1. INTRODUCTION

Seismic monitoring of structural systems constitutes an integral part of the National Earthquake Hazard Reduction Program in the United States. Recordings of the acceleration response of structures have served the scientific and engineering community well and have been useful in assessing design/analysis procedures, improving code provisions and in correlating the system response with damage. Unfortunately, there are only a few records from damaged instrumented structures to facilitate studies of the initation and progression of damage during strong shaking (e.g. Imperial County Services Building during the 1979 Imperial Valley earthquake, [Rojahn and Mork, 1981]). In the future, instrumentation programs should consider this deficiency. Jennings (1997) summarizes this view as follows: "As more records become available and understood, it seems inevitable that the process of earthquake resistant design will be increasingly, and quite appropriately,

P. Gülkan and J.G. Anderson (eds.), Directions in Strong Motion Instrumentation, 157–179.

based more and more upon records and measured properties of materials, and less and less upon empiricism and qualitative assessments of earthquake performance. This process is well along now in the design of special structures".

An instrumented structure should provide enough information to (a) reconstruct the response of the structure in sufficient detail to compare with the response predicted by mathematical models and those observed in laboratories, the goal being to improve the models, (b) make it possible to explain the reasons for any damage to the structure, and (c) to facilitate decisions to retrofit/strengthen the structural systems when warranted. In addition, a structural array should include, if physically possible, an associated free-field tri-axial accelerograph so that the interaction between soil and structure can be quantified.

Recent trends in development of performance based earthquake resistant design methods and related needs of the engineering community, as well as advances in computation, communication and data transmission capabilities, have prompted development of new approaches for structural monitoring issues and applications. In particular, (a) verification of performance based design methods and (b) needs of owners to rapidly and informedly assess functionality of a building following an event require measurement of displacement rather than or in addition to accelerations as is commonly done. Thus, new avenues in recording or computing displacement in real or near-real time are evolving. Thus, to meet the requirements for timely evaluation of damage condition of a building following an earthquake are leading the development of acquisition systems with special software that can deliver real-time or near real-time acceleration and displacement measurements.

This paper describes the past and current status of the structural instrumentation applications and new developments. The scope of the paper includes the following issues: (a) types of current building arrays and responses to be captured, (b) recent developments in instrument technology and implications, and (d) issues for the future. The scope does not include cost considerations.

2. HISTORICAL PERSPECTIVE

2.1 General Statistical Summary

In the United States, the California Division of Mines and Geology (CDMG) of the California Geologic Survey and the United States Geological Survey (USGS), manage the largest two structural instrumentation programs. Until recently, these programs have aimed to facilitate response studies in order to improve our understanding of the behavior and potential for damage to structures under the dynamic loads of earthquakes. The principal objective has been the quantitative measurement of structural response to strong and possibly damaging ground motions for purposes of improving seismic design codes and construction practices. However, to date, it has not been the objective of either instrumentation program to create a health monitoring environment for structures.

To date, the USGS has conducted a cooperative strong ground motion and structural instrumentation program with other federal and state agencies and private owners. Tables 1 and 2 summarize the current inventory and cooperative affiliations of the USGS Cooperative National Strong-Motion Program (NSMP). Within the USGS program, and unless other factors are considered and/or specific organizational choices are made apriori, the following general parameters have been considered for selecting and ranking structures for instrumentation:

1. Structural parameters: the construction material, structural system, geometry, discontinuity, and
2. Site-related parameters : severity-of-shaking on the basis of closeness to one or more of the main faults within the boundaries of the area considered (e.g. for the San Francisco Bay area, the San Andreas, Hayward, and Calaveras faults are considered).

Detailed procedures and overall description used by the USGS structural instrumentation program are described by Çelebi (2000, and 2001).

Table 1. Nationwide Distribution of USGS Cooperative Structural Instrumentation Arrays (updated 4/10/2004)

Extensively Instrumented Buildings [>6 channels]		Extensively Instrumented Bridges [>6 channels]		Extensively Instrumented Dams, Reservoirs, Pumping Plants and Power Generating Facilities [>6 channels]	
Alaska	4	California	2	Arizona	1
California	37	Oregon	10	California	21
Hawaii	1	Utah	1	Colorado	1
Missouri	2			Idaho	2
Puerto Rico	1			Montana	1
South Carolina	1			New Mexico	1
Tennessee	1			Oregon	13
Utah	2			Utah	1
Washington	2			Washington	7

Table 2. Cooperative National Strong-Motion Network of USGS [Extensively Instrumented Bldgs (> 6 channels)] (updated 4/10/2004)

Owner Agency [* Federal funds]	Stations	Recorders
Department of Veterans Affairs [*]	3	6
General Services Administration [*]	3	5
Los Angeles County	1	2
NASA-JPL [*]	8	9
University of Puerto Rico [**NSF funds]	1	5
U.S. Geological Survey [*]	31	66
Washington Dept of Natural Resources	1	1
USGS-ANSS [*]	3	8
TOTALS	51	102

On the other hand, the State of California CDMG program, which now has over 170 buildings instrumented in accordance to a predefined matrix, aims to cover a wide variety of structural systems (Huang and Shakal, 2001, and Shakal, Huang, Rojahn and Poland, 2001).

2.2 General Instrumentation Issues

2.2.1 Data Utilization

Ultimately, the types and extent of instrumentation must be tailored to how the data acquired during future earthquakes will be utilized, even though there may be more than one objective for instrumentation of a structure. Table 3 summarizes some data utilization objectives with sample references. As a recent example of data utilization, Jennings (1997) analyzed data from two buildings within close proximity (<20 km) to the epicenter of the 1994 Northridge, CA earthquake. He calculated the base shear from the records as 8 and 17 % of the weights of the buildings and the drift ratios as 0.8 and 1.6 % (exceeding code limitations). Jennings (1997) concluded: "A difference between code design values and measured earthquake responses of this magnitude – approaching a factor of ten – is not a tenable situation." Thus, recorded responses allowed the assessment of excessive drift ratios while shear forces remained reasonable.

Table 3. Sample List of Data Utilization Objectives & Sample References

GENERIC UTILIZATION
Verification of mathematical models (usually routinely performed) (*e.g.*Boroschek et al, 1990)
Comparison of design criteria vs. actual response (usually routinely performed)
Verification of new guidelines and code provisions (*e.g.*Hamburger, 1997)
Identification of structural characteristics (Period, Damping, Mode Shapes)
Verification of maximum drift ratio (*e.g.* Astaneh, 1991, Çelebi, 1993)
Torsional response/Accidental torsional response (*e.g.* Chopra, 1991, De La Llera, 1995)
Identification of repair & retrofit needs & techniques (Crosby, 1994)
SPECIFIC UTILIZATION
Identification of damage and/or inelastic behavior (*e.g.* Rojahn & Mork, 1981)
Soil-Structure Interaction Including Rocking and Radiation Damping (Çelebi, 1996, 1997)
Response of Unsymmetric Structures to Directivity of Ground Motions (*e.g.* Porter, 1996)
Responses of Structures with Emerging Technologies (base-isolation, visco-elastic dampers, and combination (Kelly and Aiken, 1991, Kelly, 1993, Çelebi, 1995)
Structure specific behavior (*e.g.* diaphragm effects, Boroschek and Mahin,1991, Çelebi, 1994)
Development of new methods of instrumentation/hardware (e.g. GPS: Çelebi et. al., 1997, 1999, 2001; wireless: Straser, 1997)
Improvement of site-specific design response spectra and attenuation curves (Boore, et. al. 1997, Campbell, 1997, Sadigh *et. al.*, 1997, Abrahamson and Silva, 1997
Associated free-field records (if available) to assess site amplification, SSI and attenuation curves(Borcherdt, 1993, 1994, 200a, 2002b, Crouse and MacGuire, 1996)
Verification of Repair/Retrofit Methods (Crosby et al, 1994, Çelebi and Liu, 1996)
Identification of Site Frequency from Building Records (Çelebi, 2003)
RECENT TRENDS TO ADVANCE UTILIZATION
Studies of response of structures to long period motions (e.g. Hall et al, 1996)
Need for new techniques to acquire/disseminate data (Straser, 1997, Çelebi, 1998, Çelebi and Sanli, 2002, Celebi and others, 2004)
Verification of Performance Based Design Criteria (future essential instrumentation work)
Near Fault Factor (more free-field stations associated with structures needed)
Comparison of strong vs weak response (Marshall, Phan and Çelebi, 1992, Çelebi, 1998)
Functionality Çelebi, 2004, Needs additional specific instrumentation planning)
Health Monitoring and other Special Purpose Verification (Heo et al, 1997)

2.2.2 Code versus Extensive Instrumentation

The most widely used code in the United States, the Uniform Building Code (UBC-1997 and prior editions), recommends, for seismic zones 3 and 4, a minimum of three accelerographs be placed in every building over six stories with an aggregate floor area of 60,000 square feet or more, and in every building over ten stories regardless of the floor area. The purpose of this requirement by the UBC was to monitor rather than to analyze the complete response modes and characteristics. UBC-code type recommended instrumentation is illustrated in Figure 1a. Following 1971 San Fernando earthquake, in 1982, in Los Angeles, the code-type requirement was reduced to one tri-axial accelerometer at the roof (or top floor) of a building meeting the aforementioned size requirements (Darragh and others, 1994). In general, code-type instrumentation is naturally being de-emphasized as a result of strong desire by the structural engineering community to gather more data from instrumented structures to perform more detailed structural response studies. Experiences from past earthquakes show that the minimum guidelines established by UBC for three tri-axial accelerographs in a building are not sufficient to perform meaningful model verifications. For example, three horizontal accelerometers are required to define the (two orthogonal translational and a torsional) horizontal motions of a floor. Rojahn and Mathiesen (1977) concluded that the predominant response of a high-rise building can be described by the participation of the first four modes of each of the three sets of modes (two translations and torsion); therefore, a minimum of 12 horizontal accelerometers would be necessary to record these modes. Instrumentation needed to provide acceptable documentation of the dominant response of a structure is addressed by Hart and Rojahn (1979) and Çelebi and others (1987). This type of instrumentation scheme is called the ideal extensive instrumentation scheme as illustrated in Figure 1b.

Specially designed instrumentation arrays are needed to understand and resolve specific response problems. For example, thorough measurements of in-plane diaphragm response require sensors in the center of the diaphragm (Figures 1c) as well as at boundary locations. Performance of base-isolated systems and effectiveness of the isolators are best captured by measuring tri-axial motions at the top and bottom of the isolators as well as the rest of the superstructure (Figure 1d). In case of base-isolated buildings, the main objective usually is to assess and quantify the effectiveness of isolators. If there is no budgetary constraint, additional sensors can be deployed between the levels above the isolator and roof to capture the behavior of intermediate floors.

2.2.3 Associated Free-Field Instrumentation

More information is required to interpret the motion of the foundation substructure relative to the ground on which it rests. This requires free-field instrumentation associated with a structure (Figure 1b). However, this is not always possible in an urban environment[1]. Engineers use free-field motions as input at the foundation level, or they obtain the motion at foundation level by convoluting the motion through assumed or determined layers of strata to base rock and deconvoluting the motion back to foundation level. Confirmation of

[1] For example, in San Francisco, California, it is not possible to find a suitable free-field location around the Transamerica building, which is extensively instrumented.

these processes requires downhole instrumentation near or directly beneath a structure. These downhole arrays will yield data on:

(1) the characteristics of ground motion at bedrock (or acceptably stiff media) at a defined distance from a source and
(2) the amplification of seismic waves in layered strata.

Downhole data from sites in the vicinity of instrumented building or other structures are especially scarce. Two new building monitoring arrays in the United States that include downhole sub-arrays are described later in the paper.

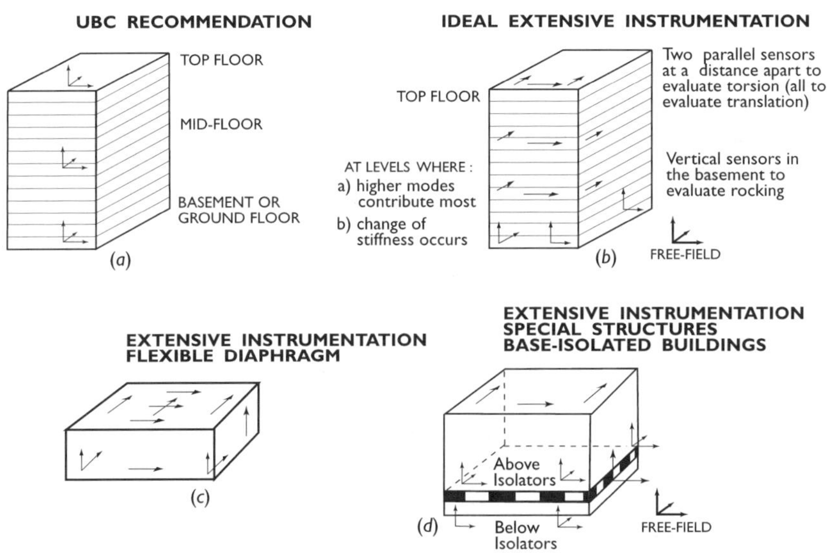

Figure 1. Typical Instrumentation Schemes

2.2.4 Record Synchronization Requirement

High-precision record synchronization must be available within a structure (and with the free-field, if applicable) if the response time histories are to be used together to reconstruct the overall behavior of the structure. Such synchronization has been achieved through extensive cabling from each of the individual sensor to the recorder. Recent technological developments enable decreasing or minimizing, and in certain cases eliminating, the use of extensive cabling. For example, the global positioning systems (GPS) is now widely used to synchronize a building instrumentation with that of a separate recorder system for the free-field; thus, eliminating cable connection between the free-field recorder and recorder within a structure. The issue here is that synchronization must be an integral part of any structure monitoring scheme whether cable or wireless transmission is the means to realize it.

2.2.5. Recording Systems, Constraints and New Developments

Until recently, commercially available recording systems have been limited to a maximum of 12-18 channels *(e.g.* analog recorder CRA-1[2] , 13 channels; the digital K-2[2], 12 channels; digital Mt. Whitney[2], 18 channels). Although multiple numbers of recording units may be used to accommodate requisite multiple-channel instrumentation systems for a structure, cost restrictions usually limit the number of channels to 12 or 18 (or multiples thereof), unless more channels are needed or special financing is available. Recently, however, with the development of PC-based data acquisition systems that utilize multiple A/D converters, several dozen channels of data can be accommodated. In such systems, the only constraints are the cost of the sensors and data transmission media required. One such system is described later in the paper.

3. SPECIAL ARRAYS – LOOKING TO THE FUTURE

3.1 Special Arrays in Los Angeles

Figure 2 shows the nine-story Millikan Library at Caltech Campus in Pasadena and the 15-story UCLA Factor Building in Los Angeles. In each building, the general objective is to thoroughly document the response of a multi-story building including the propagation of seismic waves. Another special-purpose instrumentation scheme for the twin towers at Century City, Los Angeles is shown in Figure 3. The objective of the recently-upgraded instrumentation of these two buildings is to better facilitate studies of the inter-story drift problem by means of recording the responses at several pairs of consecutive floors.

3.2 Displacement Measurement Needs and Arrays

[2] Use of commercial names or trademarks cited herein does not imply endorsement of these products by the U.S. Geological Survey.

Two important reasons are driving the recent push for developing technologies for measuring displacements in real-time or near real-time: (a) the evolution of performance-based design methods and procedures which rely on displacement as the main parameter and (b) the needs of local and state officials and prudent property owners, to establish procedures to assess the functionality of buildings and other important structures, such as lifelines, following a significant seismic event. As a result, structural engineers increasingly want the measurement of displacements during strong shaking events in order to assess drift ratios that in turn are related to performance of the structure, as schematically shown in Figure 4. On the other hand, dynamically measuring relative displacements between floors directly is very difficult and, except for tests conducted in a laboratory (e.g., using displacement transducers), has yet to be readily and feasibly achieved for a variety of real-life structures. However, recent technological developments have already made it possible to successfully develop and implement two approaches to dynamically measure real-time displacements from which drift ratios or average drift ratios can be computed. Both approaches can be used for performance evaluation of structures and can be considered as building health-monitoring applications.

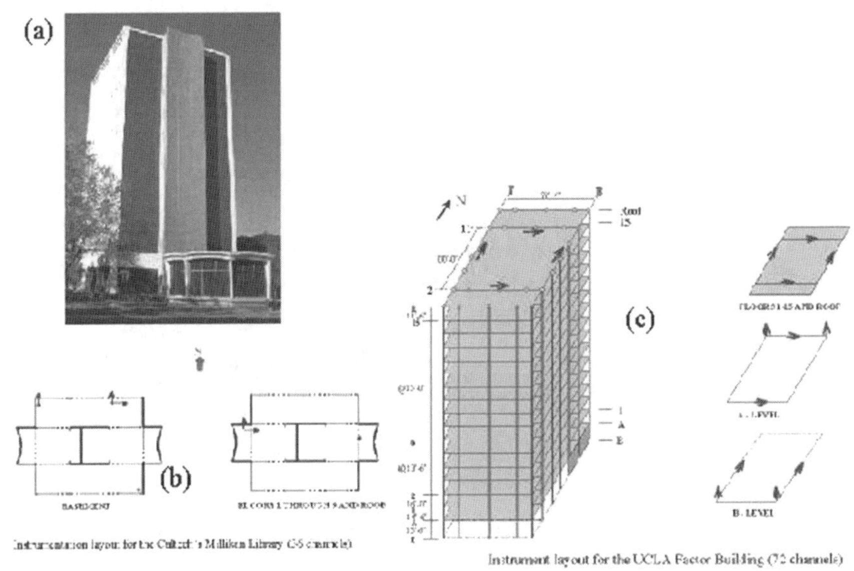

Figure 2. (a) Millikan Library (Pasadena, CA) and *(b)* its instrumentation scheme, and *(c)* Factor Building at UCLA (Los Angeles, CA) campus and its instrumentation scheme (Safak, *pers. comm. 2001*).

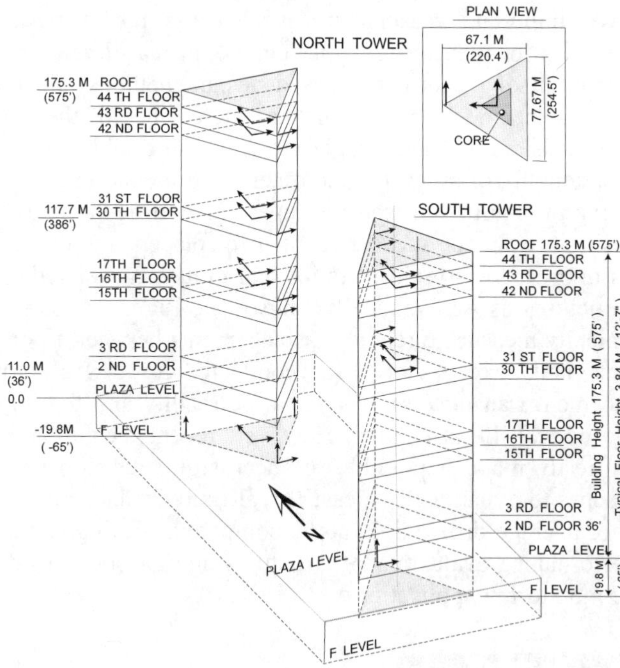

Figure 3. Twin Towers of Century City (extensively instrumented for drift studies)

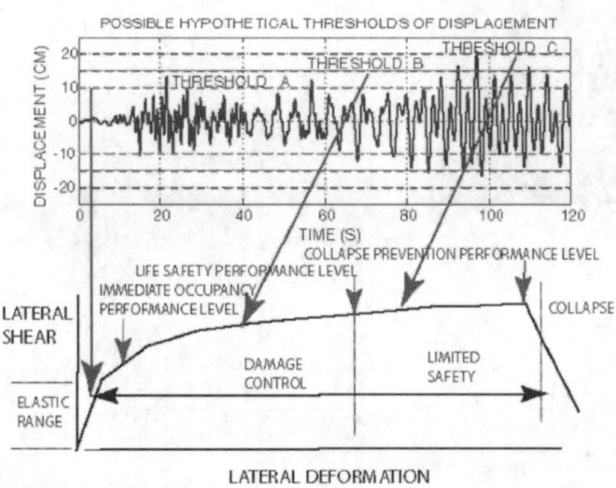

Figure 4. Hypothetical displacement time-history as related to FEMA- 274.

3.2.1. Use of GPS for Direct Measurements of Displacements

For long-period structures such as tall buildings and long-span bridges, dynamic displacement measurements using differential Global Positioning Systems (GPS) are now possible (Çelebi and Sanli 2002). However, GPS technology is limited to sampling rates of 10-20 Hz and, for buildings, measurement of displacement is possible only at the roof. The accuracy of GPS measurements is ± 1 cm horizontal and ± 2cm vertical. A schematic and photos of an application in the use of GPS to directly measure displacements is shown in Figure 5. In this particular case, two GPS units are used in order to capture both the translational and torsional response of the 34-story building in San Francisco, Ca. Furthermore, at the same locations as the GPS antennas, tri-axial accelerometers are deployed to verify the displacements measured by GPS with those obtained by double-integration of the accelerometer records. Real-time acceleration and displacement data streaming in the PC based monitoring system is shown also in Figure 5.

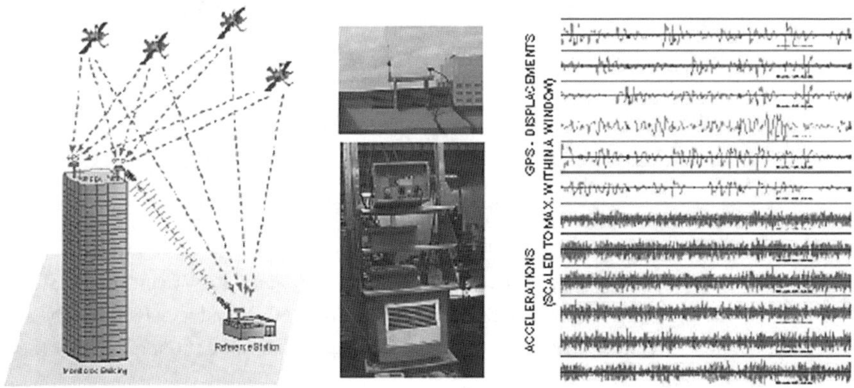

Figure 5. Special Instrumentation Using GPS and Accelerometers (San Francisco, CA.): (Left)- Schematic of thev overall system, (Center)- GPS and Radio modem antenna and the recorders connected to PC, (Right)- streaming of acceleration and displacement data in real-time.

In absence of strong shaking data from the deployed system, ambient data obtained are analyzed to infer the validity of the recorded vibration signals even though the amplitudes of both the acceleration and displacement data are very small and the data is noisy (Figure 6). The GPS displacement data is within the margin of error specified by the manufacturer (< 1 cm. horizontal).

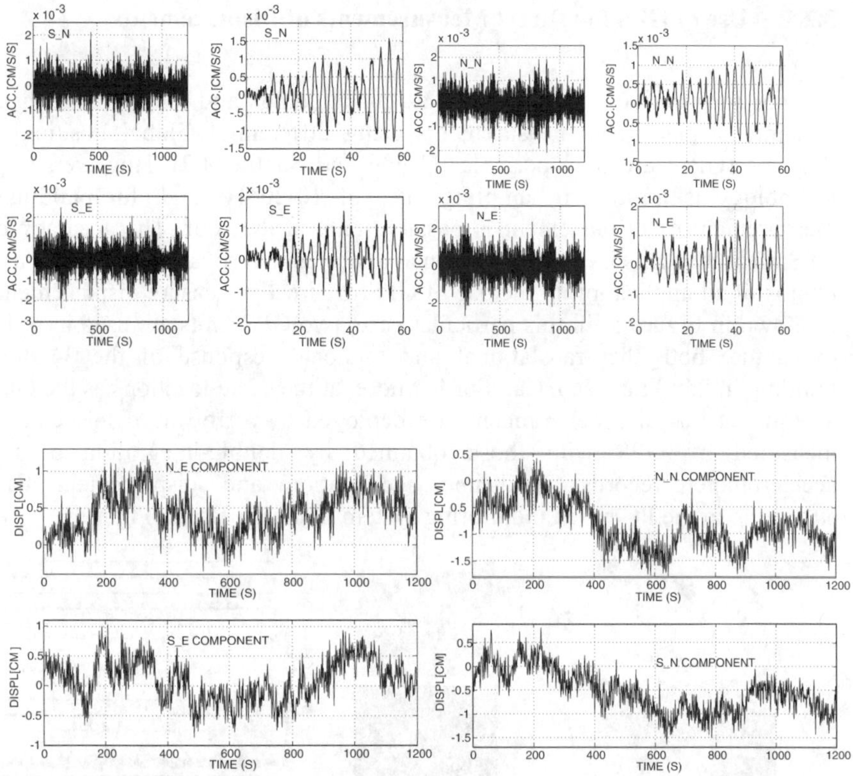

Figure 6. (Top)Remotely triggered and recorded accelerations at N (North) and S(South) locations. The figure shows pairs of 1200 second long (and 60 second window from the same) record. (Bottom) Remotely triggered and recorded displacements at N (North) and S (South) Locations.

In Figure 7, cross-spectra (Sxy) of pairs of parallel records (north-south component of north deployment [N_N] vs. north-south component of south deployment [S_N], and east-west component of north deployment [N_E] vs east-west component of south deployment [S_E]) from accelerometers are calculated. The same is repeated for the differential displacement records from GPS units. The cross-spectra (Sxy) clearly indicate a dominant frequency of 0.24-.25 Hz from both acceleration and displacement data. This frequency is within the band of expected frequency for a 34-story building. The lower peak in frequency (near ~ 0.1 Hz) seen in the cross-spectra of displacement records is due to noise, which is probably microseisms. It is expected that during larger amplitude motions with higher signal to noise ratios, such low frequency amplitudes due to noise will not be noticeable. In the acceleration data, a second frequency at 0.31 Hz is apparent. We will accept the 0.24-0.25 Hz as the fundamental translational frequency (in both directions). This is confirmed by the fact that at this frequency, the cross spectra of parallel acceleration records have a coherency of approximately unity (~ 1) and they are in-phase (0°). On the other hand, the Sxy of parallel acceleration records at 0.31 Hz also show coherency of approximately unity but they are out of phase (180°). Therefore, this frequency corresponds to a torsional mode.

For the fundamental frequency at 0.24 Hz, the displacement data exhibits a 0° phase angle; however, the coherencies are lower (~0.6-0.7) . The fact that the fundamental frequency (0.24 Hz) can be identified from the GPS displacement data, amplitudes of which are within the manufacturer specified error range, and that it can be confirmed by the acceleration data, is an indication of promise of better results when larger displacements can be recorded during strong shaking caused by earthquakes or strong winds.

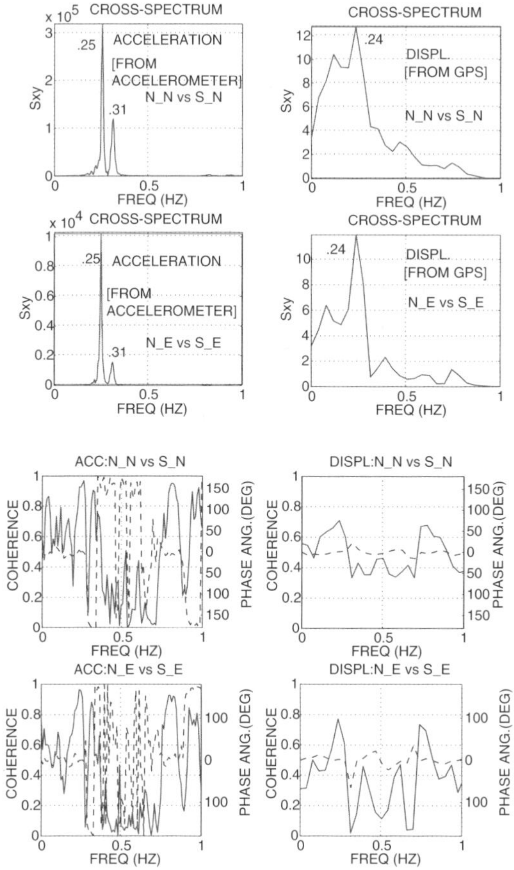

Figure 7. Cross-spectra (Sxy) and associated coherency and phase angle plots of horizontal, and parallel accelerations and displacements. [Note: In the coherency-phase angle plots, solid lines are coherency and dashed lines are phase-angle].

Since the deployment of the pioneering GPS units in San Francisco, CA, multiple other such arrays have been developed. An important array for monitoring the wind response of tall buildings in Chicago, IL have been developed by Kijewski-Correa and Kareem (2004).

3.2.2 Displacement via Real-time Double Integration

As mentioned, GPS applications are limited to sampling at $\leq$ 20 Hz, and for building monitoring, displacements measurements are possible on at the roof. This limits the application to long period structures rather than wide variety of structural systems. Therefore, the challenge is to compute displacements from recorded acceleration responses in real-time or near real-time.

A new approach in obtaining displacements in real-time is depicted in Figure 8 which also shows the distribution of accelerometers in the building - designed to provide data from several pairs of neighboring floors to facilitate drift computations. The system has a server that (a) digitizes continuous analog acceleration data, (b) pre-processes the 1000 sps digitized data with low-pass filters (herein called as the preliminarily filtered uncorrected data), (c) decimates the data to 200 sps and streams it locally (d) monitors and applies server triggering threshold criteria and locally records (with a pre-event memory) when prescribed thresholds are exceeded and (e) broadcasts the data continuously to remote users by high-speed internet.

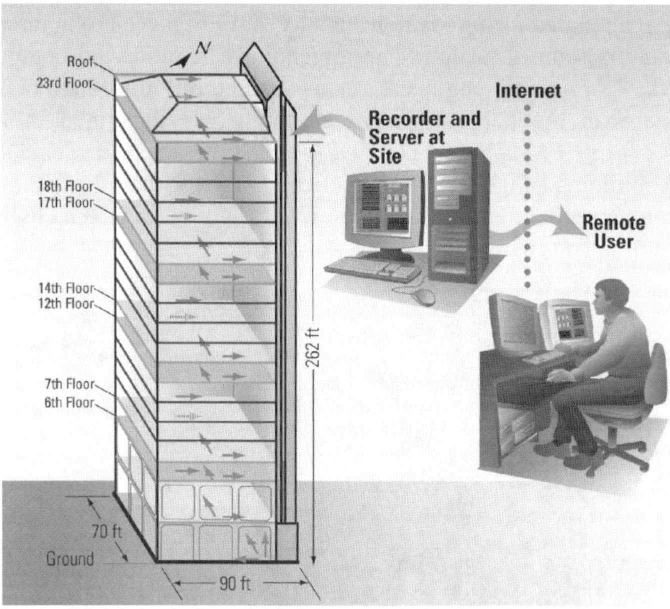

Figure 8. General schematic of data acquisition and transmittal for seismic monitoring of the building.

The broadcast streamed real-time acceleration data are acquired remotely using a "Client Software" configured to compute velocity, displacement and select number of drift ratios. Figures 9 show two PC screen snapshots of the client software display configured for 12 channels of streaming acceleration or velocity or displacement or drift ratio time series. Each paired set of acceleration response streams is displayed with a different color. The upper right shows amplitude spectra for one of the channels and is selectable by the user. It is noted that several frequencies are clearly identifiable (as discussed later in the paper). In the lower left, time series of drift ratios are shown for 6 locations, each color corresponding to the same pair of data from the window above. In order to get the drift ratios, real-time double integration of filtered acceleration data are computed. Specific filter options are built into the client software for processing of the acceleration data. To compute drift ratios, story heights, as shown in Figure 9 need to be manually entered. This figure also shows the computed pairs of displacements that are used to compute the drift ratios. Corresponding to each drift ratio, there are 4 stages of colored indicators. When only the "green" color indicator is activated, it indicates that the computed drift ratio is below the first of three specific thresholds. The thresholds of drift ratios for selected pairs of data must also be manually entered in the boxes. As drift ratios exceed the designated three thresholds, additional indicators are activated with a different color (Figure 5). The drift ratios are calculated using data from any pair of accelerometer channels oriented in the same direction. The threshold drift ratios are computed and decided by structural engineers using structural information and are compatible with the performance-based theme, as illustrated in Figure 4 (Figure C2-3 of *FEMA-274* [ATC 1997]) and summarized in Table 4 for this particular building. Figure 9 hypothetically shows that the first level of threshold is exceeded, and the client software is recording data as indicated by the illuminated red button.

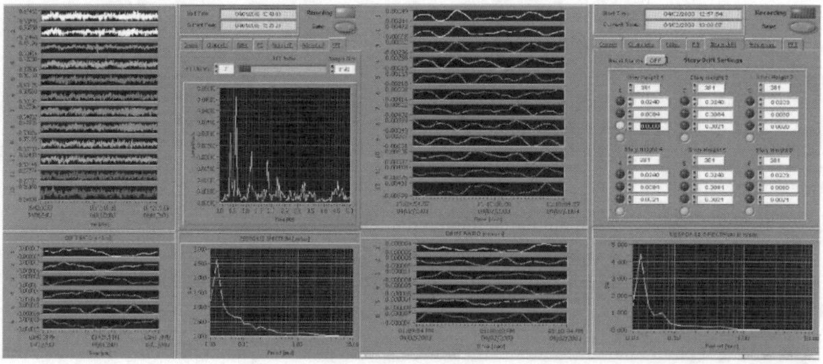

Figure 9. (Left) Screen snapshot of client software display showing acceleration streams and computed amplitude and response spectra. (Right) Screen snapshot of client software display showing 12-channel (six pairs with each pair a different color) displacement and corresponding six-drift ratio (each corresponding to the same color displacement) streams. Also shown to the upper right are alarm systems corresponding to thresholds that must be manually input. The first threshold for the first drift ratio is hypothetically exceeded to indicate the starting of the recording and change in the color of the alarm from green to yellow.

Table 4. Summary of Threshold Stages and Corresponding Drift Ratios

Threshold Stage	1	2	3
Adopted Drift Ratio	0.2%	0.8%	1.4-.2%

Thus the system can (a) be used to facilitate rapid assessment of the building integrity following an earthquake, (b) provide data in a form that is easy to correlate with known and building specific engineering parameters (e.g., drift ratio), which in turn must relate to the expected damage condition of the building, and (c) deliver the data within a relatively short time (few minutes if not in seconds), to facilitate informed decision making regarding post-earthquake building occupancy.

3.2.4 Soil-Structure Interaction Array(s)

State-of-the-art practice and analytical approaches require, when warranted, the structure-foundation system to be represented by mathematical models that include the influence of the sub-foundation media. In many cases, under a specific geotechnical environment, certain structures will respond differently than if that structure was built as a fixed based structure on a very stiff (e.g rock) site condition. This alteration of vibrational characteristics of structures due to soil-structure interaction (SSI) can be both beneficial and detrimental for their performances. To date, the engineering community is not clear about the pros and cons of SSI.

Adverse effects of SSI during the 1985 Michoacan (Mexico) earthquake were addressed by Tarquis and Roesset (1988), who showed that, in the lakebed zone of Mexico City, 400 km away from the epicenter, fundamental periods of mid-rise buildings (5-15 stories) lengthened due to SSI. Thus, such buildings were negatively affected due to SSI because the lengthening of their fundamental periods placed them in a resonating environment close to the approximately 2-second resonant period of Mexico City lakebed.

On the other hand, under different circumstances, SSI may be beneficial because it produces an environment whereby the structure escapes the severity of shaking due to shifting of its fundamental frequency. Certainly, in a basin such as that of the Los Angeles area, SSI may cause both beneficial and detrimental effects in the response of structures.

Thus, the identification of the circumstances under which SSI is beneficial or detrimental and the relevant controlling parameters is a necessity. Therefore, measurement of soil-structure interaction effects is required to fully understand the response of a major structure. This is easily accommodated along with the instrumentation schemes of the superstructure. Sensors at critical locations of the foundation are required to capture its relevant motions. Additional sensors may be needed to record the motions of the surrounding geological materials. For example, if vertical motion and rocking are expected to be significant and need to be recorded, at least three vertical accelerometers are required at the basement level (Figure 1b). In some cases; additional instrumentation (e.g. free-field accelerographs on the surface and in boreholes [downhole accelerographs]) may be required. Horizontal and vertical spatial downhole sensors will provide information on how the motions change while traveling through the media and how much it is affected by the building response. Detailed proposals for soil-structure interaction experiments resulting from a workshop are presented in USGS OFR-92-295 (Çelebi and others, 1992).

Specialized arrays that will capture SSI effects will further advance the verification of SSI effects that are currently very much limited to theoretical studies.

Two existing SSI arrays are shown in Figures 10a and b. Each of these arrays has the necessary components of sub-arrays (e.g. superstructure, foundation, surface and downhole free-field sub-arrays). Figure 10a depicts Pacific Park Plaza Building array in Emeryville, CA and Figure 10b depicts the Atwood Building in Anchorage, AK. Both building monitoring schemes are designed to capture SSI effects in addition to the traditional translational and torsional responses.

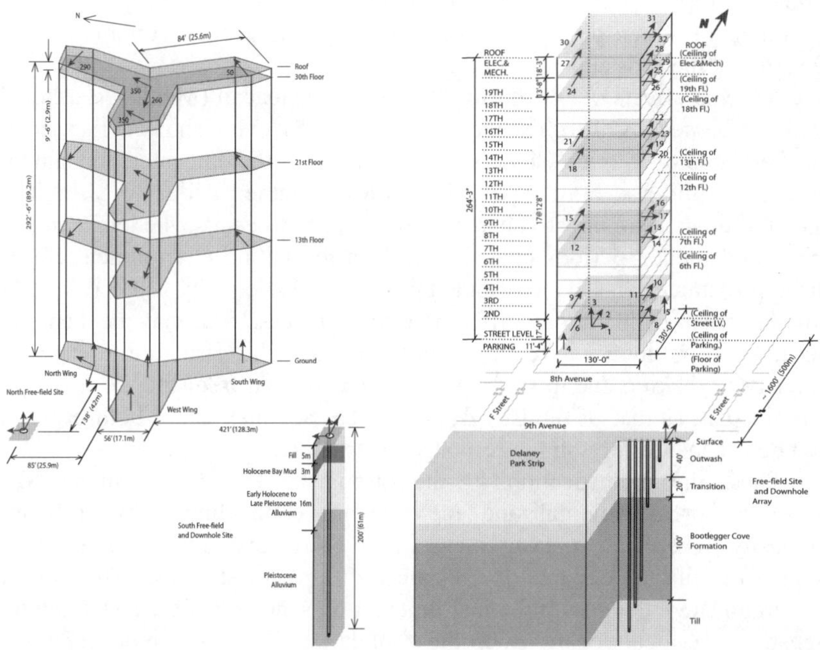

Figure 10. (Left) A three-dimensional schematic of the Pacific Park Plaza Building (Emeryville, CA) showing with integrated structure, surface and downhole sub-arrays (Note: The tri-axial downhole accelerograph was added after the 1989 Loma Prieta earthquake). (Right) A General three-dimensional schematic of the Atwood Building (Anchorage, AK) showing the general dimensions and locations of accelerometers deployed within the structure and tri-axial downhole accelerometers at free field site.

4. OUTSTANDING ISSUES AND ANSS

The following outstanding issues need to be considered in future instrumentation efforts but are not discussed in detail in this paper:

- How to better instrument to validate performance-based design procedures,
- Health monitoring needs and related cost issues for both installation and maintenance.
- Drift assessment related instrumentation needs. USGS already has 4 buildings with multiple sensors on consecutive floors that are configured to record data to assess drift ratios.
- Wireless instrumentation – is it here?
- Monitoring capability in large urban areas such as New York – in light of the September 11, 2001 event.
- Verification of specific emerging technological applications (e.g. unbonded braced system, damper systems used in new and retrofit design and construction).

There are new challenges but also new opportunities in the development of state-of-the-art seismic instrumentation projects to help better understand existing and outstanding engineering and scientific issues related to mitigation strategies in urban environs. A new initiative, the Advanced National Seismic System (ANSS), authorized by U.S. Congress and managed by the USGS, specifically aims to address such outstanding issues for seismic monitoring of structures. Development of procedures for deploying arrays related to seismic monitoring projects are currently underway while some projects are already completed (e.g. Atwood Building in Anchorage, AK).

5. CONCLUSIONS

This paper presents the current status of and methods used for the structural instrumentation program mainly by the U.S. Geological Survey but also generally in the United States. Both historical and current trends and methods used for seismic monitoring of structures are discussed in terms of utilization of data acquired by seismic monitoring. New approaches in monitoring (using GPS technology and real-time double-integration) and related data acquisition systems to meet special needs are introduced. The extent to which a structure should be instrumented to meet the code recommendations versus special needs are discussed without consideration of cost issues. Several examples of instrumented buildings are shown.

REFERENCES

1. Abrahamson, N.A. and Silva, W.J., 1997, Empirical response spectral attenuation relations for shallow crustal earthquakes, Seismological Research Letters, 68, 94—127.

2. An Assessment of Seismic Monitoring in the United States – Requirement for an Advanced National Seismic System, 1999, U.S. Geological survey, Circular 1188 , 1999

3. Applied Technology Council (ATC), 1997. NEHRP Commentary on the Guidelines for the Seismic Rehabilitation of Buildings, prepared for the Building Seismic Safety Council, published by the Federal Emergency Management Agency, FEMA 274, Washington, D.C.

4. Astaneh, A., Bonowitz, D., and Chen, C., 1991, Evaluating design provisions and actual performance of a modern high-rise steel structure, *in* Seminar on Seismological and Engineering Implications of Recent Strong-Motion Data: California Department of Conservation, Division of Mines and Geology, p. 5-1–5-10.

5. Boore, D.M., W.B. Joyner, and T.E. Fumal, 1997, Equations for estimating horizontal response spectra and peak acceleration from western North American earthquakes: A summary of recent work, Seismological Research Letters, 68,128—153.

6. Borcherdt, R.D. 1994. Estimates of site-dependent response spectra for design (Methodology and Justification), Earthquake Spectra, 10, p. 617-653.

7. Borcherdt, R.D., 1993, On the estimation of site-dependent response spectra, Proc., International Workshop on Strong-Motion Data, Menlo Park, CA, Publ. By Port and Harbor Research Institute, Japan, (Preprint of manuscript submitted as "Simplified site classes and empirical amplification factors for site-dependent code provisions", for Proc. NCEER, SEAOC, BSSC Workshop on Site Response during Earthquakes and Seismic Code Provisions, Univ. S. Calif., Los Angeles, CA, November 18 - 20, 1992), vol. 2, p 399 - 427.

8. Borcherdt, R.D., 1994, Estimates of site-dependent response spectra for design (methodology and justification), Earthquake Spectra, 10, pp. 617-653.

9. Borcherdt, R. D., 2002a, Empirical evidence for acceleration-dependent amplification factors, *Bull. Seismol. Soc. Am.*, **92**, 761-782.

10. Borcherdt, R. D., 2002b, Empirical evidence for site coefficients in building-code provisions, *Earthquake Spectra*, 18, 189-218.

11. Boroschek, R. L., Mahin, S. A., and Zeris, C., A., 1990, Seismic response and analytical modeling of three instrumented buildings, PROC., 4th U.S. National Conference on Earthquake Engineering, v.2, pp. 219-228, Palm Springs, Ca., May 20-24.

12. Boroschek, R. L., and Mahin, S., 1991, An Investigation of the Seismic Response of a Lightly-Damped Torsionally-Coupled Building, University of California, Berkeley, California, Earthquake Engineering Research Center Report 91/18, December, 291 p.

13. Campbell, K.W., 1997, Empirical near-source attenuation relationships for horizontal and vertical components of peak ground acceleration, peak ground velocity, and pseudo-absolute acceleration response spectra, Seismological Research Letters, 68, 154—179.

14. Çelebi, M., Safak, E., Brady, G., Maley, R., and Sotoudeh, V., 1987, Integrated instrumentation plan for assessing the seismic response of structures--a review of the current USGS program, ***USGS Circular 947.***

15. Çelebi, M., Bongiovanni, G., Safak, E., and Brady, G., 1989, Seismic response of a large-span roof diaphragm: Earthquake Spectra, v. 5, no. 2, p. 337-350.

16. Çelebi, M., 2001, Current practice and guidelines for USGS instrumentation of buildings including federal buildings, *in* COSMOS Proceedings, Invited Workshop on Strong-

motion Instrumentation of Buildings, Emeryville, Ca. November 2001, Cosmos Publication No: CP-2001/04.

17. Çelebi, M., 2000, Seismic Instrumentation of Buildings, U.S. Geological Survey Open-File Report 00-157.

18. Çelebi, M., 2003, "Identification of Site Frequencies from Building Records", Journal of EERI, ***Earthquake Spectra***, v.19, Issue 1, pp.1-23.

19. Celebi, M., Sanli, A., Sinclair, M., Gallant, S., and Radulescu, D., 2004, Real-Time Seismic Monitoring Needs of a Building Owner and the solution – A Cooperative Effort, in print, Journal of EERI, ***Earthquake Spectra***, v.19, Issue 1, pp.1-23.

20. Çelebi, M., and Sanli, A., 2002, GPS in Pioneering Dynamic Monitoring of Long-Period Structures, ***Earthquake Spectra,*** Journal of EERI,. Volume 18, No. 1, pages 47–61, February 2002.

21. Çelebi, M., and Safak, E., 1991, Seismic response of Transamerica Building—I, data and preliminary analysis: Journal of Structural Engineering, v. 117, no. 8, p. 2389-2404.

22. Çelebi, M. and Safak, E., 1992, Seismic response of Pacific Park Plaza—I, data and preliminary analysis: Journal of Structural Engineering, v. 118, no. 6, p. 1547-1565

23. Çelebi, M. (compiler), Lysmer, J, and Luco, E., 1992, Recommendations for a Soil-Structure Interaction Experiment Report Based on a Workshop Held at San Francisco, California on February 7, 1992), U.S. Geological Survey Open-File Report 92-295.

24. Çelebi, M., Phan, L. T., and Marshall, R. D., 1993, Dynamic characteristics of five tall buildings during strong and low-amplitude motions, Journal of Structural Design of Tall Buildings, J. Wiley, v. 2, pp. 1-15.

25. Çelebi, M., 1993, Seismic response of eccentrically braced tall building, Journal of Structural Engineering, v. 119, no. 4, p. 1188-1205.

26. Çelebi, M., 1994, Response study of a flexible building using three earthquake records, Structures Congress XII: Proceedings of papers presented at the Structures Congress '94, Atlanta, GA, April 24-28, American Society of Civil Engineers, New York, Vol. 2, 1220-1225.

27. Çelebi, M., 1995, Successful Performance of base-isolated hospital building during the 17 January 1994 Northridge earthquake, *Journal of the Structural Design of Tall Buildings*, v. 5, pp.95-109.

28. Çelebi, M., 1996, Comparison of damping in buildings under low-amplitude and strong motions, Journal of Wind Engineering and Industrial Aerodynamics, Elsevier Science, v. 59, pp. 309-323.

29. Çelebi, M. and Liu, H-P., 1996, Before and After Retrofit – Response of a Building During Ambient and Strong-motions, 8USNational Conference on Wind Eng,, The John Hopkins Univ. June 5-7.

30. Çelebi, M., Presscott, W., Stein, R., Hudnut, K., and Wilson, S., 1997, Application of GPS in Monitoring Tall Buildings in Seismic Areas, 1997a, Abstract, AGU Meeting, San Francisco, Ca., Dec.

31. Çelebi, M., 1997, Response of Olive View Hospital to Northridge and Whittier earthquakes, American Society of Civil Engineers, *Journal of Structural Engineering*, April ,v.123, no. 4, p. 389-396.

32. Çelebi, M., 1998, Performance of Building structures – A Summary, *in* The Loma Prieta, California, Earthquake of October 17, 1989 – Building Structures (M. Çelebi, *editor*), **USGS Prof. Paper 1552-C**, pp. c5-c76, January 1998.

33. Çelebi, M., Presscott, W., Stein, R., Hudnut, K., Behr, J. and Wilson, S., 1999, GPS Monitoring of Dynamic Behavior of long-Period Structures, **Earthquake Spectra**, 15,1, pp.55-66, Feb. 1999.

34. Chopra, A., and Goel, R.K., 1991, Evaluation of torsional provisions of seismic codes, **J. Struct. Eng**. ASCE, 117, 12, 3762-3782.

35. Crosby, P., Kelly, J., and Singh, J. P., 1994, Utilizing visco-elastic dampers in the seismic retrofit of a thirteen story steel framed building, ASCE Structures Congress XII, Atlanta, Ga., v. 2, 1286-1291.

36. Crouse, C. B., and McGuire, J.W., 1996, Site response studies for purpose of revising NEHRP seismic provisions, **Earthquake Spectra**, 12, 407-439.

37. Darragh, R., Cao, T., Graizer, V., Shakal, A. and Huang, M., 1994, Los Angeles code-instrumented building records from the Northridge, California earthquake of January 17, 1994: Processed Release No. 1, Report No. OSMS 94-17, California Strong Motion Instrumentation Program, California Department of Conservation, Dec. 10, 1994.

38. De La Llera, J., and Chopra, A., 1995, Understsanding of inelastic seismic behavior of symmetric-plan buildings, **Earthquake Engineering and Structural Dynamics,** 24, pp.549-572.

39. Hart, G., and Rojahn, C., 1979, A decision-theory methodology for the selection of buildings for strong-motion instrumentation: **Earthquake Engineering and Structural Dynamics**, v. 7, p. 579-586.

40. Hall, J. F., Heaton, T. H., Halling, M. W., and Wald, D. J., 1996, Near-source ground motion and its effects on flexible buildings, **Earthquake Spectr**a, v. 11, no.4, pp. 569-605.

41. Hamburger, R. O., 1997, FEMA-173 Seismic Rehabilitation Guidelines: The next step – Verification, in Proc.SMIP97 Seminar on Utilization of Strong-motion Data, California strong Motion Instrumentation Program, Div. of Mines and Geology, California Dept. of Conservation, Sacramento, Ca., 51-69.

42. Heo, G,, Wang, M. L., and Satpathi, D., 1977, Optimal transducer placement for health monitoring, Soil Dynamics and Earthquake Engineering, 16, pp-496-502.

43. Huang, M. J., and Shakal, A.F., 2001, in Proceedings, Invited Workshop on Strong-motion Instrumentation of Buildings, COSMOS, Emeryville, Ca. November 2001, Cosmos Publication No: CP-2001/04, pp. 15-19.

44. Jennings, P.C., 1997, Use if strong-motion data in earthquake resistant design, in Proc.SMIP97 Seminar on Utilization of Strong-motion Data, California strong Motion Instrumentation Program, Div. of Mines and Geology, California Dept. of Conservation, Sacramento, Ca.,1-8.

45. Kelly, J., 1993, Seismic isolation, passive energy dissipation and active control, PROC. ATC 17-1 Seminar on State of the Art and State of the Practice of Base Isolation, vol. 1, 9-22.

46. Kelly, J.M., Aiken, I.D., and Clark, P.W., 1991, Response of base-isolated structures in recent California earthquakes, *in* Seminar on Seismological and Engineering Implications of Recent Strong-Motion Data, Preprints: California Division of Mines and Geology, Strong Motion Instrumentation Program, p. 12-1--12-10.

47. Kijewski-Correa, T. and Kareem, A. (2004) "The Height of Precision: New Perspectives in Structural Monitoring," *Proceedings of Earth & Space: 9th Aerospace Division International Conference on Engineering, Construction and Operations Challenging Environments*, 7-10 March, Houston

48. Marshall, R. D., Phan, L. T., and Çelebi, M., 1992, Measurement of structural response characteristics of full-scale buildings: Comparison of results from strong-motion and ambient vibration records, NISTIR REPORT 4884, National Institute of Standards and Technology, Gaithersburg, Maryland.

49. Porter, L.D., 1996, The influence of earthquake azimuth on structural response due to strong ground shaking, *in* Eleventh World Conference on Earthquake Engineering, Acapulco, Mexico (June), (No. 1623): Elsevier Science Ltd. (CD-ROM).

50. Rojahn, C., and Matthiesen, R.B., 1977, Earthquake response and instrumentation of buildings: Journal of the Technical Councils, American Society of Civil Engineers, v. 103, no. TCI, Proceedings Paper 13393, p. 1-12.

51. Rojahn, C., and Mork, P.N., 1981, An analysis of strong-motion data from a severely damaged structure, the Imperial County Services Building, El Centro, California: U.S. Geological Survey Open-File Report 81-194.

52. Sadigh, K., C.-Y. Chang, J.A. Egan, F. Makdisi, and R.R. Youngs, 1997, Attenuation relationships for shallow crustal earthquakes based on California strong motion data, Seismological Research Letters, 68, 180—189.

53. Shakal, A. F. , Huang, M. J., Rojan, C., and Poland, C., 2001, in in Proceedings, Invited Workshop on Strong-motion Instrumentation of Buildings, COSMOS, Emeryville, Ca. November 2001, Cosmos Publication No: CP-2001/04, 5-14

54. Straser, E., 1997, Toward wireless, modular monitoring systems for civil structures, in the John A. Blume Earthquake Engineering Center Newsletter, Issue No. 2.

55. Tarquis, F., and Roesset, J., 1988, Structural response and design response spectra for the 1985 Mexico City earthquake: University of Texas, Austin, Texas, Report No. GD89-1, 208p.

56. Uniform Building Code, International Conference of Building Officials, Whittier, CA, 1970, 1976, 1979, 1982, 1985, 1988, 1991, 1994, 1997 editions.

DEVELOPMENT OF STRONG-MOTION OBSERVATION NETWORK CONSTRUCTED BY NIED

Shigeo Kinoshita
Yokohama City University, 22-2 Seto, Kanazawa-ku, Yokohama 236-0027, Japan

Abstract: This paper describes the development of strong-motion observation networks constructed by the National Research Institute for Earth Science and Disaster Prevention (NIED), Japan, in the last quarter of a century. In 1974 to 1975, we first constructed a down-hole array at the IWT station, where tri-axial negative feedback accelerometers were installed at the depths of 3,510m, 108m, and free surface, respectively. Since then NIED has continued to construct various kinds of strong-motion observation networks from 1976 to 1995 (first stage) and now has operated two nation-wide networks, K-NET and KiK-net (second stage). The aim of the present paper is to show the state-of-the-art of strong-motion observation networks in Japan.

Key words: Deep borehole, down-hole array, FCH array, Kanto borehole array, K-NET, KiK-net, KOT array, NIED, strong-motion observation network

1. STRONG-MOTION OBSERVATION NETWORK IN THE FIRST STAGE (1976-1995)

At the first stage, NIED constructed four small-scale strong-motion arrays in the Kanto region, central Japan, during the time period of 1975 to 1995, while we also improved the technical skills for array construction, the operation of network system and the interpretation of seismograms. The four strong-motion arrays are (1) the down-hole array consisted of three deep borehole stations, (2) the Fuchu (FCH) array with three-dimensional layout of seismometer, (3) the Koto (KOT) array which was composed of 20 surface stations equipped with velocity-type strong-motion seismographs, and (4) the

P. Gülkan and J.G. Anderson (eds.), Directions in Strong Motion Instrumentation, 181–196

Kanto borehole array with 27 down-holes in the Kanto region. These networks have been operated by the NIED ever since.

1.1. The Deep-borehole Stations (Construction Periods: 1974-1980)

Three deep boreholes with depths of 3,510m, 2,300m, and 2,750m were constructed at the IWT, SHM and FCH stations that are located at about 20 km apart from downtown Tokyo as marked in Fig. 1. Geologically, these deep boreholes are in depths of 600m to 800m below the top of the pre-Tertiary basement (Yamamizu et al., 1981). Our intention for constructing these three stations was to investigate how a thick sedimentary layer – basement system responding to seismic shaking and the seismic activity in and around the Kanto region as well.

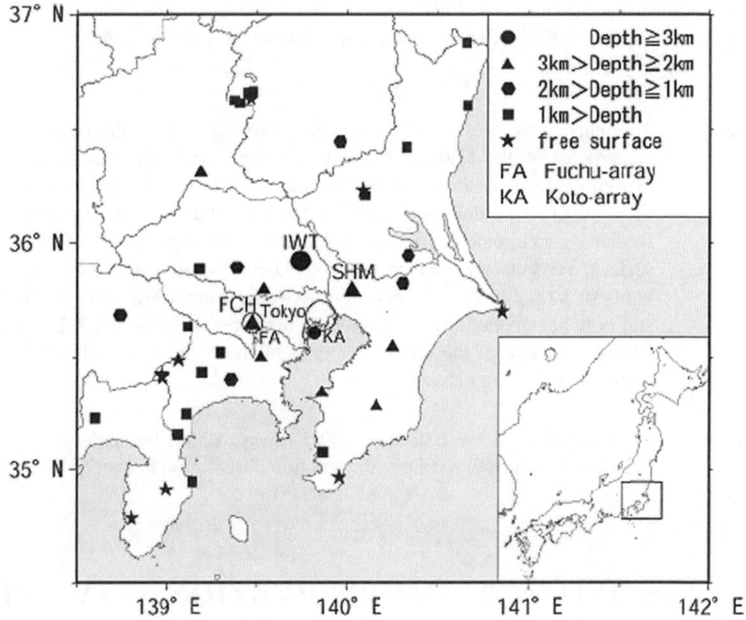

Figure 1. Layout of NIED strong-motion arrays developed in the Kanto region, Japan.

Each deep borehole station has one or two auxiliary boreholes; 108m-depth at the IWT site, 200m-depth at the SHM site, and 500m- and 200m-depths at the FCH site. At both the bottom of each auxiliary well and the free surface, a set of three-component negative feedback accelerometer was also installed. The accelerometers used at the free surface and at the bottom of auxiliary borehole are the same as one used at deep borehole except the IWT station, at which the set of three-component negative feedback accelerometer installed at the auxiliary well and the free surface has a natural frequency of

3Hz with damping factor over 300. The seismic signals received by the accelerometer at the deep borehole, the auxiliary borehole and the free surface are recorded simultaneously on the same electronic medium for each site.

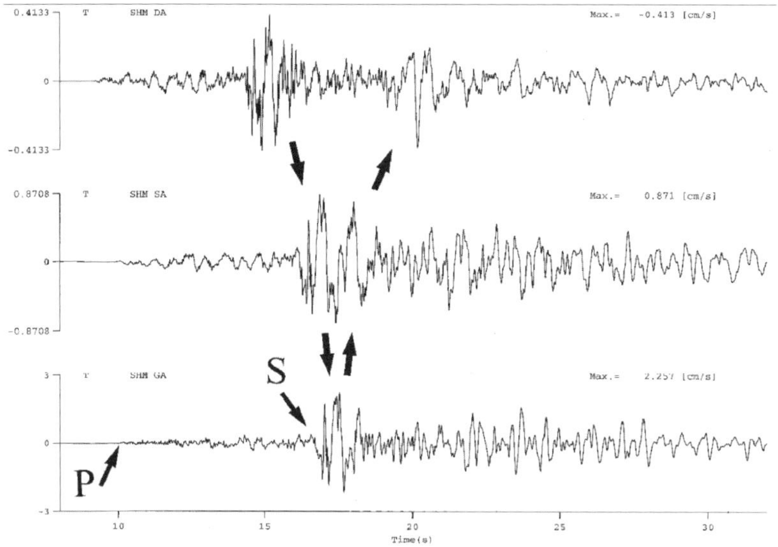

Figure 2. Transverse components of velocity recorded at the SHM site for the SE Ibaraki earthquake of February 19, 1989: (Top) 2,300m-borehole, (middle) 200m-borehole, and (bottom) surface recordings.

The same data acquisition system is used in these deep borehole stations (Takahashi and Hamada, 1975). A cylindrical capsule installed at the bottom of the deep borehole contains seismic and electronic equipment, including a set of three-component negative feedback accelerometers, amplifiers, A/D (analog to digital) converters, PCM (pulse coded modulation) transmitters, and D.C. motors for controlling the level of the seismometers. This apparatus is designed to function in the severe environment of a deep well where the temperature and pressure reach 85 deg and 35MPa, respectively. The accelerometer has a natural frequency of 450 Hz and a damping factor of 0.6 to 0.7. Signals from each accelerometer are sent into a high gain channel with a sensitivity of 60Volts/g and a low gain channel with a sensitivity of 3 Volts/g, simultaneously. The seismic signals of each channel are converted to digital data (12 bits/sample), modulated by the PCM method with a sampling frequency of 240 Hz in the capsule, and then transmitted to a surface observation room. Where the signals are demodulated and smoothed by a Butterworth low-pass filter with a corner frequency of 30 Hz and a decay rate of -18dB/octave. The final signals are stored on IC memory. Thus, this system is able to cover a high dynamic range of ground acceleration, from

0.03 to 300 gals.

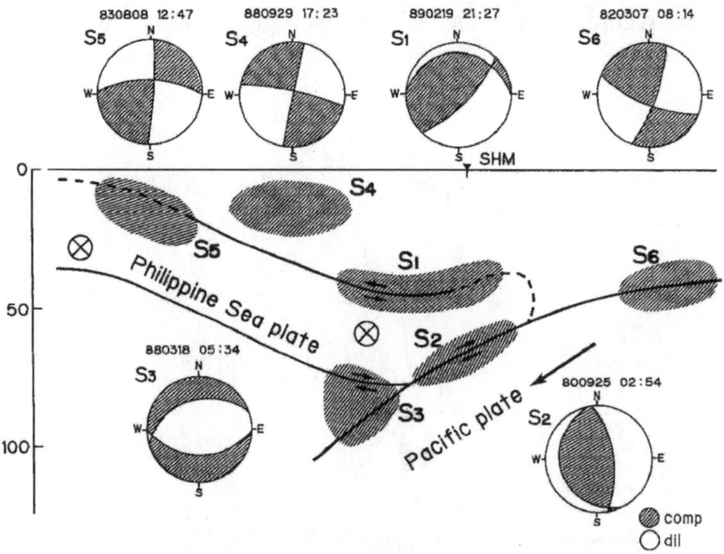

Figure 3. Source region dependence of f_{max}. The value of f_{max} for events that occurred in the S2 and S5 regions is in frequencies between 10 and 20 Hz.

From the engineering point of view, the main purpose of these down-hole arrays is to study the bedrock motion in the Kanto region and the one-dimensional earthquake response of a sedimentary layer – basement system. Fig. 2 shows an example of borehole records. These are transverse-components of velocity recorded at station SHM for the earthquake of February 19, 1989 (Ms=5.6) located in southern east Ibaraki, about 50km away from the station. Top, middle and bottom traces are recorded at the depths of 2.3km, 200m and free surface, respectively. Deep-borehole recordings which are marginally affected by site effects, contributed to study on the modeling of bedrock motion in the response spectral region and to sketch out source spectrum of earthquakes those occurred in and around the Kanto region. The sketching out source acceleration spectrum of earthquakes produced by the motion of three different plates beneath the Kanto region yielded that f_{max} is source location dependent and concluded that f_{max} is source-controlled (Kinoshita, 1992). The result shows that the events in the terminal zone of the Philippine Sea plate where the plate is colliding with the Pacific plate and the North American plate, S2 and S5 zones in Fig. 3, respectively, have lower f_{max} than that of the other events. The value of f_{max} for events that occurred in these zones is in frequencies between 10 and 20Hz. On the contrary, for other events, f_{max} is over 25Hz. The study on the modeling of bedrock motion requires at least two different models to explain the response spectrum of deep borehole data (Kinoshita, 1984). One model is constructed for events that occurred in and around the volcanic front and the other is for events that occurred in the fore-arc region of the volcanic front. These models for response spectrum were used as the input motion for estimating site amplification factor in the Kanto region (Kinoshita et al, 1986).

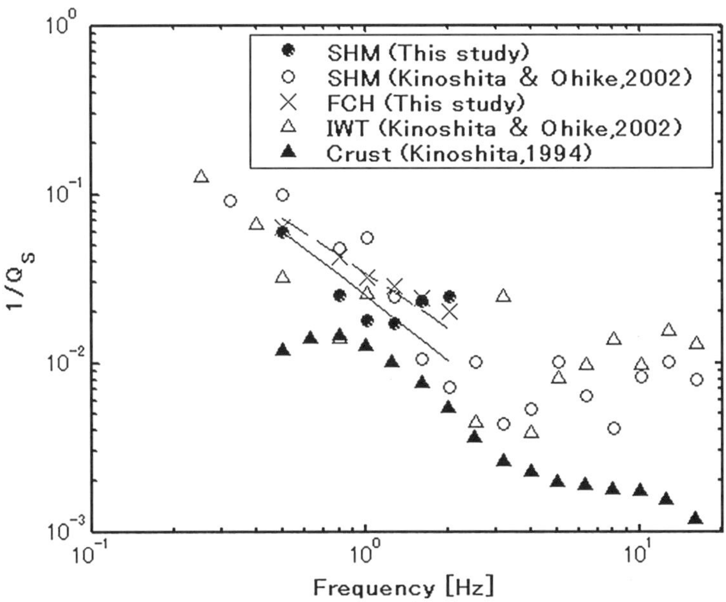

Figure 4. The estimates of 1/Qs in a sedimentary layer – basement system. The estimates of 1/Qs in the crust of the Southern Kanto region are shown by solid triangles.

In experimental studies on the determination of $1/Q$ for body waves traveling in a sedimentary layer – basement system, several methods were proposed. A unique method for the estimation of $1/Q$ is probably to make a comparison between incident phase and its surface-reflected one on the same seismogram recorded at the bottom of deep borehole which is drilled in the pre-Tertiary basement (Kinoshita, 1983). Such a source-canceling method uses waveform recorded by the same seismometer, hence there is no need for instrument compensation and it is applicable to coherent body waves. A recent study obtained by applying this method to the deep borehole seismograms recorded at the FCH and SHM stations yielded the following results:

$$1/Q_P(f) \approx 0.05 f^{-1} \text{ for } f = 0.5-5\,\text{Hz}$$

$$1/Q_S(f) \approx 0.03 f^{-1} \text{ for } f = 0.5-2\,\text{Hz}.$$

The ratio of $Q_S(f)$ to $Q_P(f)$ is about 1.5 in the frequency range 0.5 to 2Hz. Frequency dependence of $1/Q_S(f)$ was also estimated by using the spectral ratio of the S-waves at free surface to deep-borehole recordings in different area as shown in Fig. 4 (Kinoshita and Ohike, 2002b) for the wide frequency range from 0.25 to 16 Hz.

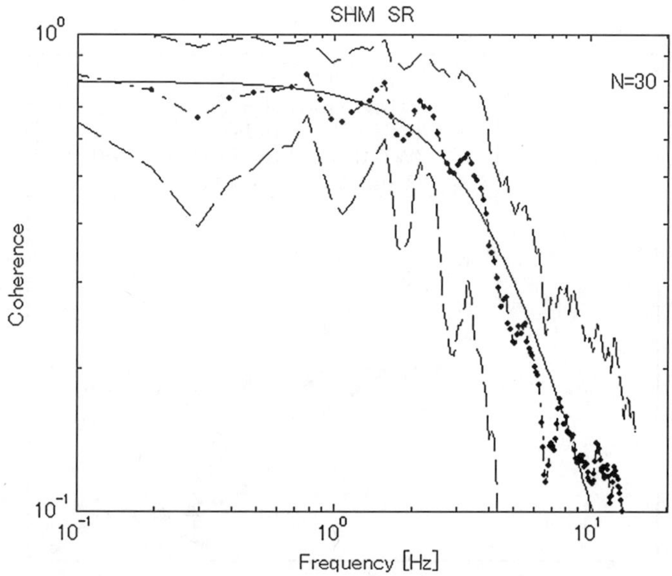

Figure 5. Estimated coherence at the SHM station (dotted line). Solid line is the model coherence of ω^{-2} -type.

A key point for the estimation of $1/Q$ in a sedimentary layer - basement system is a high-frequency bound where body waves are able to propagate coherently in the system in its lower frequencies. We can investigate the coherency characteristics of body waves by making a comparison between direct body waves simultaneously recorded at deep-borehole and surface sites, and between incidence phase and its surface-reflected one on the same deep-borehole seismogram. This means that we can find the coherency characteristics of body waves that travel one-way distance or two-way distance between the levels of deep borehole and surface by using down-hole array recordings. Fig. 5 shows the average coherence estimated by using deep-borehole seismograms recorded at the SHM station. Estimated coherence has the characteristics of ω^{-2} -type:

$$\gamma^2(f) = \frac{\gamma_0^2}{1+(f/f_c)^2}.$$

Thus, the coherence of body waves traveling in a sedimentary layer – basement system, has a corner frequency f_c that decreases as increasing in travel distance or travel time. However, there is a critical corner frequency that becomes a constant independent of travel distance. In the case of S waves, the critical corner frequency is about 4Hz. This means that S waves in the pre-Tertiary basement is already incoherent in high frequencies, that is, the coherence of S waves is less than 80% of γ_0^2 in frequencies higher than a one-half of critical corner frequency.

1.2. The FCH Array (Construction Periods: 1980-1982)

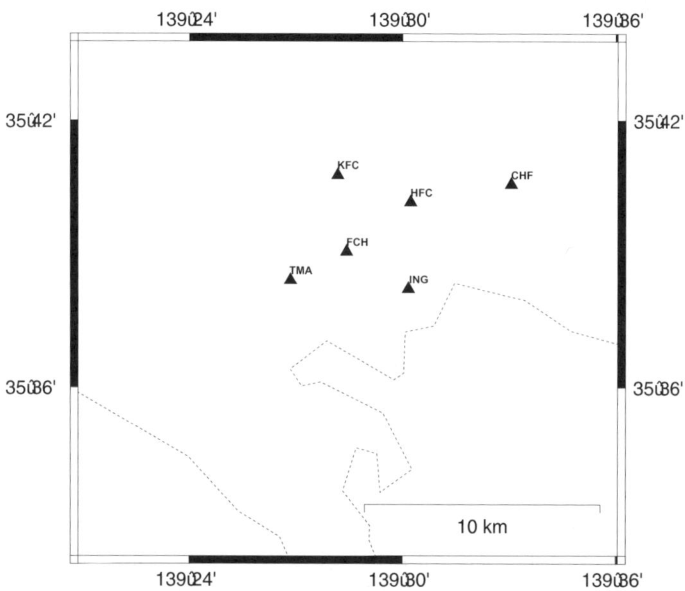

Figure 6. Layout of FCH strong-motion array. Four boreholes are drilled at the center site, FCH. Three-component negative feedback accelerometer is installed at the bottom of each borehole. Six surface sites are equipped with three-component velocity seismometers.

The above-mentioned studies on down-hole array recordings are based on the one-dimensional propagation of body waves in a sedimentary layer - basement system. We need two-dimensional layout of stations in order to investigate the one-dimensional propagation of body waves. Thus, three-dimensional layout of stations is also required for investigating the two-dimensional propagation of body waves. The FCH array was constructed for this purpose, although its array size is compact. The FCH strong-motion array is located in a suburban area west of Tokyo as shown in Fig. 1. At the center of the array (site FCH), the thickness of a sedimentary layer is 2km, which overlies on the pre-Tertiary basement. The closest edge of the sedimentary basin located less than 20km to the west. The NIED has been operating this strong-motion array since 1983 to investigate the two-dimensional response of a sedimentary layer – basement system during earthquakes (Kinoshita, 1985; Phillips et al., 1993). Four sets of three-component negative feedback borehole accelerometer were installed in station FCH at depths of 0, 150 (moved from 200m on 19 March 1986), 500, and 2,750m, respectively. Only the deepest borehole was drilled in the pre-Tertiary basement. The accelerometers have a natural frequency of 450 Hz and the damping factor of 0.6 to 0.7. Six sets of three-component negative feedback velocity seismometers were installed at the surface sites as shown in Fig. 6. Velocity response is flat in the frequency range of 0.01 to 100Hz. Dynamic range is 96dB (16 bits) and sampling rate is 200Hz for surface recording system. Station interval for the surface sites is roughly 3 km.

1.3. The KOT (Tokyo Lowlands) Array (Construction Periods: 1983-1987)

Down-hole arrays including a deep borehole and the FCH array were constructed for the investigation of body waves. On the other hand, the KOT array was constructed, intentionally, to study surface waves, in particular, basin-induced surface waves (Kinoshita et al., 1992). The KOT strong-motion array is located in the Tokyo lowlands near downtown Tokyo (Fig. 1) and consists of 20 surface stations as shown in Fig. 7. The predominant period of Love waves in this region is about 8 seconds. Thus, the Koto array used negative-feedback velocity seismometer that is possible to record, effectively, long-period surface waves comparing to accelerometers. The diameter of the KOT array is roughly 10 km, a one-half wavelength of the fundamental mode of basin-induced Love wave in this region. Each site in the KOT array is equipped with the same three-component set of negative-feedback velocity meter, VSE-11/12, with the flat response in the frequency range of 0.01 Hz to 100Hz. The measurable velocity of VSE-11/12 velocity seismometer is 200 cm/s at its maximum. The KOT array used a 24-bit type recording system with the sampling rate of 200 Hz.

Typical seismograms recorded by the KOT array are shown in Fig. 8, which were transverse-components obtained for the Near Izu-Ohsima earthquake of February 20 1990. The surface magnitude of this earthquake

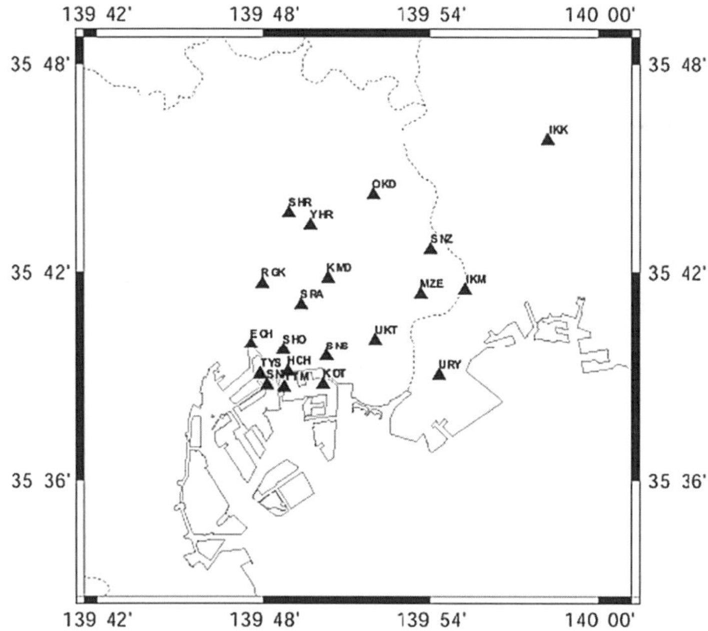

Figure 7. Layout of KOT strong-motion array.

is 6.6 and the epicenter distance is about 150km. As appeared from the array recordings, basin-induced surface waves (Love waves) have the duration of about 5 minutes. Based on frequency-wave number analysis, Kinoshita (2003) discriminated three types of Love waves (LQ1, LQ2 and LQ3 in Fig. 8). The character of each phase is shown in Fig. 9. Two wave-number spectra shown in Fig. 9 are estimated for different portions of array recordings by a moving 40-second time window. The start and end time of each window are 105s-145s and 175s-215s for Fig. 9(a) and Fig. 9(b), respectively. The results show that the fundamental structure of basin-induced Love waves appearing in the Tokyo metropolitan area consists of three phases of surface waves. The last phase named LQ3 phase, in particular, is a basin-induced Love wave emanated from the opposite side of source. By using the estimated propagation velocity and the approaching azimuth of LQ3 phase, we can elucidate the existence of LQ3 wave in the time domain. Fig. 10 shows the original seismogram recorded at the KMD station in the KOT array and the beam-formed LQ3 wave at the same station. We can make sure the clear existence of LQ3 wave in the time domain from the beam-formed seismogram.

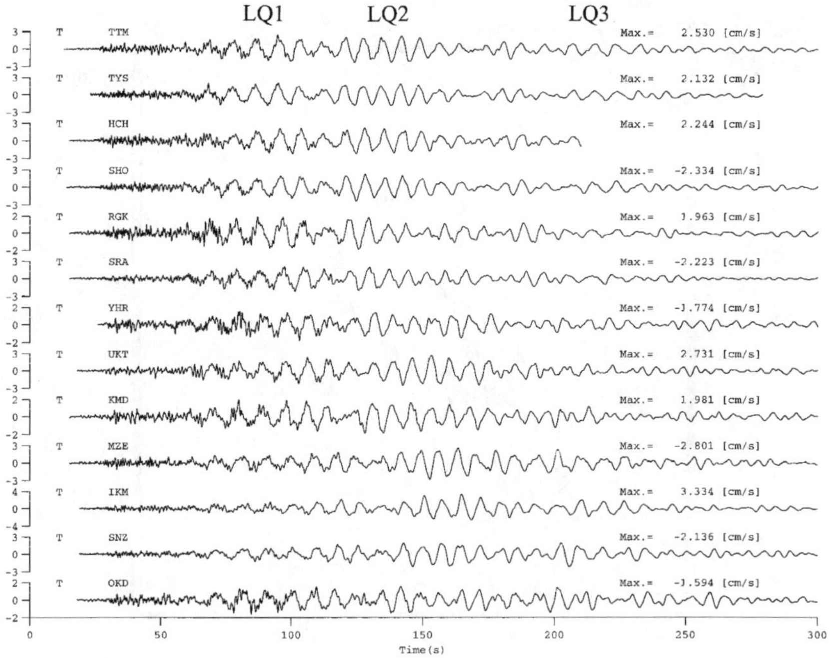

Figure 8. Transverse-components of velocity recorded at the KOT strong-motion array for the near Izu-Oshima island earthquake of February 20, 1990.

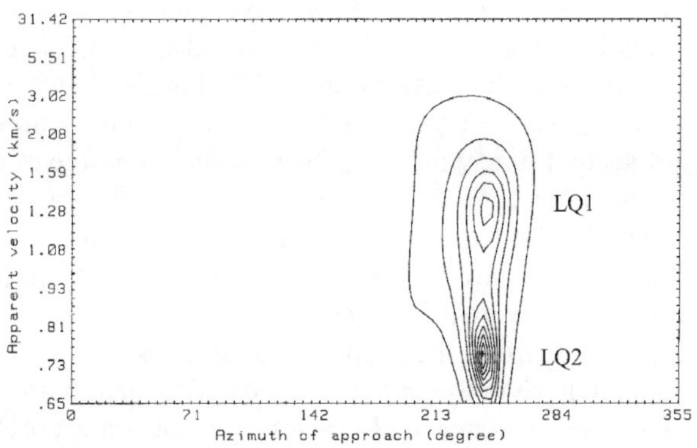

Figure 9(a). Wave-number spectrum at 7.5s (105s-145s).

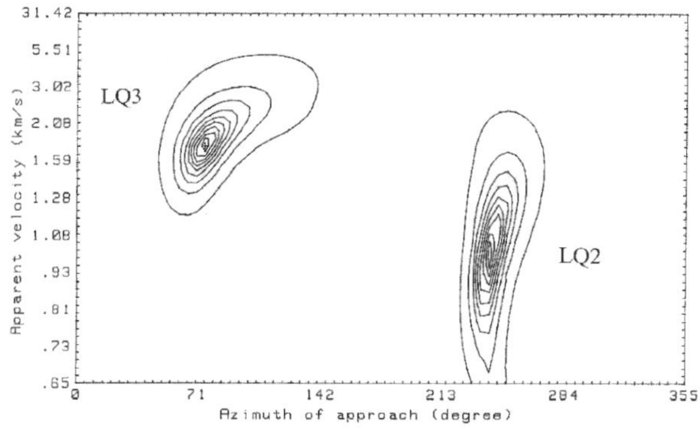

Figure 9(b). Wave-number spectrum at 7.5s (175s-215s).

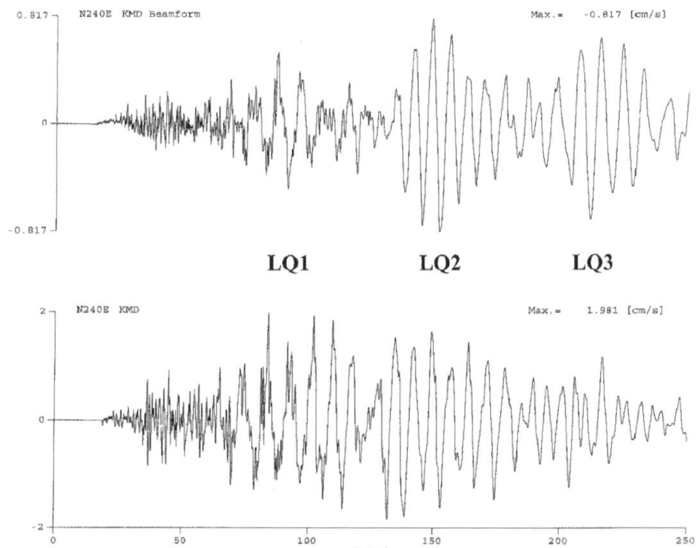

Figure 10. Beam-forming of transverse component of velocity recorded at the KOT array for the near Izu-Oshima island earthquake of February 20, 1990. (Top) Beam-formed surface wave at the KMD site and (bottom) original wave at the same site.

1.4. The Kanto Borehole Array (Construction Periods: 1991-1995)

On the contrary to the purpose for the FCH and KOT arrays, the borehole array deployed in the whole Kanto region is to study path effects and source characteristics. In 1991 to 1995, the NIED constructed 12 boreholes with depths of 1km to 2km in the Kanto region. At the bottom of each borehole, a set of three-component negative feedback accelerometer with a natural frequency of 450Hz and the damping factor of 0.6 to 0.7 was installed. Overall characteristic of recording system used in these borehole sites has the flat response in the frequency range of DC to 30Hz. In addition, borehole stations with depths of less than 200m and surface stations on the outcrop of hard rock areas surrounding the Kanto plane were constructed as well (Fig. 1).

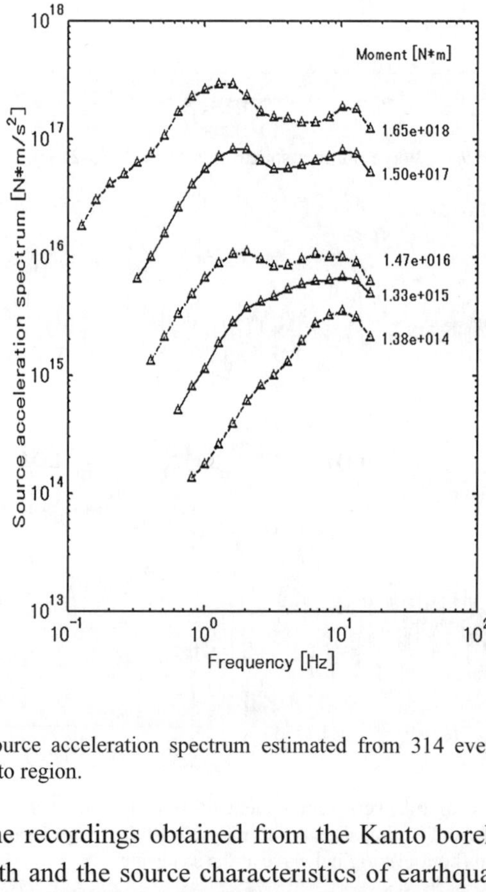

Figure 11. Source acceleration spectrum estimated from 314 events that occurred in and around the Kanto region.

Using the recordings obtained from the Kanto borehole array, $1/Q_S(f)$ along the path and the source characteristics of earthquakes that occurred in and around the Kanto region were investigated. The resultant $1/Q_S(f)$ is shown in Fig. 4 by solid triangles (Kinoshita, 1994) and has a peaked structure in a frequency range of 0.5 to 16Hz on the assumption that geometrical spreading is the reciprocal of hypocenter distance. In particular, $1/Q_S(f)$ has a peak of about 0.01 at 0.8 Hz and decreases with increasing frequency, having a knee structure at around 8Hz, which characterizes the path effect between the network and the Philippine Sea subducting-slab. Source acceleration spectrum was also estimated by using the Kanto borehole array. Estimated acceleration source spectrum is different from that of the standard ω^{-2} source model and shows two characteristic bumps (Kinoshita and Ohike, 2002a). Fig. 11 shows acceleration source spectrum which is estimated by using 314 events that occurred in and around the Kanto region (Kinoshita, 2003). The resultant spectra show two bumps: One is in frequencies higher than 5Hz and the other one is in the low-frequency band around the corner frequency. The bump in the low-frequency band moves toward the low-frequency side with increasing seismic moment. This effect appears to be the result of larger-scale heterogeneity on the earthquake fault.

2. STRONG-MOTION OBSERVATION NETWORK IN THE SECOND STAGE (1995-NOW)

The second stage of strong-motion observation is constructing nation-wide. The occasion for constructing nation-wide networks was the 1995 Kobe earthquake and two networks had been constructed under three principles (Fig. 12): (1) systematic observation network with the same seismograph on the free surface or installed at the bottom of borehole drilled in the basement, (2) a uniform station-to-station distance of about 25km (K-NET) or 50km (KiK-net), and (3) complete data release through the Internet.

2.1. K-NET (Construction Periods: 1995-1996)

This nation-wide strong-motion network had been constructed just after the Kobe earthquake (Kinoshita, 1998) and has been releasing data through the Internet since June 1996 (http://www.k-net.bosai.go.jp). Now the number of stations is 1,034 including the KOT array. At each surface site, a set of three-component negative feedback accelerometer with a natural frequency of 450Hz and the damping factor of 0.6 to 0.7 was installed. Sampling rate is 100Hz and a 24-bits type recording system with the maximum acceleration of 2g was used. The renewal of strong-motion seismograph instead of the K-NET95 type (Kinoshita, 1998) has been promoting since March 2004. New seismograph is designed with the aim of semi-online use of strong-motion data. Also, the new strong-motion seismograph adopted a set of three-component negative feedback accelerometer with a natural frequency of 100 Hz and the damping factor 0.6 to 0.7. A 24-bits type recording system is also adopted.

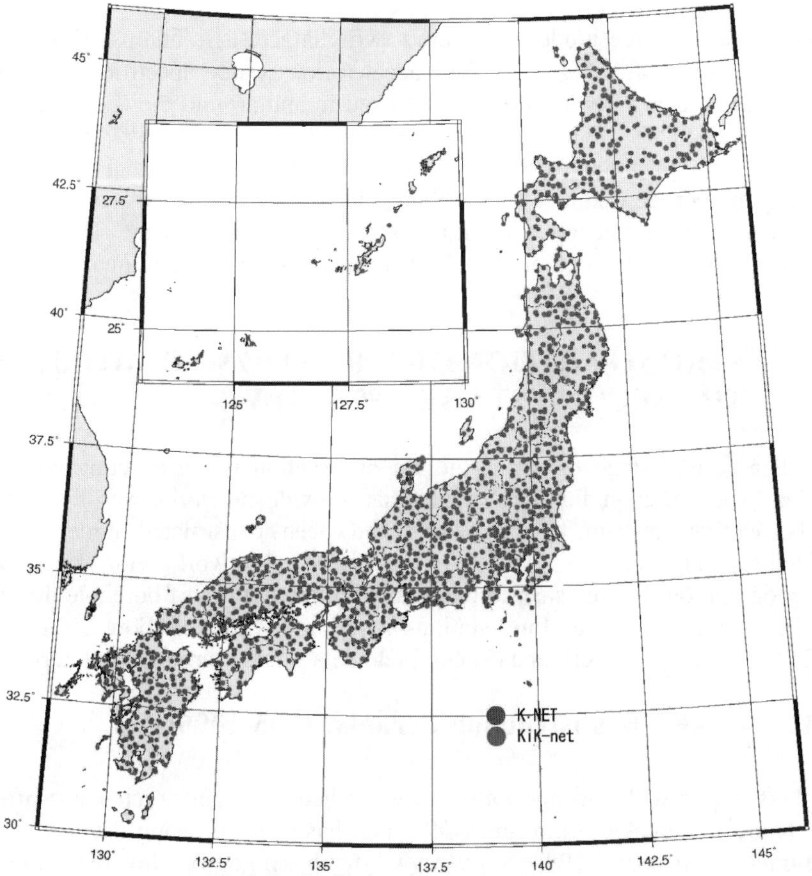

Figure 12. Layout of K-NET and KiK-net stations.

As an example of the application of K-NET data, Fig. 13 shows the site amplification map of the K-NET sites in the Kanto region. Site amplifications for S waves were estimated by using a spectral inversion method, where the SHM 2.3km-depth borehole site was used as the reference. This figure is the result of 0.5Hz-band and shows the relation between the thickness of a sedimentary layer and site amplification. For longer period bands, the regions showing high site amplification were concordant with the regions of thick sediments (Kinoshita and Ogue, 2002).

2.2. KiK-Net (Construction Periods: 1997-2001)

KiK-net stands for Kiban (basement) K-NET. It is the twin of the Hi-net seismic network over Japan. This network consists of 559 sites. Each site has a borehole with depths of 100 to 200m. However, in major city such as Osaka or Sendai city, the boreholes reach depths over 1km. In each KiK-net site, two sets of three-component negative feedback accelerometer with a natural frequency of 450Hz and the damping factor of 0.6 to 0.7 were installed at the bottom of borehole and the free surface. Sampling rate is 200Hz and a 24-bits type recording is used with the maximum acceleration of 2g. The data from KiK-net have been releasing on the Internet (http://www.kik-net.bosai.go.jp).

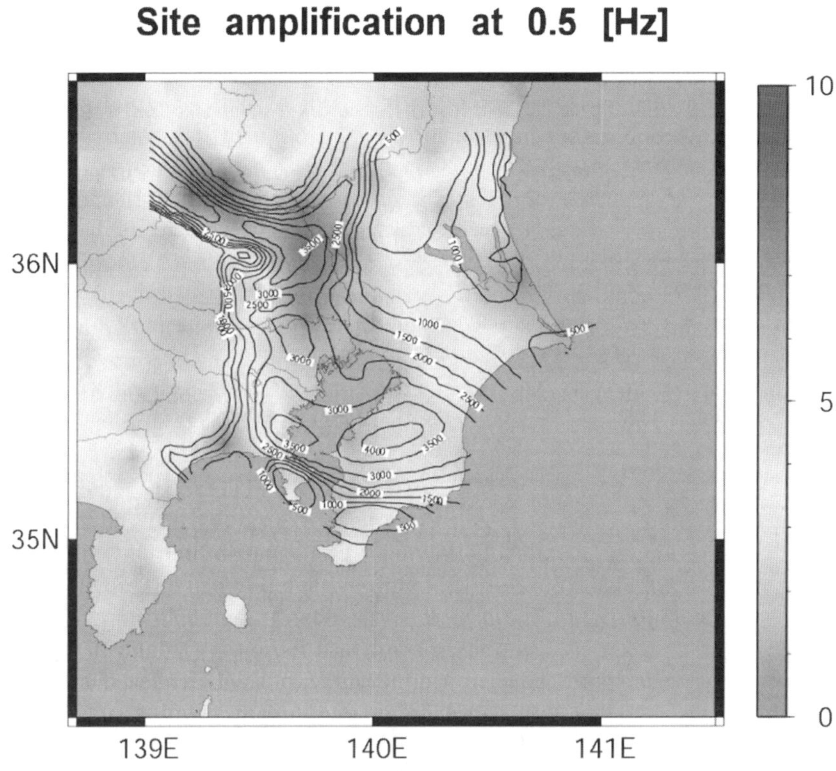

Figure13 Site amplification map at 0.5Hz-band

The K-NET covers urban areas in Japan. Thus, the K-NET is used for studying site effects, mostly. On the contrary, the KiK-net sites were constructed in the region where the effects of surface layer on seismograms are minor. Hence, the KiK-net is used for studying path effects and source characteristics, basically.

3. CONCLUDING REMARKS

This paper described the history of NIED strong-motion observation network for the last quarter of a century. The process toward the construction of the nation-wide networks started from the construction of single station gives us various chances to learn various technical skills and knowledge about strong-motion seismographs, borehole drilling, and the design and operation of observation network. Also, interpretation of seismograms on the-job-training plays an important role to promote the construction work.

REFERENCES

1. Kinoshita, S., 1983, A study for damping characteristics of surface layers, *Proc. Japan Soc. Civil Eng.* **330**: 15-25 (in Japanese).
2. Kinoshita, S., 1984, Spectral characteristics for bedrock motion in the Tokyo metropolitan area, *Proc. Japan Soc. Civil Eng.* **344**: 89-94.
3. Kinoshita, S., 1985, Propagation of total reflected plane SH pulses in a dipping layer, *Zisin* **38**: 597-608 (in Japanese).
4. Kinoshita, S., Mikoshiba, T., and Hoshino, T., 1986, Estimation of the average amplification characteristics of a sedimentary later for short period S waves, *Zisin* **39**: 67-80 (in Japanese).
5. Kinoshita, S., 1992, Local characteristics of the $f_{\max}$ of bedrock motion in the Tokyo metropolitan area, Japan, *J. Phys. Earth* **40**: 487-515.
6. Kinoshita, S., Fujiwara, H., Mikoshiba, T., and Hoshino, T., 1992, Secondary Love waves observed by a strong-motion array in the Tokyo lowlands, Japan, *J. Phys. Earth* **40**: 99-116.
7. Kinoshita, S., 1994, Frequency-dependent attenuation of shear waves in the crust of the Kanto area, Japan, *Bull. Seism. Soc. Am.* **84**: 1387-1396.
8. Kinoshita, S., 1998, Kyoshin net (K-NET), *Seism. Res. Let.* **69**:309-332.
9. Kinoshita, S., and Ohike, M., 2002a, Scaling relations of earthquakes that occurred in the upper part of the Philippine Sea plate beneath the Kanto region, Japan, estimated by means of borehole recordings, *Bull. Seism. Soc. Am.* **92**: 611-624.
10. Kinoshita, S., and Ohike, M., 2002b, Attenuation characteristics of S waves in a sedimentary layer – basement system in the Kanto region, Japan, for a frequency range of 0.5 to 16Hz, *Zisin* **55**: 19-31 (in Japanese).
11. Kinoshita, S., and Ogue, Y., 2002, Site amplification of K-NET sites in the Kanto region, central Japan, in: *Seismotectonics in Convergent Plate Boundary*, Fujinawa, Y., and Yoshida, A., eds. TERRAPUB, Tokyo, pp. 393-405.
12. Kinoshita, S., 2003, Local event seismograms, *Technical Note of the Natl. Res. Inst. for Earth Science and Disaster Prevention* **240**: 1-190 (in Japanese).
13. Phillips, W. S., Kinoshita, S., and Fujiwara, H., 1993, Basin-induced Love waves observed using the strong-motion array at Fuchu, Japan, *Bull. Seism. Soc. Am.* **83**: 64-84.
14. Takahashi, H., and Hamada, K., 1975, Deep borehole observation of the earth activities around Tokyo, *PAGEOPH* **113**: 311-320.
15. Yamamizu, F., Takahashi, H., Goto, N., and Ohta, Y., 1981, Shear waves velocities in deep soil deposits, *Zisin* **34**: 465-479 (in Japanese).

DENSE STRONG-MOTION ARRAY IN YOKOHAMA, JAPAN, AND ITS USE FOR DISASTER MANAGEMENT

Saburoh Midorikawa
Center for Urban Earthquake Engineering, Tokyo Institute of Technology

Abstract: This paper describes the dense strong-motion array in the city of Yokohama, Japan. The strong-motion array, which consists of 150 stations in the area of about 400 km^2, has been originally planned for disaster management of the city. Based on the array which can report the ground motion intensity data immediately after an event, the real-time seismic hazard and risk assessment system is developed for emergency response. The data from the array are also used for ground motion studies which contribute to seismic hazard assessment of the city. In order to improve earthquake awareness and preparedness of citizens, seismic hazard and risk maps for scenario earthquakes are prepared.

Key words: Strong-motion array, Ground motion studies, Disaster management, Real-time damage assessment, Hazard and risk map, Yokohama

1. INTRODUCTION

The 1995 Kobe, Japan earthquake gave a great impact on disaster management strategy. Just after the earthquake, few immediate emergency response was taken by the national and local governments, which might allow the spread of the disaster. This is due to lack of prompt information on the disaster. The mayor of the city of Yokohama seriously realized the importance of the earthquake information, and he decided to develop the dense strong-motion array for disaster management with suggestion of a seismologist, the late Prof. M. Kikuchi who is one of the pioneers of real-time seismology. The strong-motion array was originally planned for disaster management of the city; however the data from the array have been used not only for disaster management, but also for seismological and engineering

P. Gülkan and J.G. Anderson (eds.), Directions in Strong Motion Instrumentation, 197–208.

researches. This paper describes the dense strong-motion array and its use for disaster management.

2. DENSE STRONG-MOTION ARRAY IN YOKOHAMA

The city of Yokohama is the second largest city in Japan whose population and area are 3.5 millions and 433 km^2, respectively. As shown in Fig. 1, accelerographs are installed for ground shaking monitoring at 150 free field stations, which satisfy 2 km spacing between stations. The requirement of the 2 km spacing arises from the width of the narrow disaster belt zone observed in the 1995 Kobe earthquake. A high precision digital accelerograph is used to record weak to very strong ground motion. Ground acceleration up to 2000 cm/s^2 is registered with 18 bits A/D resolution at 200 Hz sampling. In addition to the accelerographs at ground surface, borehole accelerometers are installed at 9 stations for liquefaction hazard assessment.

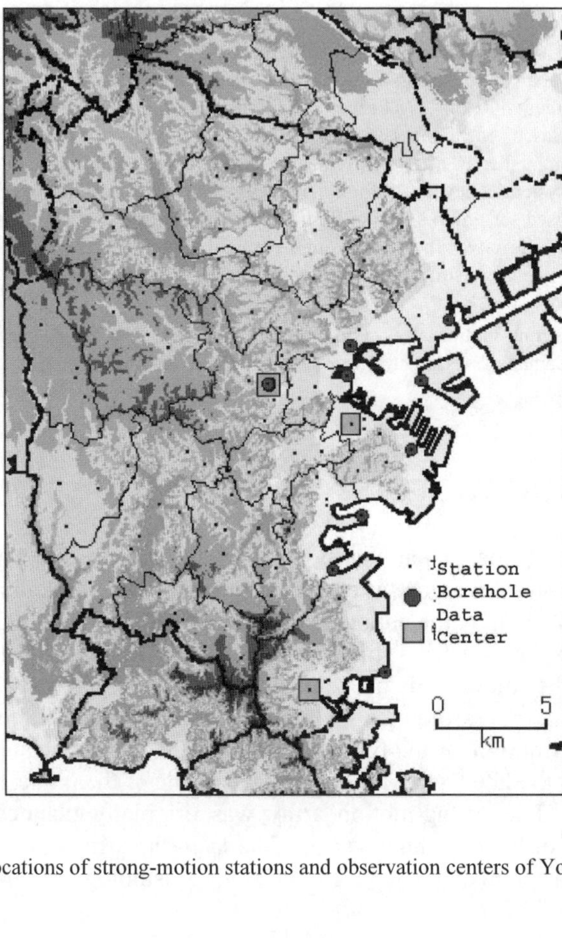

Figure 1. Locations of strong-motion stations and observation centers of Yokohama array.

These stations are connected to three observation centers, the disaster preparedness office of the city hall, the fire department office of the city and Yokohama City University, by the high-speed and higher-priority telephone line. At 18 stations, the backup communication system by satellite is available. The full operation of the monitoring system started on May, 1997. When the accelerograph is triggered by an earthquake, the station computes ground-motion parameters such as the instrumental seismic intensity, peak amplitudes, predominant frequency, total power, duration and response spectral amplitudes. These parameters are automatically reported to the centers.

On receiving more than 10 reports in a prescribed time interval, the centers activate alert systems. The seismic intensity data is immediately conveyed to the city officials by the pager. The intensity map of the city is drawn within a few minutes after the earthquake, and is open to the public through the Internet (http: www.city.yokohama.jp/me/bousai/eq/index.html) and local cable TV. The map is utilized as the most early information for disaster management. The seismic intensity data at 36 stations are also reported to the Japan Meteorological Agency. The data are put together with those from the other sources and are transmitted by nationwide broadcasts.

3. GROUND MOTION STUDIES USING ARRAY DATA

While the main objective of installation of the strong motion array is disaster management, the data is open to researchers at universities and private companies for ground motion studies. The studies may be divided into three categories; a) local site amplification studies, b) wave propagation studies and c) deep underground prospecting studies.

a) local site amplification studies

Large variation has been observed in ground motion records at 150 stations. Figure 2 shows comparison of velocity time histories of a horizontal component. The variation of peak ground velocities is about ten times. This suggests large effects of local site conditions in the city. Torii et al. (1998) show that the local site effects observed in the records can be quantitatively evaluated by the multiple reflections of SH-waves in soil deposits. Several techniques for site amplification evaluation, such as the H/V spectral ratio technique, have been verified by using the records (Tsuboi et al., 2001; Rodriguez and Midorikawa, 2003).

b) wave propagation studies

As the stations are close to each other, the records provide unique opportunity to examine characteristics of wave propagation. Kohketsu and Kikuchi (2000) and Miura and Midorikawa (2001) show the propagation of long-period ground motion in the Kanto basin using the records of a local earthquake observed at the Yokohama array and surrounding stations as shown in Fig. 3. Dispersion observed in the records suggests that fundamental mode Love waves are dominant in long-period ground motion (Miura and Midorikawa, 2001).

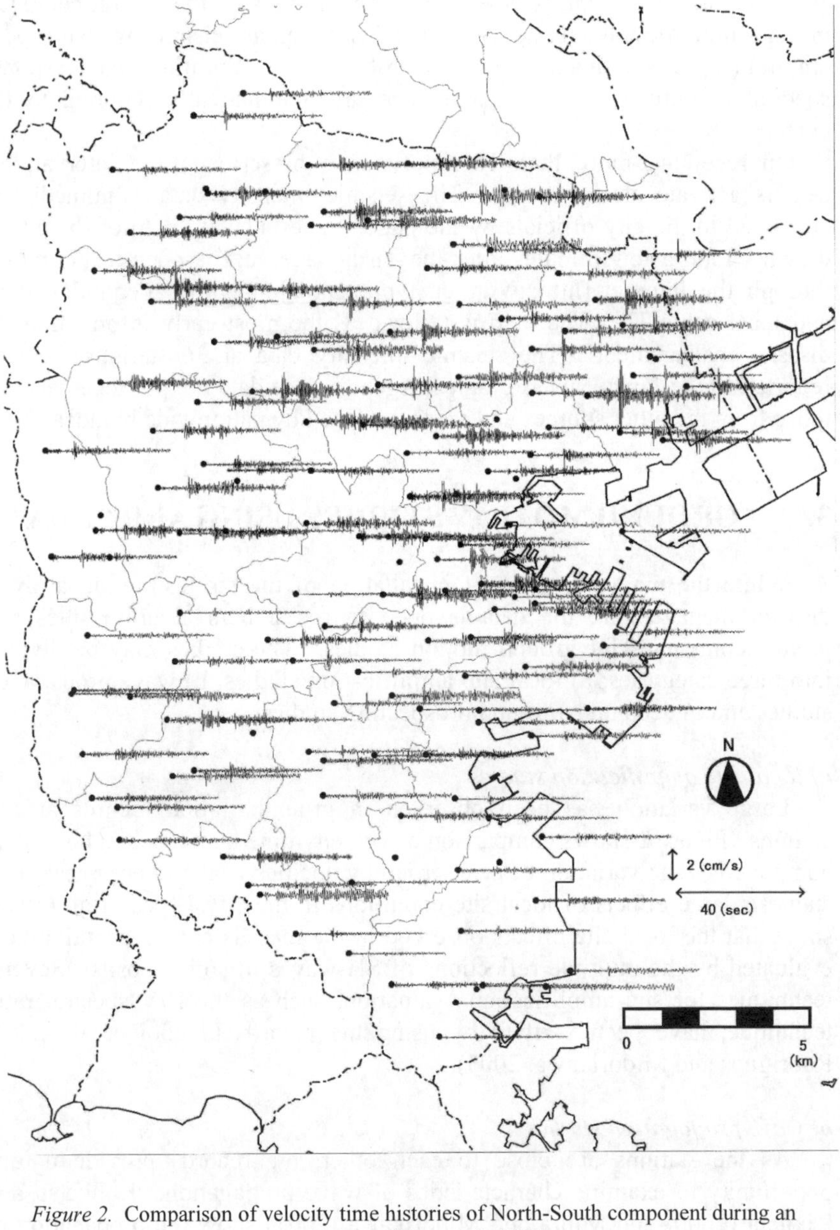

Figure 2. Comparison of velocity time histories of North-South component during an earthquake of August 9, 1997.

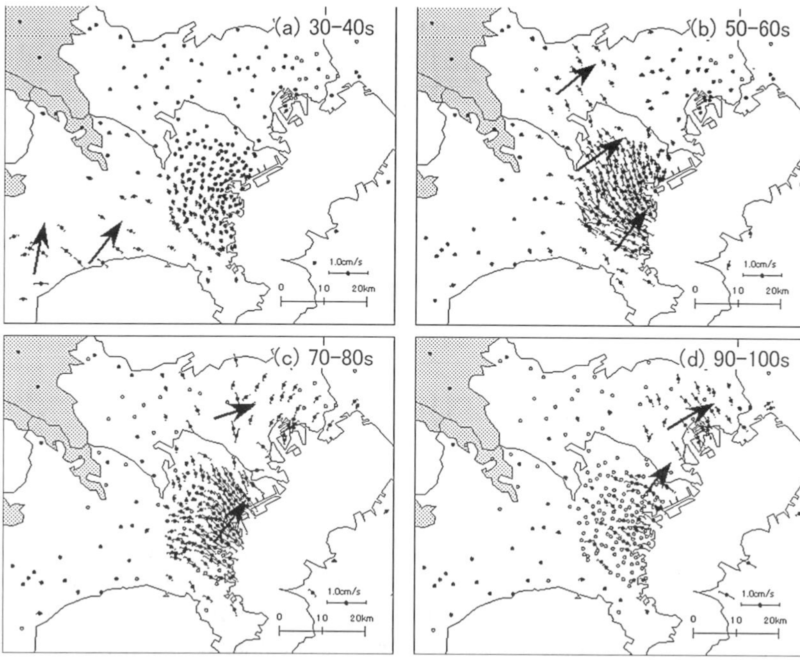

Figure 3. Distribution of principal axes calculated from horizontal band-pass filtered velocity components for the period of 8 seconds (Miura and Midorikawa, 2001).

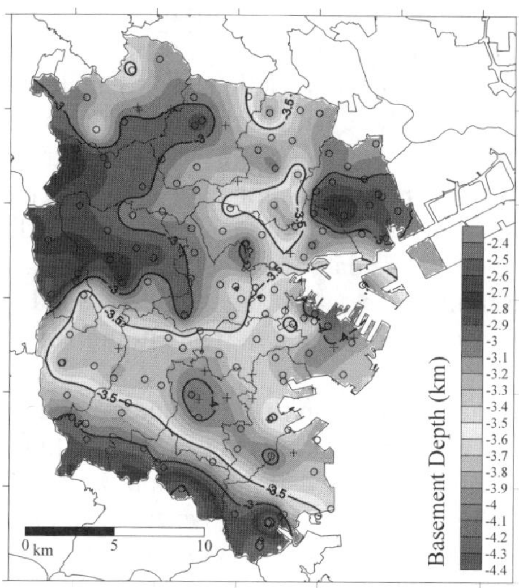

Figure 4. Distribution of estimated depth to basement rock (Miura and Midorikawa, 2001).

c) deep underground prospecting studies

It is important for strong motion prediction to understand the deep underground structure down to the seismic bedrock. The delay of the travel time of P- and S-waves in the records is used to estimate thickness of sediment layers (City of Yokohama, 2001). The time delay of PS converted waves to direct P waves is also used to model the 3-D deep underground structure in the city (Miura and Midorikawa, 2001). As shown in Fig. 4, the results indicate that the depth to the seismic bedrock varies from 2.5 to 4.0 km in the city and that observed long-period motion is amplified in the area with thicker sediments.

4. REAL-TIME SEISMIC HAZARD AND RISK ASSESSMENT SYSTEM FOR EMERGENCY RESPONSE

In order to grasp a scale of the disaster and its spatial distribution immediately after the event, the ground motion data from the stations are used for the real-time seismic hazard and risk assessment. The assessed items are ground motion, liquefaction and building damage. The various GIS (Geographic Information System) data which have been built by the different departments of the city are gathered and compiled for the assessment. The operation of the assessment system started in June, 1998.

In the mapping of ground motion hazard, the city is divided into 170,000 elements with the mesh size of 50 x 50 meters. The fine mesh system is adopted because of strong spatial variation of the geologic conditions in the city. More than 15,000 borehole data in the city are compiled. The P- and S-loggings down to the sediment rock are conducted at all the 150 strong motion stations. The soil profile sections are drawn from the borehole data and the geomorphological information, and the soil condition of the city is classified into 268 types. The data from borings or measurements of S-wave velocity are used to build up the generalized underground model for each soil type. The amplification factor of the corresponding soil type is assigned to each element.

To eliminate the site effects from the observed ground motion, the bedrock motion spectrum at each station is estimated by dividing the observed response spectrum by the computed amplification factor. The ground motion spectrum at each element of the fine mesh system is computed by multiplying the bedrock motion spectrum at the neighboring station by the amplification factor at the element (Midorikawa and Abe, 2000). Thus, the very detailed ground shaking map is constructed by using the observed ground motion data and the soil amplification factors. Figure 5 shows an example of the detailed ground shaking map constructed for a local earthquake on May 3, 1998.

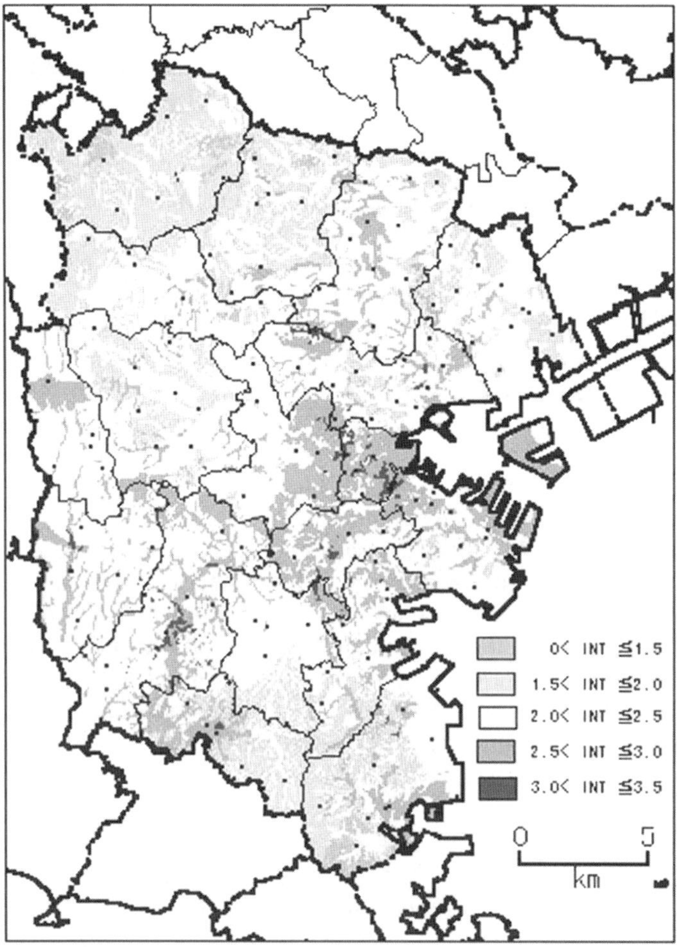

Figure 5. 50m-mesh ground shaking map constructed from observed ground motion data and soil amplification factors.

Liquefaction hazard is also assessed. In this assessment, a dynamic load induced in a soil element during an earthquake is computed not by the record at surface, but by the record of the borehole accelerometer at sediment rock. The liquefaction resistance of soil is calculated by the method adopted in the specifications for highway bridges (Japan Road Association, 1996). The liquefaction potential for each element of the fine mesh system is evaluated by comparison of the liquefaction resistance with the dynamic load. The shaking and liquefaction maps are drawn within ten minutes after the event.

Although there are many commercial buildings in the city, the majority of the buildings in the city are wooden houses. Generally, they are more vulnerable to strong shaking than non-wooden buildings such as reinforced concrete buildings. Accordingly, the assessment of the seismic risk of wooden houses is included in the system. The GIS database of about 450,000 wooden houses is built up by the data matching of the fixed asset tax inventory and the digital base map. The dynamic vibration characteristics of each house are assigned according to the types of structure, use and roof and the construction age. The response of the house is computed by using the response spectrum of ground motion, to determine the degree of the damage. The distribution of the wooden-house damage is mapped within twenty minutes after the event. The damage map is displayed with other information such as locations of hospitals, refuges and major roads for emergency transportation. Thus, the hazard and risk maps are created within twenty minutes after the earthquake, so that the maps can be used for selecting strategy of emergency response activities.

The damage information gathering system is constructed to overlay the actual damage information which is gradually obtained after the event, on the estimated damage map. The city has an agreement with about 500 civil construction firms. When the seismic intensity larger than 5 in the Japanese scale (approximately 7 in the M.M. scale) is observed, they will immediately survey the damage to road structures on the 93 major roads, and report the damage to the 18 civil construction branch offices of the city. The damage information reported to the 18 branch offices is transmitted to the city hall in the form of the GIS data by the intranet GIS system. The damage GIS data such as the damage of the road and restriction of the traffic are overlaid on the estimated damage map in order to up-to-date the damage information.

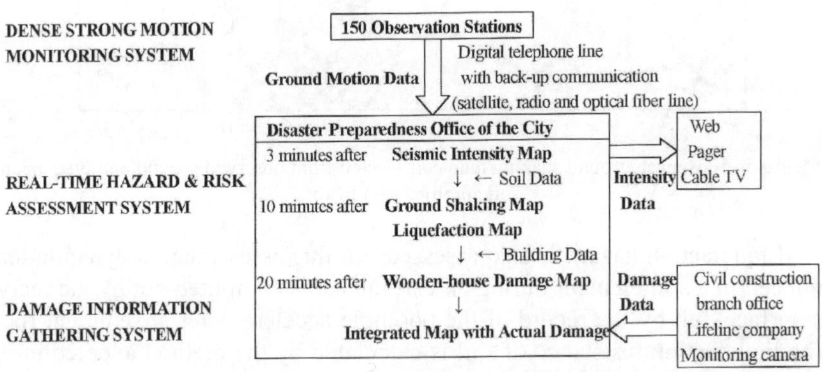

Figure 6. Structure of system for REal-time Assessment of earthquake Disaster in Yokohama (READY system).

The three systems, such as the dense strong motion monitoring, real-time seismic hazard and risk assessment and damage gathering systems, compose the system for REal-time Assessment of earthquake Disaster in Yokohama

(READY system), as shown in Fig. 6. The city conducts earthquake drills twice a year on the days of the Kanto earthquake (September 1) and of the Kobe earthquake (January 17). At the drill, the simulated results from the real-time system for a scenario earthquake are used to make the drill more practical and realistic.

5. SEISMIC HAZARD AND RISK MAPPING FOR PREPAREDNESS

At emergency, the real-time system mentioned above should be useful for disaster mitigation. Also, at ordinary times, disaster mitigation activities such as preparedness by citizens should be promoted. In order to improve earthquake awareness of citizens, the seismic hazard map is effective. When a citizen look the seismic hazard map, he thinks that a) Is my house in the red zone ? and b) If my house is in the red zone, what should I do ? (Olshansky, 2000). To reply his questions, the map should be the micro-scale one. To meet the demand of citizens, the city published “Yokohama City Shake Map” in June, 2001 (City of Yokohama, 2001).

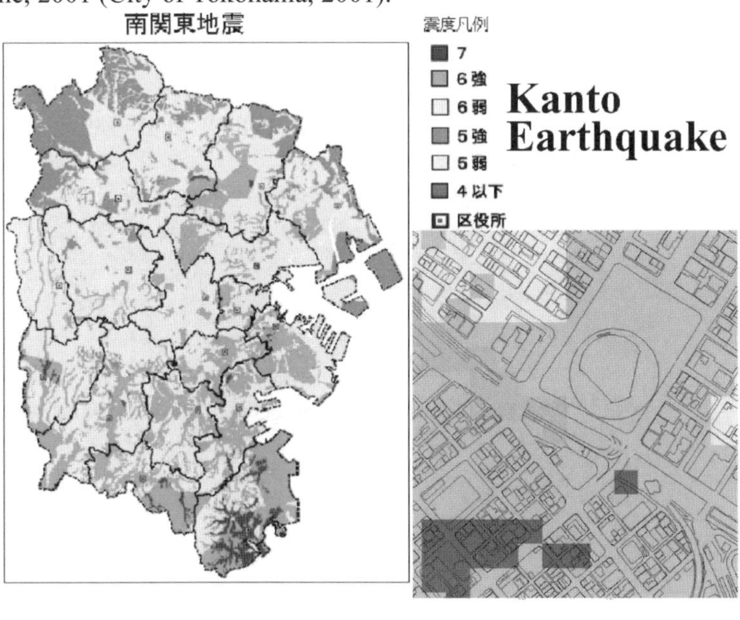

Figure 7. Yokohama Shake Map (left: whole map, :right: close-up map).

Considering the detailed subsoil data constructed for the real-time system and the deep underground structure estimated from the observed records, the deterministic ground shaking maps are computed by the hybrid simulation technique for several scenario earthquakes including the recurrence of the 1923 Kanto earthquake. Figure 7 shows an example of the shake map. As the 50 meter mesh system is used, the map is so detailed that the citizen can identify the seismic hazard at his house and may have stronger interests in earthquake disaster and its mitigation. To verify the computed map, the intensity questionnaire survey of local earthquakes (Matsuda et al., 2000) and the damage distribution survey of the 1923 Kanto earthquake (Midorikawa, 2002) are conducted.

More than 100,000 copies of the map were printed and distributed to citizens. As the results, the numbers of applicants for seismic performance appraisal of wooden house and for seismic retrofitting subsidies by the city increased to about 2000 and 150 in 2001, respectively. These numbers are twice of those in the previous year. This is an evidence of improvement of earthquake awareness of citizens by the map.

If we focus our objectives on improvement of citizen's emergency preparedness, the detailed hazard map alone does not provide sufficient information. Information maps for earthquake emergency preparedness should be developed. The information map consists of three kinds of maps, such as a risk map, evacuation aid map, and fire fighting and rescue aid map (Ishizawa and Midorikawa, 2002). The city produced the prototype of the maps in 2003, as shown in Fig. 8. The maps include the items which are related to evacuation actions, such as evacuation place, road width, slope failure and building collapse potential. The complete maps will be provided on the web GIS system by January 2005.

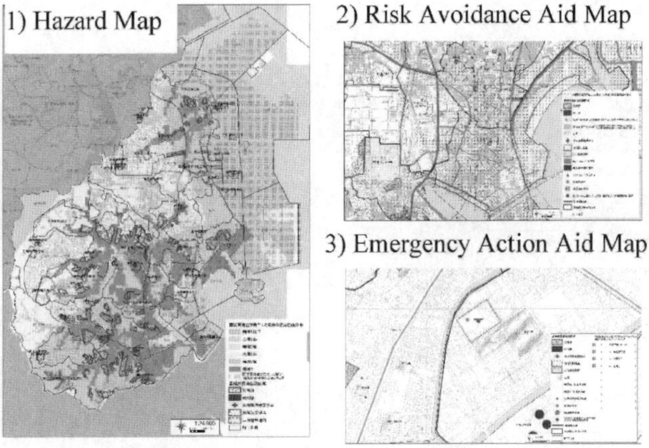

Figure 8. Prototype of Information Map for Earthquake Emergency Preparedness.

6. CONCLUDING REMARKS

The city of Yokohama installed a dense strong-motion array for earthquake disaster management. As shown in Fig. 9, the data from the array has been used in several aspects;

1. For research which contribute to seismic hazard assessment of the city, studies on effects of near-surface geology and deep sediments,
2. For emergency response, the prompt report of seismic intensity in the city and the real-time seismic hazard and risk assessment,
3. For preparedness, seismic hazard and risk mapping.

Further use of the array data will contribute to construct more safety and resilient society to earthquakes.

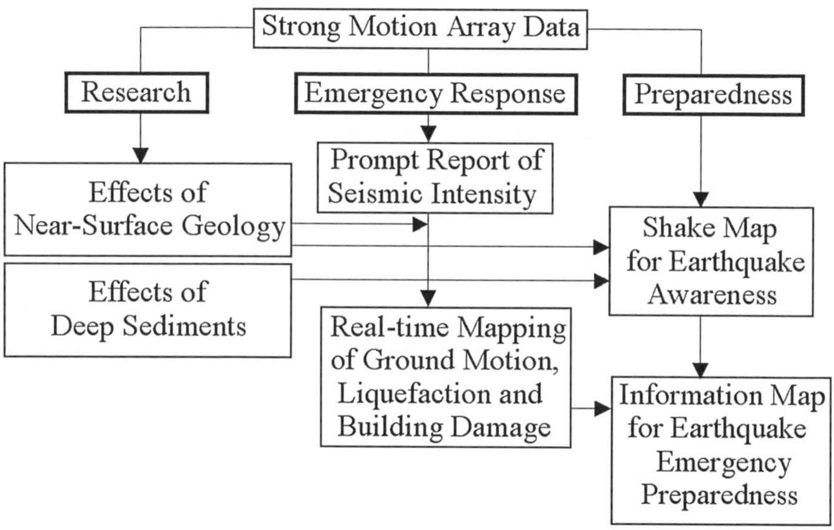

Figure 9. Use of Strong Motion Array for Disaster Management in Yokohama.

REFERENCES

1. City of Yokohama, *Yokohama City Shake Map*, Yokohama City Brochure B-BE152, 2001.
2. Ishizawa E., Midorikawa S. A guide map for earthquake disaster management in order to improve citizen's disaster consciousness. Proceedings of the Eleventh Japan Earthquake Engineering Symposium; 2002 November 20 – 22; Tokyo: Japan Association for Earthquake Engineering, 2002.
3. Japan Road Association, *Specifications for Highway Bridges, Part V Earthquake Resistant Design*, 1996 (in Japanese).

4. Koketsu K., Kikuchi M. Propagation of seismic ground motion in the Kanto basin, Japan. Science 2000; 288:1237-1239
5. Matsuda Y., Abe S., Inui S., Midorikawa S., Shinbo H., Questionnaire seismic intensity survey using the Internet in Yokohama. Proceedings of the Annual Conference of the Institute of Social Safety Science 2000; 10: 47-48 (in Japanese with English abstract).
6. Midorikawa S. Importance of damage data from destructive earthquakes for seismic microzoning – damage distribution during the 1923 Kanto, Japan, earthquake -. Annals of Geophysics 2002; 45:769-778.
7. Midorikawa S., Abe S. Real-time assessment of earthquake disaster based on dense strong-motion network. Proc. of the Twelfth World Conf. on Earthquake Engineering; 2000 January 30 – February 4 ; Auckland, New Zealand: New Zealand Society for Earthquake Engineering, 2000.
8. Miura H., Midorikawa S. Effects of 3-D deep underground structure on characteristics of rather long-period ground motion. J. Seism. Soc. Jpn. 2001,;54:381-395 (in Japanese with English abstract)
9. Olshansky R.B. Is my house in the red zone? local communities and seismic hazard mapping. Proceedings of the Sixth International Conference on Seismic Zonation; 2000 November 12 – 15; Palm Springs: Earthquake Engineering Research Institute, 2000.
10. Rodriguez V., Midorikawa S. Comparison of spectral ratio techniques for estimation of site effects using earthquake motions recorded at surface and in boreholes, and microtremor data, Earthquake Engineering & Structural Dynamics 2003; 32:1691-1714
11. Torii M., Abe S., Suzuki M., Shinbo Y., Aoki T., Saito S., Saito M., Midorikawa S. Investigation for earthquake disaster mitigation using the dense strong motion network in Yokohama, Japan, *The Effects of Surface Geology on Seismic Motion*, Balkema, 1998.
12. Tsuboi S., Saito M., Ishihara Y. Verification of horizontal-to-vertical spectral-ratio technique for estimation of site response using borehole seismographs. Bull. Seism. Soc. Am. 2001; 91:499-510

THE COSMOS VIRTUAL DATA CENTER

Open dissemination of worldwide strong motion data

Ralph Archuleta, Jamison Steidl and Melinda Squibb
Institute for Crustal Studies, University of California, Santa Barbara, 91306, USA

Abstract The COSMOS VDC is a comprehensive, unrestricted, on-line, interactive strong ground-motion search engine for engineers, seismologists, and other earthquake professionals which implements a variety of search interfaces and allows users to preview data and configure design spectra overlays on response spectra. Users may download data files from data providers' servers transparently through this web portal: **http://db.cosmos-eq.org**.

Keywords: accelerograms, database, strong motion, earthquake, response spectra

1. INTRODUCTION

The Consortium of Organizations for Strong Motion Observation Systems (COSMOS) Virtual Data Center (VDC) is a comprehensive, unrestricted, on-line, interactive strong ground-motion search engine for earthquake engineers and scientists, emergency response and recovery agencies and officials, and other earthquake professionals. With the increasing deployment of digital data recorders with high dynamic range such as those available through CISN, ANSS, and KNET, we can expect a continuing increase in the rate at which strong ground motion data is accumulated. However, the full impact of the strong-motion data on public safety depends on effective and efficient access to these data by engineers and scientists. With the goal of providing practicing earthquake professionals and public officials efficient, routine access to strong motion data as part of their practices, COSMOS has overseen the initial development of the VDC (http://db.cosmos-eq.org/). Small engineering practices and individual researchers now have, through this free, web-based resource, the same access to legacy and current data as large companies and university consortia.

P. Gülkan and J.G. Anderson (eds.), Directions in Strong Motion Instrumentation, 209–222.

To be of greatest use to this wide range of users, the data must be retrievable based on each user's needs. To assist the user in utilizing an increasing number of accelerograms and other data products and the array of supporting information, the COSMOS VDC allows the user to search for and select the data most appropriate for a particular project or application. By permitting searches through all of the available data and metadata collected by the various recording networks, the user knows that important data are not overlooked. Moreover, users may search the large dataset for specific characteristics, view data in a geographical perspective, preview records, compare recorded data with design spectra, and retrieve data and metadata. The COSMOS VDC is the only strong-motion center that is attempting to provide on-line access to all freely available, strong-motion data, worldwide.

2. BACKGROUND OF THE COSMOS VDC

A prototype for the COSMOS VDC was first developed at UCSB within the framework of the Southern California Earthquake Center (SCEC) activities, as the SCEC Strong-Motion Database (SMDB). Expanding upon that prototype, with enhancements for ease of use, accessibility and record keeping, the COSMOS VDC was developed in 1999 as part of the COSMOS mission. The COSMOS VDC staff now continually updates the database with new metadata, preview plots and URLs to data files; maintains legacy data and researches and develops mechanisms for effective, efficient, flexible data search, preview and delivery. The basic mission is to provide a one-stop shopping site for finding and downloading strong motion data.

In that the accelerograms and other data products are stored and maintained by the organization that collected the data, the COSMOS VDC is a 'virtual' data center. The VDC provides a means of searching and previewing strong motion data and metadata, but the accelerograms, when downloaded, come directly from the agencies that collected the data. This ensures that quality control rests with the organizations that best know the nature of the data, and that the user is getting the most recent version of the data. In some cases, where a data provider has no resources for data archival, the data provider may elect to store data at the COSMOS VDC.

Table 1. Increase in data available through the COSMS VDC from 2001 to 2004.

	Sept. 2001	Dec. 2004	% Increase
Earthquakes	199	486	144
Stations	1,744	2,880	65
Accelerograms	11,537	23,111	100

As shown in Table 1, as of December 2004 the COSMOS VDC contains the searchable metadata for 486 earthquakes, 2,880 stations, and 23,111 traces. In the last few years substantial data sets representing earthquakes with magnitude greater than 5.0 have been added from ChiChi, Taiwan; all New Zealand records from 1966-1999; the NSF-funded Guerrero array on the Pacific coast of Mexico, as well as smaller but seismically important data sets from Central Asia, Turkey and India. The reputation of the COSMOS VDC is now so well established, even abroad, that the Institute of Geological and Nuclear Sciences Ltd, New Zealand, and the Department of Earthquake Engineering, Indian Institute of Technology, Roorkee, India requested that we include their data for distribution.

The large legacy dataset distributed by the NGDC is currently being added to the data center. The COSMOS VDC continues to incorporate all data available from the USGS and CISN with magnitude greater than 5.0 in highly seismic areas and greater than 4.5 in areas of low seismicity. The VDC also incorporates all data from the K-Net and Kik-Net Japanese networks with magnitude greater than 5.0, depth less than 100km and with a pga of at least 0.1g. Recent data sets from these sources include the 2004 Parkfield and Niigata earthquakes and aftershocks. We have also recently received an offer to include all past and future strong-motion records from the Geological Survey of Canada and are in negotiations with the European Strong-Motion Database to link to their dataset through the COSMOS VDC. Thus the COSMOS VDC can be expected to continue its exceptional expansion.

3. COSMOS VDC USAGE

The COSMOS VDC is the portal of choice for access to strong-motion data for many U.S. as well as international users. A Google search on COSMOS returns hundreds of external references from such diverse sources as the ANSS, USGS NSMP, California Integrated Seismic Network, Northern California Earthquake Data Center, Earthquake Engineering Research Institute, MCEER, California Geological Survey, Southern California Earthquake Center, Pacific Northwest Seismological Network, Caltech, Boston College, Institution of Structural Engineers, International Seismological Centre (UK), Geonet (New Zealand), the National Strong-Motion Program of Turkey, the Sahand Earthquake Research Center (Iran), Civil Engineering Portal (Greece) and the SRRC, Dankook University (Korea).

The effect of the ease and flexibility of data discovery, the longevity and stability of the COSMOS VDC website, as well as a recently instituted New Event Notice, is demonstrated in the steady increase in usage of the COSMOS VDC (Figure 1).

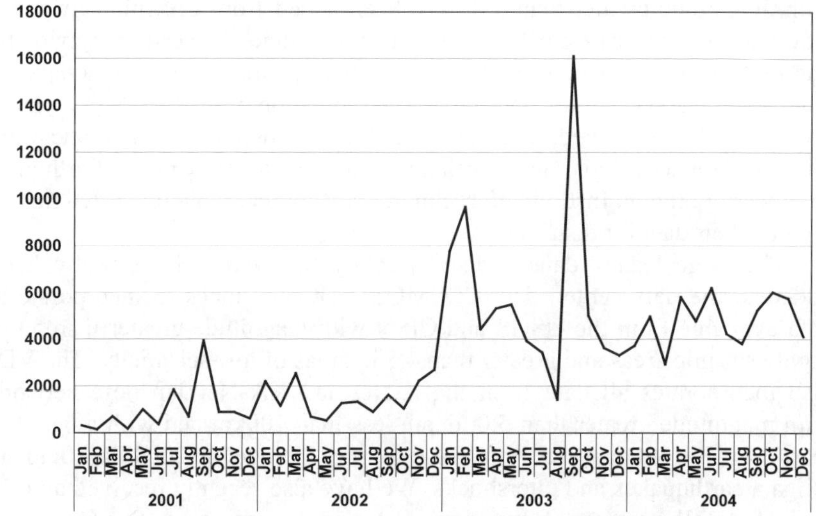

Figure 1. Files selected for downloading from the COSMOS VDC, 2001-2004.

The number of files downloaded in 2004, by data provider, is shown in Figure 2. The first eight are U.S. sources, the remaining are from international sources. Although USGS and CGS files have consistently been the most popular downloads, an appreciable percentage of the whole have been

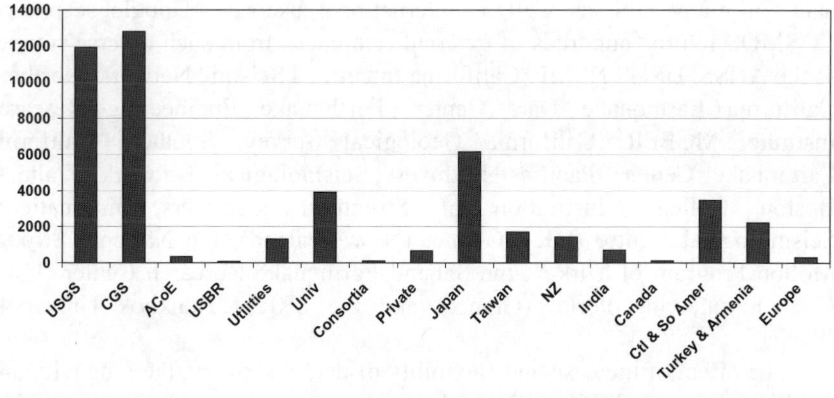

Figure 2. Number of files selected for downloading Jan 2004 - Oct. 2004, by data provider

provided by international sources.

The breadth of the VDC user community is demonstrated in the pie chart distribution shown in Figure 3, including practicing engineers (.net and .com domains), research engineers and seismologists (.edu domain), as well as the international community. This broad community of VDC users and the steady increase in the number of web pages accessed and data files downloaded over the last three years are all testaments to the usefulness and the impact that the VDC is having on the earthquake engineering community.

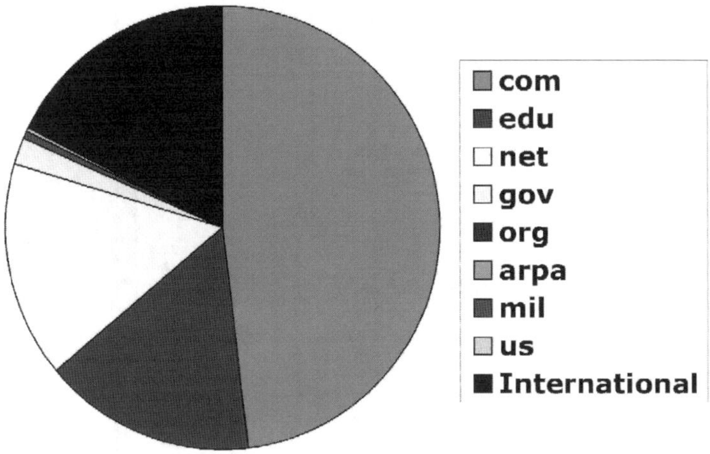

Figure 3. Distribution of web pages accessed Sept. 2001-Dec. 2004. Data is grouped by domain for U.S. sources.

4. VDC METHODS OF OPERATION

4.1 Metadata Acquisition

COSMOS maintains agreements with many data providers, including government agencies, universities, and private owners of data, to distribute their data through the VDC. These data files are pulled to the COSMOS VDC to generate plots, and to extract and insert the associated metadata into the database, together with the URL from which the data files may be retrieved. These metadata and the fully relational database in which they reside enable the extensive search capabilities of the COSMOS VDC. The process of metadata extraction and publication is shown in Figure 4.

The structure of the relational database for the metadata is represented in Figure 5. The COSMOS VDC has a more complex table structure than earlier

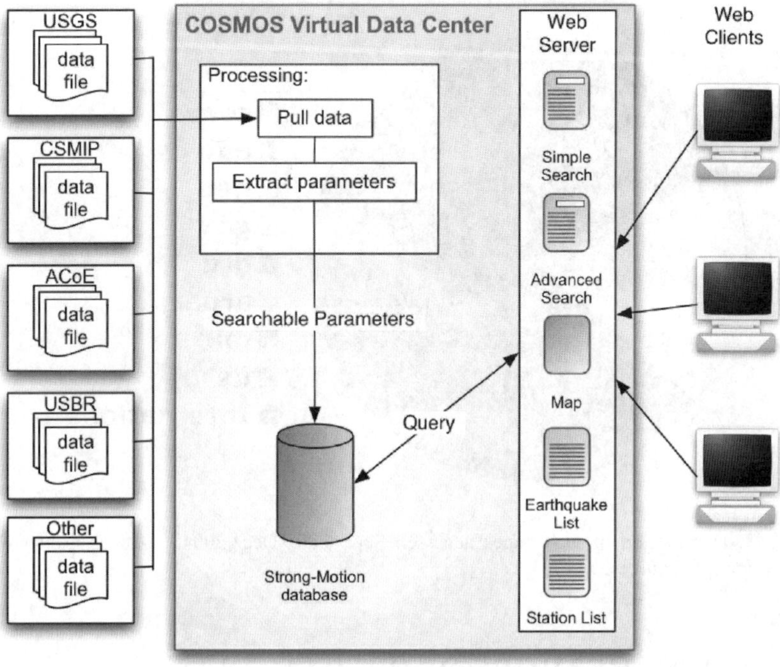

Figure 4. COSMOS VDC Metadata Acquisition and Publication.

Strong-Motion databases to increase the efficiency of searches, storage and data delivery; and to facilitate the representation of more complex data. For example, the separation of instrument from station tables allows complex sites, such as dams or geotechnical arrays to be represented in the COSMOS VDC. The format table will facilitate the automation of record format conversion. The Trace Table contains fields for the URLs of multiple data products: acceleration, velocity, displacement, Fourier and response spectra, while the dataset table allows a single URL for a zipped archive of all files associated with an event.

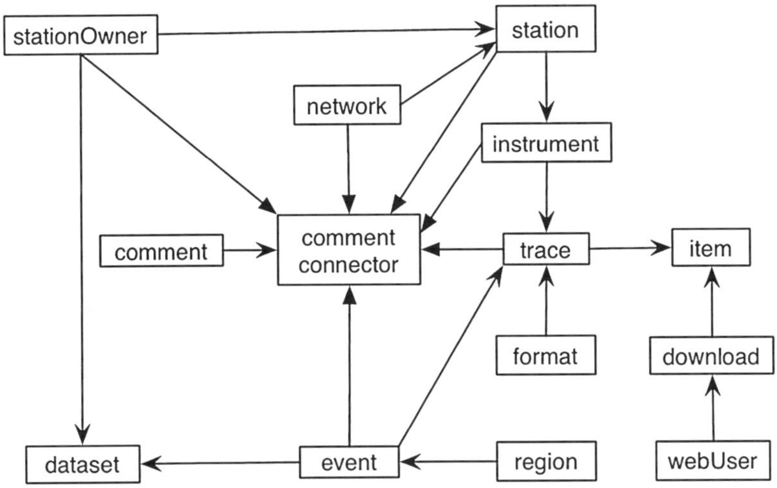

Figure 5. The Structure of the COSMOS VDC relational database.

4.2 Search Capabilities

4.2.1 The Map Interface

The map interface, shown in Figure 6, presents a world map displaying all earthquakes and stations for which data records are available through the COSMOS VDC. Users may customize the map by entering latitude and longitude ranges to view, or click on a blank area of the map to zoom in. For maps on which fewer than 50 earthquakes appear, a list of earthquakes appears at the bottom of the page. Users can use adjacent buttons to highlight the earthquake and reporting stations for that earthquake. Each earthquake symbol on this map displays name and date information in the status bar when the user moves the mouse across the symbol; and when clicked, generates an Event page with data for all stations reporting that earthquake.

The Event page displays:

1) Information about the earthquake itself, including location, magnitudes, mechanism, and references;
2) Links to another, localized map showing all of the stations which recorded this earthquake;
3) A table of these stations showing pga, pgv, hypocentral distance, component orientation, station name, location, owner, processor and data products available for each station;

4) Buttons to add either all data on the page or the individual traces to a 'shopping cart' of downloadable records;
5) Links to plots of the data for acceleration and response spectra.

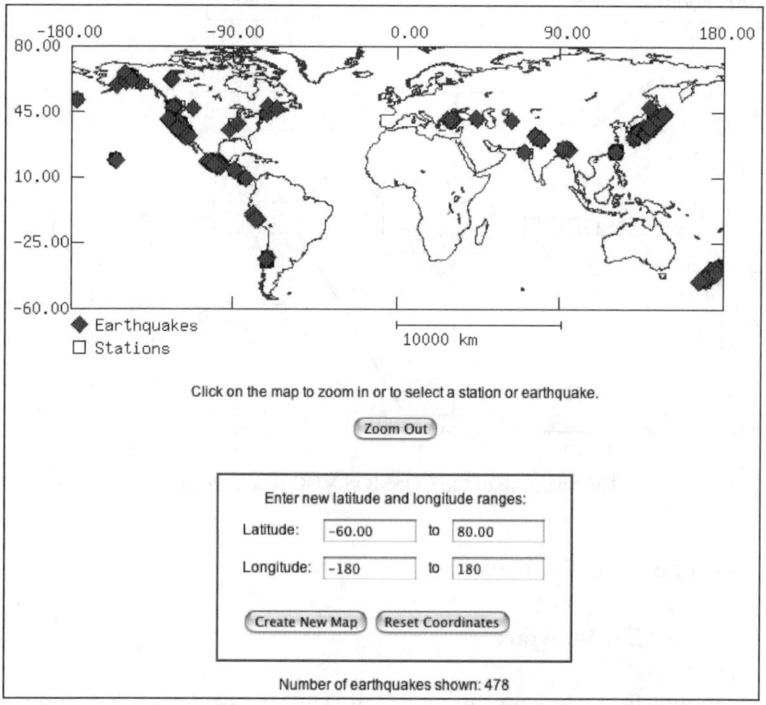

Figure 6. A world map showing the location of the earthquakes contained in the VDC. The user can zoom to a smaller region by clicking and dragging on a rectangle on the map, or setting longitude and latitude ranges.

When the user zooms in to a closer view of a map, the display may display faults as well as the information in the larger scale map, as shown in Figure 7. Each station symbol on these maps displays station name in the status bar when the user moves the mouse over the symbol; and when clicked, generates a web page with data for all records associated with that station.

The station page displays:

1) Information on the station's location, owner and geology, and the owner's web page for the station, if available;
2) Links to another, localized map showing all of the earthquakes recorded by this station;
3) A table of these earthquakes showing earthquake name and date, and pga, pgv, hypocentral distance and component orientation for this station;
4) Buttons to add either all data on this page or the individual traces to a 'shopping cart' of downloadable records;
5) Links to acceleration and response spectra plots of the data.

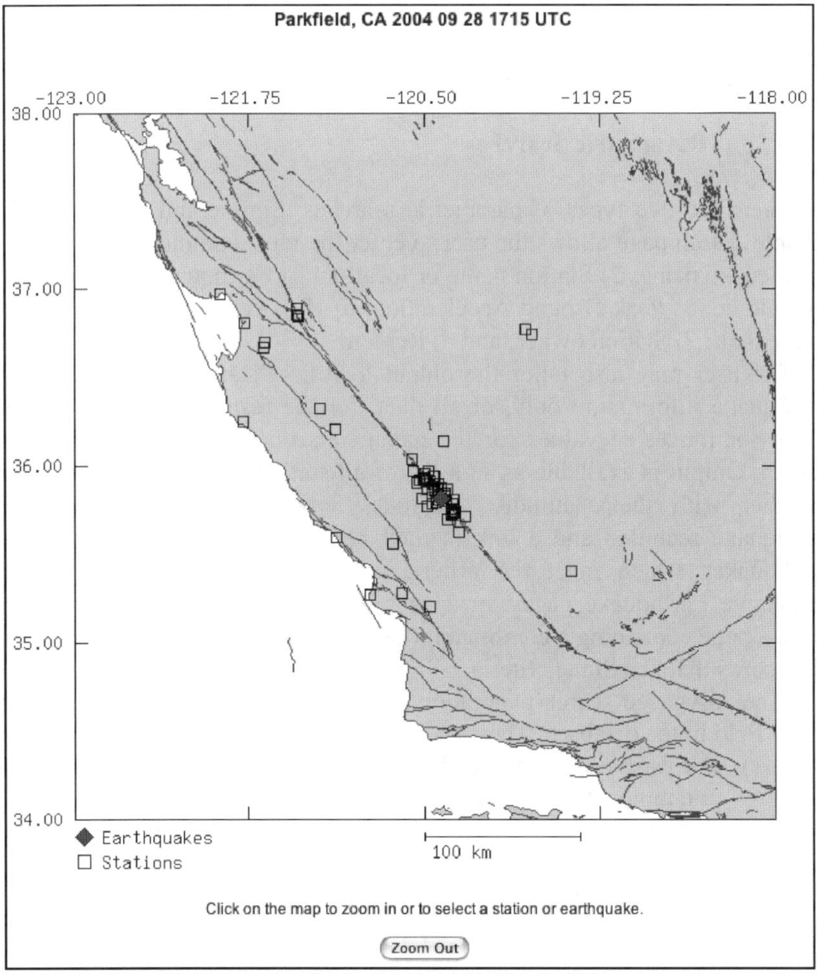

Figure 7. Map display for the Parkfield, California 2004 earthquake.

4.2.2 Earthquake and Station-Based Searches

The Earthquakes page lists all earthquakes available through the COSMOS VDC by region, with a drop-down list of regions at the top of the page for quicker navigation to the region of interest. The listing for each earthquake consists of name, date, preferred magnitude, number of stations available for this earthquake and the agencies providing the data, as well as linking to the full Earthquake page for this earthquake, as described in the section on the Map interface. Where the data provider makes a zip archive of all records for the earthquake available, there is a link to that archive.

The Stations page lists all stations available through the COSMOS VDC by region, also with a drop-down list of regions at the top of the page. For each station the listing consists of owner and station name. The user selects a station and clicks a button to display the full Station page as described in the section on the Map interface.

4.2.3 Parametric Searches

There are two types of parametric searches, Simple and Advanced. The Simple search page allows the user to enter the most common parameters: 1) Earthquake name; 2) Station name or location; 3) Earthquake mechanism; 4) Magnitude; 5) Peak Ground Acceleration; 6) Hypocentral or closest distance to the fault; 7) Station owner; and 8) Region.

The user may also tailor the output to reflect station information only, earthquake information only, or all data, for the result set. The user enters values or ranges of values for the parameters on which he/she chooses to search. Output is available as 1) a list of all stations which meet the criteria together with their latitude, longitude, hypocentral distance from the earthquake recorded and a link to their respective Station pages; or 2) all earthquakes which meet the criteria, with their date and link to their respective Earthquake pages or, 3) if 'all data' was selected as the output type, a page containing a combination of earthquake and station page data for all records that match the criteria.

The Advanced search page allows the user to query and recover almost every field in the database. These include:

1) For earthquakes: name, date, longitude, latitude, depth, magnitudes and moment, mechanism, strike, rake and dip;
2) Region;
3) For station: name, owner's code, location, address, geology, and s-wave velocity for the top 30m;
4) For station owner: name, acronym, contact person, ftp site, web site, and address; for the network: name, acronym, owner, and website;

5) For the instrument: location, type, serial number, latitude and longitude;
6) For the trace: the url of the uncorrected and corrected acceleration, velocity, displacement, fourier spectrum and response spectrum files; epicentral, hypocentral, and closest distance to the fault; azimuth, component vertical and horizontal offset; pga, pgv, and 5% damped response spectrum at 0.1 sec, 0.3 sec, 1 sec and 3 sec periods.

The user chooses which fields to retrieve or search on and, on the subsequent page, enters the value or ranges of values for the fields to search on. The output is available in three forms: 1) a combined station-earthquake page like that returned by the simple search; 2) an html table listing all parameters selected for the result set; 3) a tab-delimited text page which contains all parameters selected on the Advanced search page for each trace returned in the result set. This last may be saved as a rtf file and imported into a spreadsheet file or database.

4.3 Preview and Download Options

4.3.1 Data Preview

Each html search result page, whether reached via a map, parameter search or list interface, presents several options for further inquiry: a link to a new map that corresponds to the stations and earthquakes in the result set, and a link for each station in the result set to plots of acceleration for that station.

Each search result page also contains a link to both logarithmic and linear plots of response spectral acceleration, as shown in Figure 8, at damping values provided by the data processor. The user may select the units in which acceleration is displayed, scale the response spectra and enter parameters to define the shape of the design response spectra. The COSMOS VDC currently provides response spectra plots only for data providers that offer that data product, but plans to process all remaining unprocessed data, adhering to guidelines developed by the COSMOS Working Group.

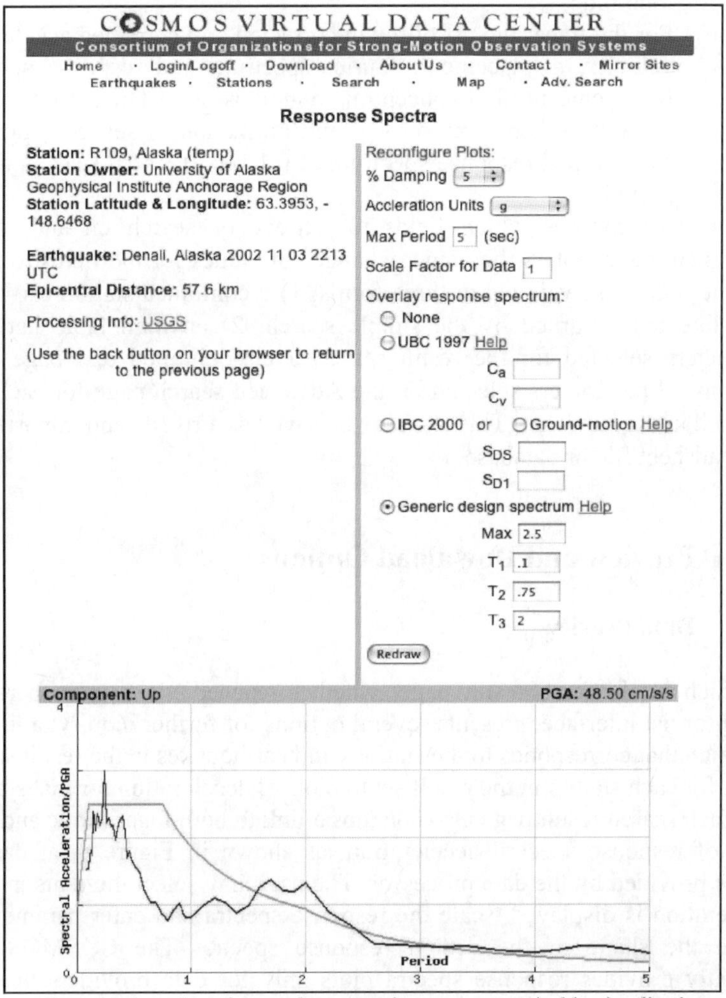

Figure 8. Sample web page demonstrating user-customizable visualization methods for response spectra with design spectra overlay.

4.3.2 Parameter and Data File Downloads

There are two methods of downloading metadata parameters from the COSMOS VDC database, both following an Advanced Search. After selecting data, the user is given the option of displaying all user-selected fields as a 1) tab-delimited file, or 2) as an html table, as described above.

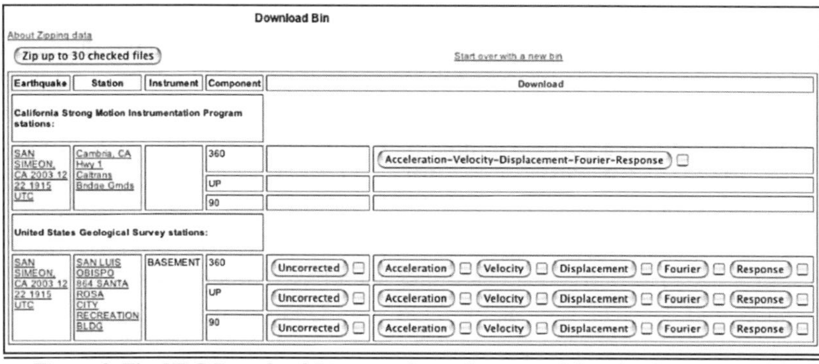

Figure 9. The VDC Download Bin page shows the types of download options and the range of data products available. The data provider is listed for each station.

There are three methods for downloading data files through the COSMOS VDC:

1) on the Event page, many events are accompanied by a link to a zipped archive of files for that event;
2) on the Download Bin page, users may select individual files for downloading;
3) on the Download Bin page, users may select up to 30 individual files and receive them as a single zipped file, accompanied by a log of files included.

After selecting the files to download from a search results page and entering his email address, the user is directed to the Download Bin. The user can see on the Download Bin page how files are configured by the arrangement of buttons. For example, in Figure 9, the single file from CSMIP contains the data for all three components, as well as combining the acceleration, velocity, displacement, Fourier and response spectra in one file, while the data from the USGS station is broken into individual files for each component and data product. The user may click on one of these buttons to download the files individually, or click on the checkbox next to the button to add to a zipped set. Once up to 30 files have been selected to zip, the user clicks on the 'Zip up to 30 checked files' button to retrieve and zip the files. Data files are pulled via FTP or http calls from the data provider at the time of the user's request. Thus, users always receive the most recent copy of data available from the data provider.

5. FUTURE ENHANCEMENTS OF THE VDC

In the near future, the VDC plans make all data available through the VDC in a standard format, the COSMOS text format, which was developed by the multi-agency COSMOS Strong Motion Programs Board. The data will also be available in a corresponding standards-based XML format. Research is currently in progress to ensure ISO and GML compliance in the XML format. In addition, the VDC will make available web-deployed tools for visualization of the data in these formats and conversion to legacy formats. The VDC is also actively developing web-based interactions with related databases and datasets, such as the COSMOS Geotechnical VDC.

ACKNOWLEDGMENTS

The COSMOS VDC gratefully acknowledges the COSMOS leadership, in particular, Carl Stepp, Bruce Bolt and James Davis, for their continuing support. The COMOS VDC would not exist without the initial multi-year support of the Southern California Earthquake Center. Of course, the VDC would not exist if it were not for the dozens of agencies that have expended their financial and personnel resources to collect strong-motion data. We will always be in their debt. The COSMOS VDC is funded by NSF, Civil and Mechanical Systems (CMS-0201264), with additional funds from COSMOS, USGS and CGS. This is an ICS contribution.

REFERENCES

1. PEER Strong Motion Database.,Pacific Earthquake Engineering Research Center, http://peer.berkeley.edu/smcat/.
2. Strongmo Database System. Lamont-Doherty Earthquake Observatory, Columbia University, http://www.ldeo.columbia.edu/res/data/nceer/strongmo.
3. Worldwide Strong Motion Data Archive. National Geophysical Data Center (NGDC), http://www.ngdc.noaa.gov/seg/hazard/strong.shtml.
4. COSMOS Geotechnical VDC, COSMOS/PEER Lifelines Project 2L02. http://geoinfo.usc.edu/gvdc/.
5. Geography Markup language (GML), Open Geospatial Consortium (OGC), OpenGIS Specifications, http://www.opengeospatial.org/specs/.
6. ISO 19115:2003 Geographic Information – Metadata. ISO/TC 211 Publications, Geographic Information/Geomatics, International Organization for Standards (ISO), http://www.isotc211.org/.

SITE-DEPENDENT GROUND-MOTION DATA RECORDED BY GERMAN TASKFORCE IN TURKEY

Jochen Schwarz,[1] Clemens Ende[2] and Dominik H. Lang[1]
[1]*Bauhaus-University Weimar, Earthquake Damage Analysis Center, Marienstr. 7, D-99421 Weimar, Germany*
schwarz@bauing.uni-weimar.de, lang@uni-weimar.de
[2]*seismotec GmbH, Beratende Ingenieure, Th.-Müntzer-Str. 35, D-99423 Weimar, Germany*
ende@seismotec.de

Abstract: Within the western segment of the Northanatolian fault zone two strong damaging earthquakes occurred on August 17 (Ms 7.8) and November 12, 1999 (Ms 7.4). Immediately after both earthquakes seismic stations were temporarily installed within the disaster-struck area. The recording stations provided a unique and comprehensive data base of near-field aftershocks, establishing the basis for elaborating subsoil-dependent attenuation laws for peak ground acceleration as well as spectral accelerations.In addition to the aftershock records of the strong motion stations of German TaskForce for Earthquakes 145 triaxial acceleration records of main- and aftershocks (rock 6, stiff 36, soft 103) from the 1999 Turkey earthquakes provided by the Kandilli Observatory and Earthquake Research Institute (KOERI) and the General Directorate of Disaster Affairs (AFET) could be implied. Attenuation functions were elaborated for different data sets considering the impact of data composition, lower magnitude bound, source depth and subsoil conditions. Predicted peak ground accelerations (PGA) and horizontal spectral accelerations are remarkably lower than those from attenuation functions mainly derived from European strong motion records.The different data sets lead to similar spectra. For shorter and longer periods data set of German TaskForce (being dominated by near-field records of lower magnitude) is slightly underestimating the results. Nevertheless, the results can be applied for a realistic and site-dependent prediction of design spectra and probabilistic seismic hazard assessment.

Key words: Strong motion data, peak ground acceleration, spectral acceleration, aftershock data, attenuation functions, experimental site classification

P. Gülkan and J.G. Anderson (eds.), Directions in Strong Motion Instrumentation, 223–239

1. INTRODUCTION

Several field missions were carried out by the reconnaissance teams of German TaskForce for Earthquakes after strong damaging earthquakes in Turkey since the year 1992 [12]. Strong ground motion recordings were realized by network of temporarily installed strong motion recorders during the field surveys to Turkey in 1998 (Adana/Ceyhan), 1999 (Izmit/Kocaeli and Duzce/Bolu), and 2002 (Sultandagi/Afyon). Particularly both missions in 1999 provided a unique and comprehensive data base of recorded aftershocks [9]. Figure 1 exemplarily illustrates the locations of strong motion accelerographs of German TaskForce for Earthquakes (GTFE) being installed immediately after the Izmit/Kocaeli earthquake on August 17, 1999.

In addition to these reconnaissance missions, post-earthquake investigations were initiated within the provinces Adana, Kocaeli, Sakarya, Bolu, and Afyon (in October 2000 and September 2002) in order to gain more insight into the effect of local site conditions. Microtremor measurements were conducted at sites of the previously installed strong motion accelerographs as well as at locations of evident concentration or remarkable scatter of structural damage. By the application of spectral H/V-ratio method on microtremor data, a first classification of the recording sites (into soft soil, stiff soil, rock-type conditions) could be performed ([13],[14]).

Results of these different field studies provide the basis for elaborating subsoil-dependent attenuation laws for peak ground acceleration as well as spectral accelerations. It has to be noticed that the amount of analyzed aftershock data has been increased due to the source and magnitude detections by the seismology group of German TaskForce (GFZ Potsdam). At the present state of elaboration the aftershock data base consists of 538 triaxial acceleration records (cf. Table 1).

Since the year 2000 different attenuation functions for a more less (in)complete data set were elaborated. It can be concluded that the general tendency of results has experienced only slight changes. This especially holds true for the strategic question (of further TaskForce missions) whether a data base which is predominated by near-source aftershocks of small to moderate magnitudes can supply useful results for ground motion prediction in terms of seismic hazard assessment or design purposes.

2. STRONG MOTION DATA BASE(S)

For the derivation of attenuation laws different data bases were applied. As a whole five different data bases (DB) are considered.

Data base I: GTFE (cf. Figure 2a)

Only strong motion data from the Izmit/Kocaeli aftershock series (August-September 1999) recorded by the network of German TaskForce for Earth-

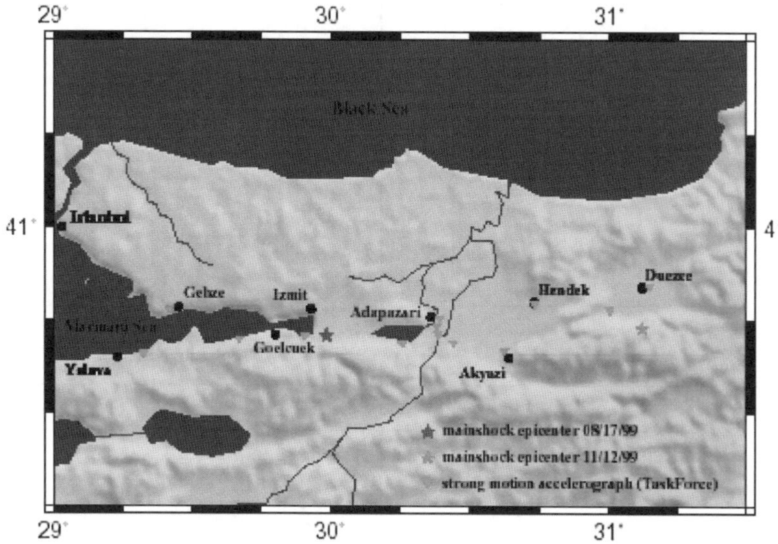

Figure 1. Sites of temporary stations of German TaskForce for Earthquakes (GTFE)

quakes (GTFE) are included within this data base which can be regarded as unique with respect to its consistent quality and the level of quality assurance:

- The whole data are *uniformly elaborated.* The strong motion recording instruments are well-calibrated and maintained (technical aspect). The site of each station was selected by trained engineers, thus ensuring seismic recordings on free-field site conditions not affected by effects coming from soil-structure interaction, building response, or strong noise perturbations.

- Both, locations and magnitudes of the events are *uniformly determined* by seismologist from GeoForschungsZentrum Potsdam [11]. Aftershock localization represents the processing state of April 2004.

- Due to the temporarily installed seismic network, *predominantly near-field* data are recorded, cf. Figure 2a.

- Subsoil conditions at the recording stations are *uniformly classified*, cf. Section 3.

Data base DB I (GTFE) consists of 538 triaxial acceleration records (rock 53, stiff 52, soft 433), with a mean magnitude $M_L = 2.8$ and a mean epicentral distance of 10 km (cf. Table 1).

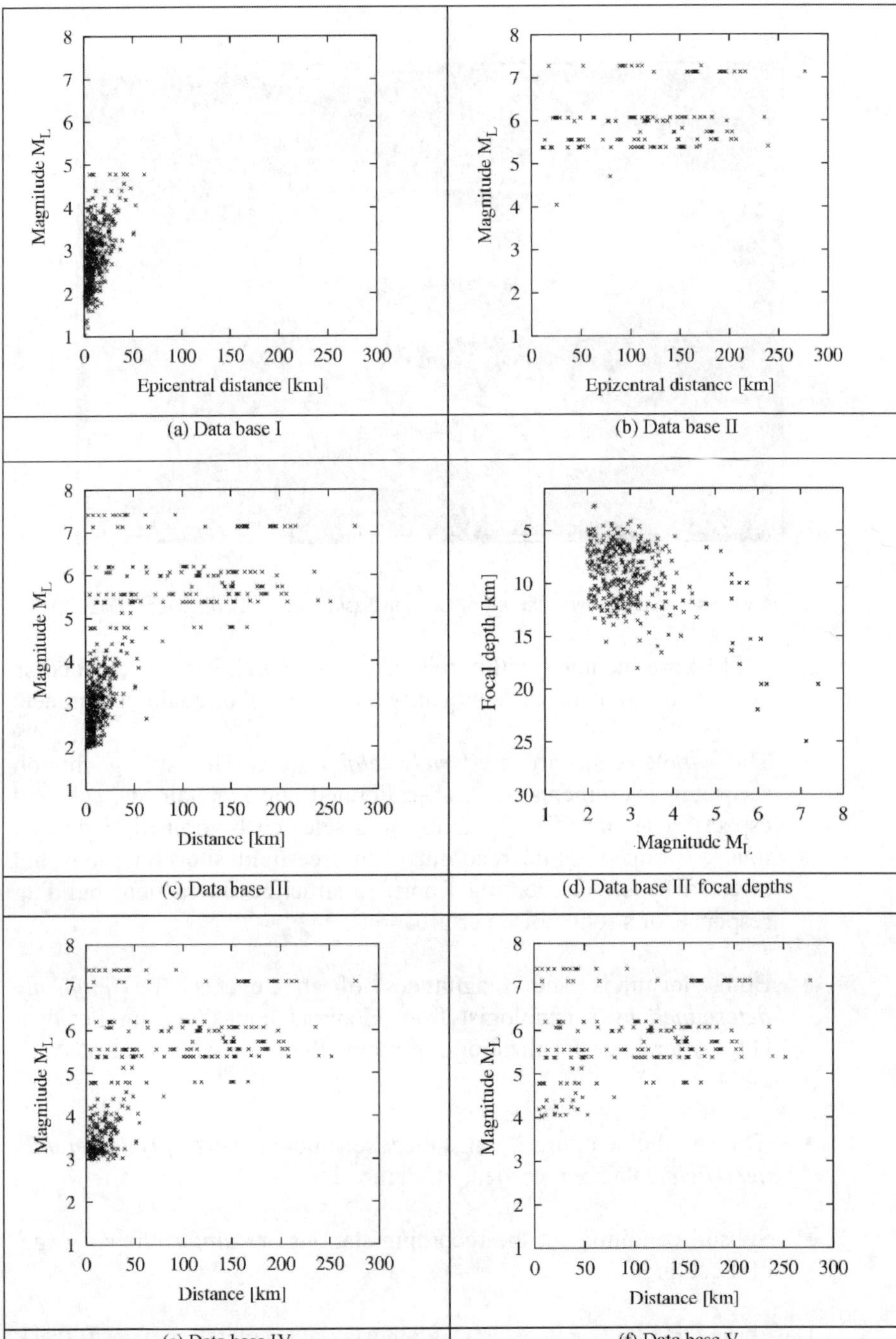

Figure 2. Magnitude distance relationships of different data sets applied

Data base II: KOERI (cf. Figure 2b)

In order to check the sensitivity of attenuation laws with respect to the composition of the considered data set the study should be repeated with an enlarged strong motion data base taking into account recordings from other Turkish earthquakes. In addition to the aftershock records of the strong motion stations of German TaskForce for Earthquakes (GTFE, [14]), a small data set of main- and aftershock recordings from the 1999 Turkey earthquakes provided by the Kandilli Observatory [8] could be implied. It has to be emphasized that the near epicentral region of the August 17, 1999 Izmit/Kocaeli earthquake was only scarcely equipped with strong motion stations by KOERI.

Data base DB II used in this study comprises 145 triaxial acceleration records (rock 6, stiff 36, soft 103) with a mean magnitude $M_L = 6.0$ and a mean epicentral distance of 117 km (cf. Table 1).

Attenuation functions according to the combined data base (I + II) are compared to those of data base DB I alone. The magnitude-distance composition of both data sets is illustrated in Figures 2a and 2b. Both data sets are quite different in terms of magnitude and distance range. While the data set of German Task Force (DB I) only consists of small magnitude aftershocks (at the present state of elaboration: $M_L \leq 4.9$) being recorded in epicentral distances $R_{epi} \leq 70$ km, the KOERI dataset (DB II) is featured by larger magnitudes ($M_L = 4.8 - 7.2$) and by a distance range of about 10 - 250 km.

Data base III*: GTFE, KOERI + AFET

Data base (I+II) was extended by further recordings from AFET-stations [4] given with fault distance and meanwhile localized TaskForce data.

To avoid the use of defective time histories, all acceleration records were visually controlled and, if necessary, sorted out. Furthermore every recording was put through a zero order baseline correction.

Data bases III, IV and V are simply derived from data base III* while introducing a lower magnitude bound as criteria for the data elimination (Figures 2c, 2e, and 2f). Thus data bases follow from DB III* by truncating records with magnitude $M_L \leq 2.0$, 3.0, and 4.0, respectively.

The truncation leads to a drastic reduction of the number of events as well as of the number of records. Information on the data composition can be taken from Table 2 and Figure 2.

The weighted center of the data bases (expressed by the mean magnitude and distances) is shifted from lower magnitude and near-field data (DB I + II or m/r) to larger and distant records (DB IV, DB V or M/R). There is no doubt that this aspect has to be recognized while interpreting the results.

Table 1. Basic characteristics of data bases I and II, and their combination (I+II)

Data base	*Magnitude range*	*Mean magnitude*	*Mean epicentral distance [km]*	*Number of records*	*Soil conditions*		
					rock	*stiff*	*soft*
I	$M_L \geq 0.8$	2.8	10	538	53	52	433
II	$M_L \geq 4.0$	6.0	117	145	6	36	103
I+II	$M_L \geq 0.8$	3.5	34	693	59	88	536

Table 2. Basic characteristics of data bases III, IV, and V

Data base	*Magnitude range*	*Mean magnitude*	*Mean distance [km]*	*Number of events*	*Number of records*
III	$M_L \geq 2.0$	3.8	38	394	665
IV	$M_L \geq 3.0$	4.7	60	148	386
V	$M_L \geq 4.0$	6.0	99	29	205

3. SUBSOIL CLASSIFICATION OF STRONG MOTION SITES

In order to elaborate attenuation relations for different subsoil conditions, and moreover, to compare own findings with the results of other attenuation functions, a uniform site classification has to be ensured. On the basis of available information and own estimations, strong motion recording stations of German TaskForce for Earthquakes (GTFE), Kandilli Observatory Istanbul (KOERI), and Istanbul Technical University (ITU) could be categorized according to a conventional site classification using the shear wave velocity V_s of the uppermost 30 m (100 ft) of soil layers ($V_{s,30}$) as the main criterion. For most of the recording sites a conventional site classification has already been performed. In addition to the coarse classification into rock, stiff soil, and soft soil conditions, stations of KOERI (2002) were arranged according to the scheme of the 1997 Uniform Building Code [16], cf. Table 3. As only minor information on the geological conditions at the strong motion stations of German TaskForce was available, a site classification based on instrumental investigations was applied. The applied *Method of an Experimental Seismic Site Assessment* [10] correlates instrumental results with those of theoretical investigations for generated model subsoil profiles. Transfer functions of the sites were elaborated by applying spectral H/V-ratio method on microtremors. Once a so-called site-specific subsoil class could be allocated according to the coincidence between theoretical and experimental results, a calibration of the best-fitting model profile by variation of stiffness and/or sediment depth can be realized.

The classification of the recording sites was done by comparing the shapes of spectral H/V-ratios on microtremors (Figure 3b) with theoretical

H/V-ratios as well as with theoretical transfer functions determined for idealized subsoil profiles according to the German earthquake code DIN 4149 [5], cf. Figure 3a.

This approach is based on a differentiation between geological subsoil classes (A, B, C) representing the total thickness of the overlying stratum, and subsoil types (1-rock, 2-stiff soil, 3-soft soil) for the uppermost 25 m of soil layers leading to a combination of relevant site situations (cf. Table 3). Geological subsoil class A can be characterized by missing or overlying sediments with maximum thicknesses of 25 m, whereas class C is described by deep, mostly Quaternary alluvial layers reaching depths between 100 and 1000 m. Geological subsoil class B is representing the transition zones between classes A and C. The classification of temporary strong motion stations of GTFE and other recording stations can be taken from Lang [10] while explaining the basic steps of this hybrid technique in more detail. The successfully applied procedure seems to be a first step towards a more detailed elaboration of spectral as well as ground acceleration attenuation laws taking into account the total depth of sedimentary layers.

The conventional site classification leads to a subdivision of strong motion data bases I (GTFE) and II (KOERI) as given by Table 1. It becomes obvious that the different data bases are dominated by recordings on soft soil sites ($V_{s,30} \leq 360\ m/sec$).

Table 3. Assignment of subsoil classes

Classification	*Shear wave velocity*	*DIN 4149*	*UBC-97*
rock	$V_s^* \geq 800\ m/sec$ $V_s^+ = 760 - 1500\ m/sec$	A1 (A2)	B
stiff soil	$V_s^* = 350 - 800\ m/sec$ $V_s^+ = 360 - 760\ m/sec$	(A2) B2 C2	C
soft soil	$V_s^* \leq 350\ m/sec$ $V_s^+ = 180 - 360\ m/sec$	A3 B3 C3	D

* benchmarks which are not constituted in DIN 4149 (blueprint Feb. 2002) [5]; V_s^* complies with the average shear wave velocity of the uppermost 25 m.

\+ V_s^+ complies with the average shear wave velocity of the uppermost 30 m.

4. REGRESSION MODEL

Check of lognormal distribution. After the selection and compilation of time histories absolute response spectra with a damping of 5 percent were calculated for both horizontal components (and). Furthermore, the peak ground accelerations (PGA) of each time history were determined.

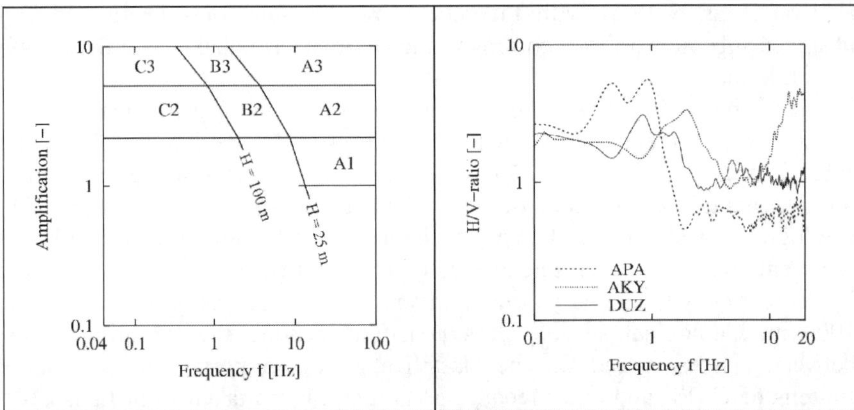

(a) The different areas mark the possible locations of predominant peaks of the H/V-ratio

(b) Subsoil class B3 (soft soil with sedimentary thickness of about 25-100 m)

Figure 3. Spectral H/V-ratios on microtremors allocated into site-specific subsoil classes of German earthquake code DIN 4149 [5]

A precondition for the regression analysis is that the data set (acceleration) at each discrete period is following a lognormal distribution. While the empirical data of peak ground acceleration (T = 0.01 sec) reasonably fit the lognormal distribution function calculated by the mean and standard deviation, however, notable differences appear with increasing period, which might be explained by the combination of different data sets [14]. Both data bases I and II individually regarded show quite regular lognormal distributions.

Magnitudes. The magnitudes of the recordings of GTFE (data base I) are determined as local magnitudes M_L [11]. To convert local magnitudes M_L into moment magnitudes M_w as well as into surface wave magnitudes M_s the following equations (1) and (2), referring to Grünthal and Wahlström [7] and Ambraseys [1], respectively, are used:

$$M_w = 0.9(M_L + 0.722)^2 + 1.65 \tag{1}$$

$$M_s = 1.33M_L - 1.733 \tag{2}$$

Distance parameter. For the strong motion records of data bases I and II, the epicentral distance R_e will be applied. For the other data bases the shortest distance between the station and the fault (fault distance R_f) is taken using available distance parameters from Turkish institutions [8].

Focal Depth. The focal depth is more or less documented for all data offering the opportunity to study the impact of this parameter on peak ground acceleration (PGA) and horizontal spectral acceleration (SA). For this purpose data base III is considered, only. Figure 2d is illustrating the composition of this data base with respect to the local magnitude M_L and focal depth h. Earthquakes are concentrated in a range between 5 and 15 km which should be representative for many other seismic regions (e.g. Central Europe).

Regression. The attenuation relations elaborated in this study are based on the model given by equation (3), which is identical to the one published by Ambraseys et al. [3]. Both peak ground acceleration and spectral acceleration (maximum of both components in units of gravity, g) depend on the local magnitude M_L, distance R, and the local subsoil conditions.

Regression analyses were performed to determine the coefficients C_1, C_2 C_4, and σ. These coefficients are estimated by varying h_i to minimize the standard deviation σ. The coefficients C_r, C_a, and C_s, considering the subsoil conditions (rock, stiff soil, and soft soil, respectively) are computed by the mean residues of each subsoil class (cf. Ambraseys et al. [3]).

$$log(Y)=C_1+C_2M_L+C_4\ log(R)+C_rS_r+C_aS_a+C_sS_s+\sigma P \tag{3}$$

with:

- $R=\sqrt{D^2+h_O^2}$, D - epicentral distance for data bases I and I + II and fault distances for DB III - V in [km].
- P = 0 and 1 for 50 % and 84 % fractiles, respectively.
- Y = PGA or spectral acceleration SA within a period range between T = 0.05 - 2.0 sec.

The coefficients of equation (3) are determined on the basis of the maximum horizontal component for PGA and 5 % damped response spectra. Different regression types were checked in the frame of accompanying investigations.

Table 4. Regression coefficients for horizontal peak ground accelerations for data bases I and (I + II)

Data base	C_1	C_2	C_4	σ	h_0	C_r	C_a	C_s
I	-2.7742	0.4479	-1.0628	0.2859	2.75	-0.1006	-0.1364	0.0287
I+II	-3.0815	0.5161	-0.9501	0.3193	2.00	-0.1620	-0.1078	0.0355

5. PREDICTION OF PEAK GROUND ACCELERATION IN DEPENDENCE ON SUBSOIL CONDITIONS

The calculated coefficients for the prediction of the horizontal peak ground acceleration according to equation (3) are given in dependence on the subsoil conditions (Table 4).

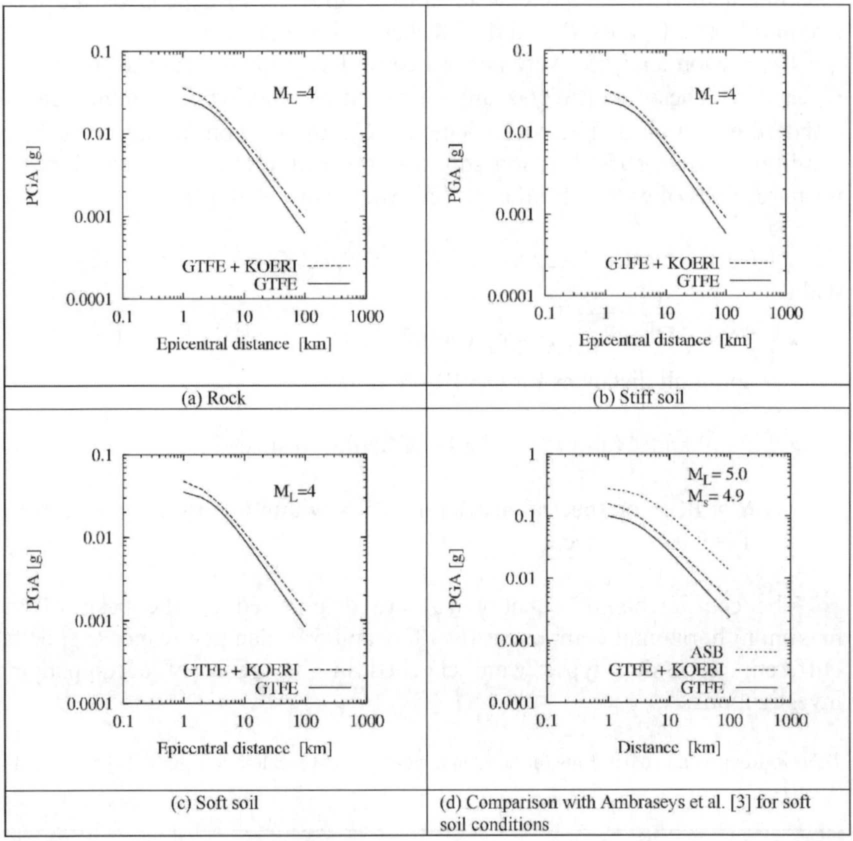

Figure 4. Attenuation of peak ground acceleration for data base I (GTFE) and I + II (GTFE + KOERI) for different subsoil conditions: a) rock, b) stiff soil, c) soft soil, and d) comparison with Ambraseys et al. [3] for soft soil.

The proposed attenuation relations for the horizontal peak ground acceleration derived from the data bases I and (I + II) are compared in Figure 4a to 4c for a magnitude $M_L = 4.0$. Data base (I+II) leads to higher PGA. The attenuation is of similar form. Only slight differences occur between the three different subsoil classes. Peak ground accelerations following from the attenuation functions by Ambraseys et al. [3] are given by Figure 4d for soft soil conditions and a local magnitude M_L = 5.0 (curve ASB). M_L was converted into M_L using equation (2). It becomes evident that the attenuation function of Ambraseys (mainly derived from European strong motion records) result in accelerations which are twice as high as the presented one.

This conclusion coincides with the outcome of similar studies on Turkish strong motion data recently presented by Gülkan and Kalkan [6]. The data base of these authors contains the mainshock records from damaging earthquakes in Turkey covering a period of about 20 years. Question remain whether the existing attenuation functions are not suited for Turkish earthquake regions, and vice versa, whether the predicted attenuation relationships are in a manner particular that for engineering design purposes further proposals have to be considered, too. This general difference requires further consideration and might be explained by the homogeneity and composition of the TaskForce data set (dominated by near-field data of aftershocks). It should be stressed that due to the quality of data elaboration and data processing a remarkable level of reliability has to be attested to the derived results.

To provoke the discussion it should be clarified whether the attenuation relationships derived from purely regional data are more representative for the regarded region than those derived from a mixture of in-homogeneous records.

6. PREDICTION OF SPECTRAL ACCELERATION

The calculated coefficients for the prediction of the horizontal spectral acceleration according to equation (3) are given in dependence on the subsoil conditions and for selected periods in Table 5. Horizontal response spectra for a magnitude M_L = 4.5 and epicentral distance R_e = 20 km derived from the data bases I and (I + II) are compared in Figures 5a to 5c. Figure 5d shows spectra for magnitudes M_L = 4.5 and 5.5 for soft soil conditions. Since the amount of records on soft soil sites dominates the entire data set, attenuation curves for this class induce more reliable results. Further investigations considering subsoil conditions should mainly focus on rock and stiff soil recordings.

Similarly to Figure 4d, spectra from the attenuation relationship of Ambraseys et al. [3] are determined. Again equation (2) was used to convert magnitudes M_L into M_L.

As it could be stated for the peak ground acceleration data base I and data base I + II (including strong motion data from higher magnitudes and more distant earthquakes) lead to similar spectra. At shorter and longer periods data base I is slightly underestimating the results. This tendency can be observed for all subsoil classes (rock, stiff soil, soft soil).

The attenuation functions of Ambraseys (curves ASB) result in spectral ordinates which are twice as high as the presented one.

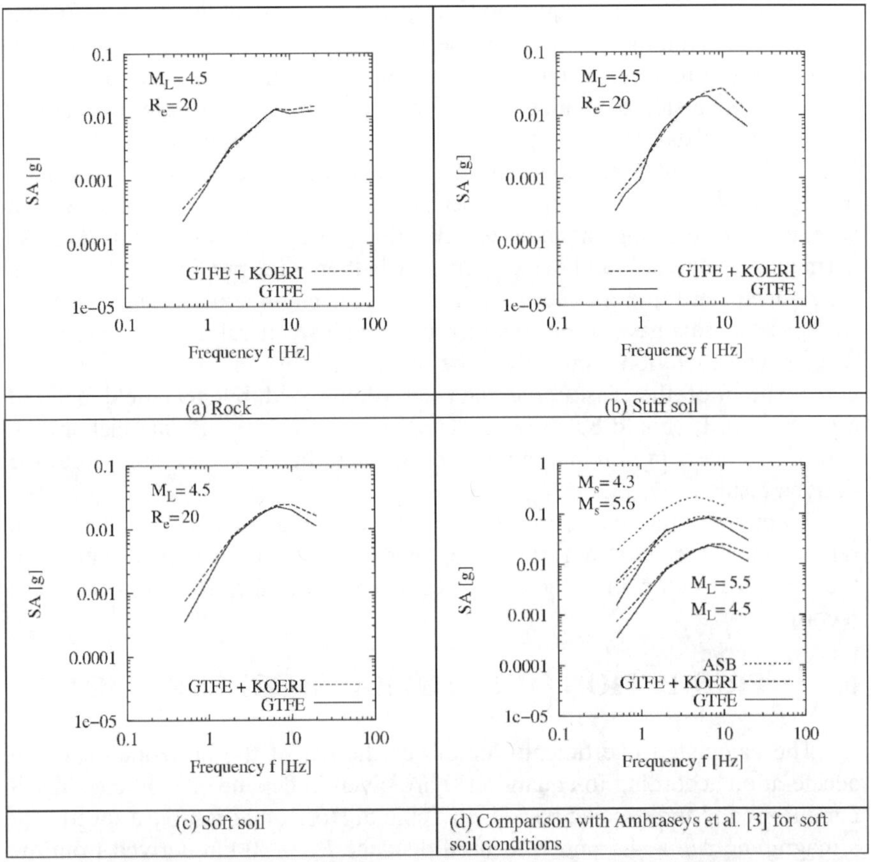

(a) Rock

(b) Stiff soil

(c) Soft soil

(d) Comparison with Ambraseys et al. [3] for soft soil conditions

Figure 5. Attenuation laws for horizontal spectral acceleration (epicentral distance = 20 km) and for different subsoil conditions: a) rock, b) stiff soil, c) soft soil, and d) comparison with Ambraseys et al. [3] (ASB) for soft soil

7. RESULTS OF PARAMETER STUDIES

Impact of subsoil conditions. As it was already mentioned, there are only slight differences between the three subsoil classes (cf. Figures 4a to 4c, Figures 5a to 5c). Prediction coefficients C_r (rock), C_a (stiff soil), and C_s (soft soil) in Tables 4 and 5 indicate that both, peak ground accelerations as well as spectral accelerations are higher on soft soil than on stiff soil. At higher frequencies spectral acceleration for rock might be higher than the corresponding soil values. This is especially true for data base I (cf. spectral ordinates at frequencies f > 10 Hz in Figures 5a to 5c).

Table 5. Prediction coefficients for horizontal spectral acceleration at selected periods

Period T [sec]	C_1	C_2	C_4	σ	h_0	C_r	C_a	C_s
0.05	-2.6398	0.4876	-1.0518	0.3387	2.00	-0.0165	-0.1371	0.0243
0.10	-2.4436	0.5058	-1.1295	0.3314	4.50	-0.2417	-0.0723	0.0385
0.15	-2.7300	0.5602	-1.0855	0.3478	6.00	-0.2286	0.0110	0.0233
0.20	-3.0872	0.5874	-0.9720	0.3720	6.00	-0.2796	0.0159	0.0282
0.25	-3.3142	0.6111	-0.9336	0.3896	6.75	-0.3370	0.0322	0.0424
0.50	-4.4357	0.7504	-0.8318	0.3869	2.75	-0.3552	-0.1134	0.0577
0.75	-5.0358	0.7826	-0.7261	0.3884	1.75	-0.3537	-0.1237	0.0592
1.00	-5.3878	0.7981	-0.6757	0.3960	1.50	-0.3544	-0.1257	0.0596
1.50	-5.7566	0.7982	-0.6280	0.4036	1.00	-0.2929	-0.1314	0.0538
2.00	-5.9185	0.7851	-0.6137	0.4041	0.75	-0.2719	-0.1336	0.0519

Impact of lower magnitude bound. Studies on the impact of the lower magnitude bound are related to the cardinal questions whether low magnitude but near-field data can be applied for engineering design purposes and to which extent and in which direction results might be influenced. The answers to this are of twofold importance. The first aspect targets the strategic content of TaskForce missions and here, in particular, the relevance of strong motion measurements, which under best circumstances will start three days after a mainshock. A second aspect, in general, concerns regions of lower or moderate seismicity (e.g., Central Europe) and the sense of a strong motion network which over the years will supply only low magnitude recordings. It has to be clarified if and where attenuation relationships derived from such a data set can be applied or extended by other (higher magnitude) data. It still remains an unsolved problem whether results from such studies alone should be applied for zoning maps.

Within this study data base V ($M_L \geq 4.0$) is taken as reference condition. To illustrate the impact of the lower magnitude bound the ratio between respective spectral accelerations is plotted as a function of period T and distance R. The ratio between spectral accelerations from DB III ($M_L \geq 4.0$) and DB V, as well as from DB IV ($M_L \geq 3.0$) and DB V are calculated for the magnitudes $M_L = 4.0$ (Figures 6 and 7), $M_L = 5.0$ (Figures 8 and 9), and $M_L = 6.0$ (Figures 10 and 11). The color of isolines indicates the effect of data reduction. Truncating the lower magnitude in general leads to higher PGA, and further to lower spectral accelerations in the near-field range ($R <$ 10 km) at periods shorter than 0.8 seconds. It depends on the deaggregation of seismic hazard and the general design philosophy to artificially cut the lower magnitude data. In Central Europe, damage due to earthquakes can only be provoked by near events. In such a case the neglect of lower magnitude data would underestimate

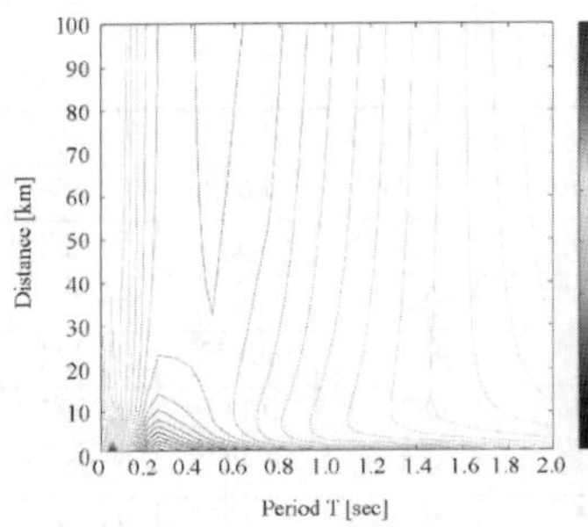

Figure 6. Ratio DB III to V for $M_L = 4.0$

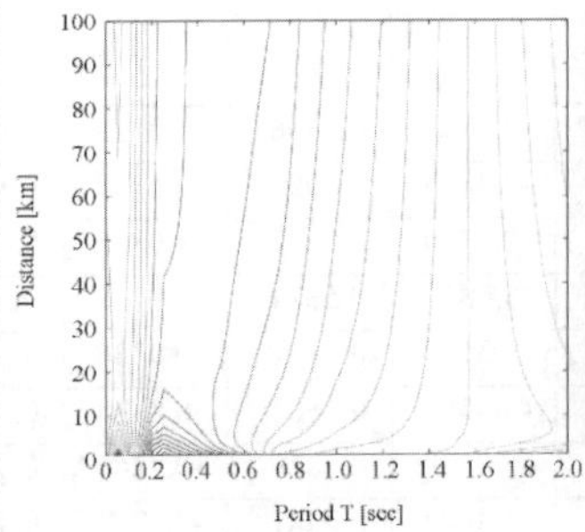

Figure 7. Ratio DB IV to V for $M_L = 4.0$

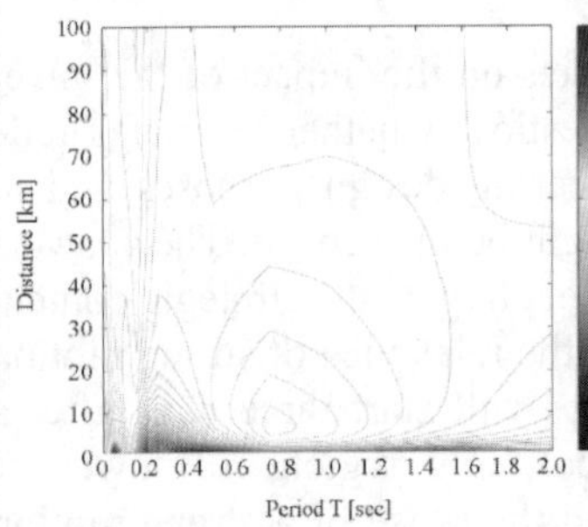

Figure 8. Ratio DB III to V for $M_L = 5.0$

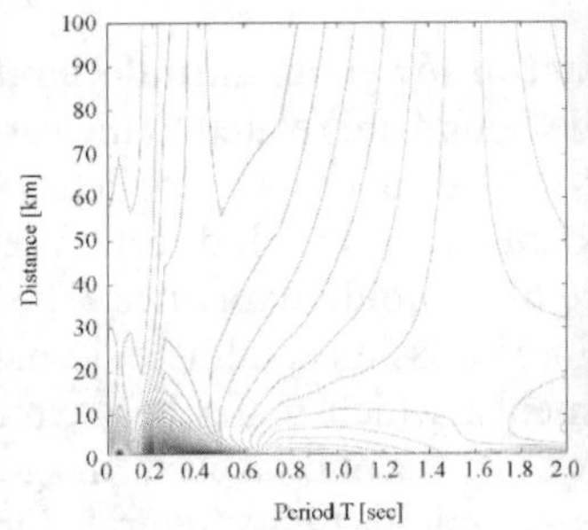

Figure 9. Ratio DB IV to V for $M_L = 5.0$

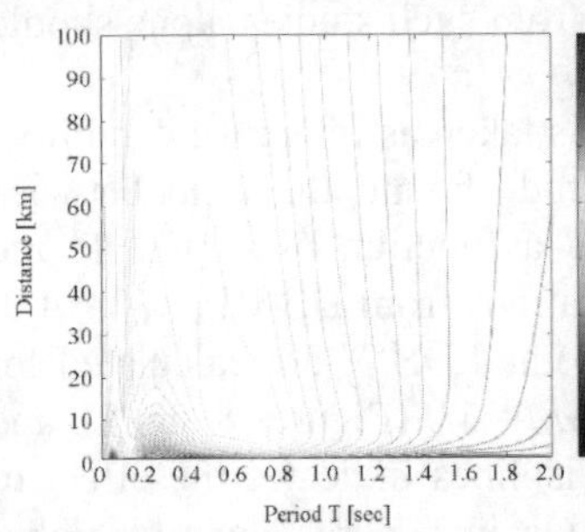

Figure 10. Ratio DB III to V for $M_L = 6.0$

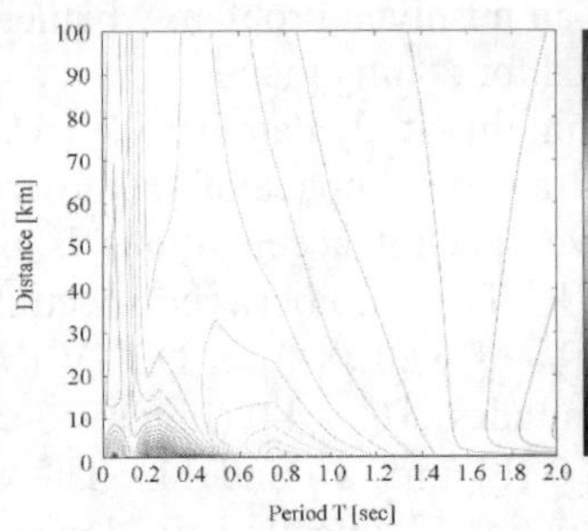

Figure 11. Ratio DB IV to V for $M_L = 6.0$

the predictions. Furthermore, it has to be considered that during the life time of buildings these events have a certain level of probability and, therefore, may affect the structure. It should be concluded from the study that these attenuation functions can be applied for the prediction of realistic and site-dependent design spectra, as well as for probabilistic seismic hazard assessment.

The elaborated data base may serve as a basis for supplementing and repeating the regression procedure. Nevertheless, the study should be continued and repeated with respect to a more refined classification of subsoil conditions.

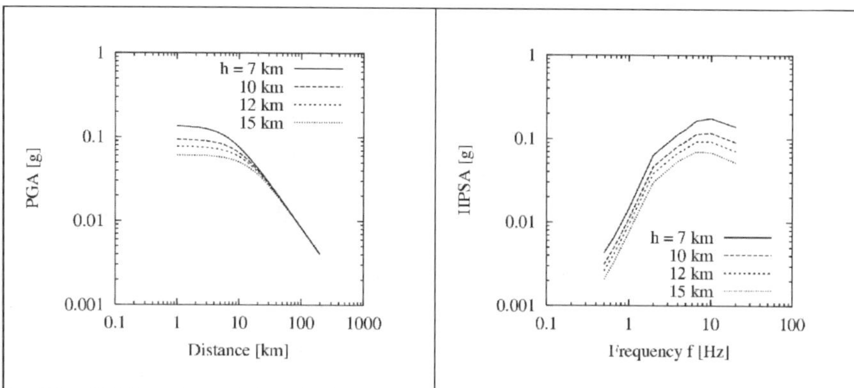

(a) Attenuation of *PGA* for magnitude $M_L = 5$ and focal depths ranging from 7 to 15 km

(b) Depth-dependent spectra for magnitude $M_L = 5$ and distance $R_f = 5$ km

Figure 12. Source depth-dependent attenuation laws of horizontal peak ground acceleration and spectral acceleration for all data (subsoil independent)

Impact of focal depth. Results of parametric studies with respect to focal depth are given by Figure 12. Data base III without any specification of subsoil conditions is considered. The regression analysis is repeated with the modification that in equation (3) for h_0 the real depth of each earthquake is used. Regression coefficients can be taken from [15].

Results in Figures 12a and 12b are related to the horizontal peak ground acceleration (PGA) for local magnitude $M_L = 5.0$ and to the acceleration response spectra for $R_f = 5$ km, respectively, and refer to focal depths between 7 and 15 km. It should be noticed that a comparable investigation was performed by Ambraseys [2] for peak ground acceleration only. Both studies lead to similar results and support the conclusion that the focal depth-independent prediction equations overestimate the ground motion and spectral ordinates in the near-field range (distance $R < 20$ km) if the focal depth exceeds a value of $h = 10$ km.

8. CONCLUSIONS

With the recorded aftershock data base it was possible to generalize strong motion recordings in terms of attenuation coefficients for peak ground acceleration as well as spectral accelerations considering different subsoil conditions (rock, stiff soil, soft soil). The derived attenuation relations strongly vary with the applied data set.

Reliability of the results only exists within the limits of the recorded data (in terms of magnitude and epicentral distance), and strongly depends on the amount of available events. This effect is considered by combining different data bases and reducing the data amount through the variation of a lower magnitude bound. As a whole, the regression procedure was repeated for five different data bases. Additionally, the extended data base with a wider focal depth-magnitude distribution is studied.

In order to check the sensitivity of attenuation laws with respect to the composition of the considered data set the study might be repeated with an enlarged strong motion data base taking into account recordings from previous Turkish earthquakes. Nevertheless, the studied data bases react quite stable on variations of the data composition.

The investigations show that derived peak ground accelerations for the study area (Northwestern Turkey) are remarkably lower than those predicted on the basis of the European data set (Ambraseys et al. [3]).

Furthermore, only slight differences between attenuation relations predicted from soft and stiff recording sites can be observed, supporting the authors' opinion that the severity of ground motion is less responsible for the high extent of building damage than the vulnerability of the building stock.

It should be concluded from the study that attenuation functions can be applied for the prediction of realistic and site-dependent design spectra as well as for probabilistic seismic hazard assessment.

The study is intended to contribute to the general problems of strong motion attenuation relations concerning the appropriate data composition and future directions of ground motion studies where an enormous quantity of studies could not really improve the quality of attenuation functions with respect to the scatter and uncertainty.

ACKNOWLEDGMENTS

The authors would like to thank Polat Gülkan for the invitation to the workshop and NATO for providing funds.

Missions of German TaskForce for Earthquakes are sponsored by Hannover Rückversicherungs AG (Hannover Re), GeoForschungsZentrum Potsdam, and Earthquake Damage Analysis Center at Bauhaus-University Weimar.

REFERENCES

1. Ambraseys, N.N., 1990, Uniform magnitude re-evaluation of European earthquakes associated with strong-motion records, *Earthquake Eng. and Structural Dyn.* **19**: 1–20.
2. Ambraseys, N.N., 1995, The prediction of earthquake peak ground acceleration in Europe, *Earthquake Eng. and Structural Dyn.* **24**: 467–490.
3. Ambraseys, N.N., Simpson, K.A., and Bommer, J.J., 1996, Prediction of horizontal response spectra in Europe, *Earthquake Eng. and Structural Dyn.* **25**: 371–400.
4. Ambraseys, N.N., Smit, P., Sigbjörnsson, R., Suhadolc, P., and Margaris, B., "Internet-Site for European Strong-Motion Data *http://www.isesd.cv.ic.ac.uk/*," EVR1-CT-1999-40008, European Commission, Directorate-General XII, Environmental and Climate Programme, Bruxelles/Belgium.
5. DIN 4149 - Bauten in deutschen Erdbebengebieten. Deutsches Institut für Normung Berlin (National code provision), 2002.
6. Gülkan, P. and Kalkan, E., 2002, Attenuation modeling of recent earthquakes in Turkey, *Journal of Seismology* **6**: 397–409.
7. Grünthal, G. and Wahlström, R., 2001, Sensitivity of parameters for probabilistic seismic hazard analysis using a logic tree approach, *Journal of Earthquake Eng.* **5**: 309–328.
8. Kandilli Observatory and Earthquake Research Institute (KOERI, 2002), "Internet-Site *http://www.koeri.boun.edu.tr/defaulteng.htm*," Istanbul/Turkey.
9. Lang, D.H., Schwarz, J., and Habenberger, J., 2000, Strong-motion data recorded during missions of German TaskForce, 17th General Assembly of the European Seismological Commission, Lisbon/Portugal.
10. Lang, D.H., 2004, Damage potential of seismic ground motion considering local site effects, *scientif ic technical reports* **1**, 297 pp., Earthquake Damage Analysis Center, Bauhaus-University Weimar.
11. Milkereit, C., Lokalisierung von Nachbeben, Technical report, SABOnet Group, 1999-2000.
12. Schwarz, J., Lang, D.H., and Raschke, M., 2000, Lessons from recent earthquakes – Field missions of German TaskForce, *Proceedings* of the 12th World Conference on Earthquake Engineering, Auckland/NZ.
13. Schwarz, J., Ende, C., Habenberger, J., Lang, D.H., and Raschke, M., 2002, Engineering analysis of strong-motion data recorded during German TaskForce Missions to Turkey (1998-2000), 27th General Assembly of the European Geophysical Society (EGS), Nice/France.
14. Schwarz, J., Ende, C., Habenberger, J., Lang, D.H., Baumbach, M., Grosser, H., Milkereit, C., Karakisa, S., and Zünbül, S., 2002, Horizontal and vertical response spectra on the basis of strong-motion recordings from the 1999 Turkey earthquakes, 18th General Assembly of the European Seismological Commission (ESC), Genoa/Italy.
15. Schwarz, J. and Ende, C., 2004, Spektrale Abnahmebeziehungen zur Bestimmung seismischer Bemessungsgrössen auf Grundlage von Starkbebenregistrierungen in der Türkei, *Schriften* der Bauhaus-Universität Weimar **116**: 91–104.
16. International Conference of Building Officials (ICBO), 1997 Uniform Building Code (1997-UBC), National code provision, 492 pp., Whittier/United States, 1997.

OBSERVATION AND PREDICTION OF STRONG GROUND MOTION IN CHINA

TAO Xiaxin[1,2] ZHOU Zhenghua[2] ZHOU Yongnian[2] WANG Haiyun[2]
1, Harbin Institute of Technology, China
2, Institute of Engineering Mechanics, China Earthquake Administration

Abstract: Development and current status of the strong ground motion measurement and interpretation in China are described.

Keywords: ground motion prediction, seismogenic source, slip, Green Function method, finite fault model, Tangshan earthquake

1. INTRODUCTION

In this paper, the evolution and the state-of-the-art of instrumentation of strong ground motion in China are summarized briefly, as well as a new project on instrumentation of strong ground motion is introduced. In review of prediction of strong ground motion in practice, finite fault model is emphasized for near field, and a random source model is introduced after a talk on synthesis procedures of ground motion.

2. STRONG GROUND MOTION OBSERVATION NETWORK IN CHINA [1]

The strong ground motion observation in China was initiated by the effort of the Institute of Engineering Mechanics (IEM), in 1962 after an Ms 6.1 earthquake induced by Xingfengjiang Reservoir. The first mobile strong motion observation was carried out in the epicenter area of the Xingtai Earthquake (Ms7.2) in 1966 after the main shock. Subsequently, the first groups of permanent strong motion station were deployed in Xingtai and Shijiazhuang area in Hebei Province. The permanent instrumentation networks of strong motion observation were gradually built up in some seismic zones, especially in the North China, after Tonghai Earthquake (Ms7.7) in 1970, Haicheng Earthquake (Ms7.3) in 1975 and Tangshan

P. Gülkan and J.G. Anderson (eds.), Directions in Strong Motion Instrumentation, 241–254.

Earthquake (Ms7.8) in 1976, mainly with the instruments manufactured in China. In 1980's, a series of international cooperative projects with USA, Japan, Russia and Australia were worked out, and some instruments including the most advanced digital accelerographs were also imported.

Since 1980's, the free-field observation has been in the majority, while motions on some typical structures such as hydrological structure and high-rise frame structures have been also observed. In the recent years, the observed objects have been extended to nuclear power station, and arrays on local site effect and canyon topography. With digital accelerographs manufactured home and abroad, the first telemetric array of rapid seismic intensity was established around Beijing in 1990's, which consists of 80 dial-up strong motion stations and 72 real-time satellite transmission strong motion stations. Some digital dense arrays, such as Xiangtang site effect array, structural array of Disaster Prevention Building of China Earthquake Administration and topography effect array at Xiluodu Dam site, were also built in this period.

Up to 2002, there are totally 488 strong motion stations in the networks in China mainland as listed in Table1 (from www.iem.net.cn/1d/qztzflb.htm), but those in Taiwan is not included in the table and also in the following text for the well known reason.

Table 1 Strong motion stations in China

Installation Type	Observed Object	Number of Stations	Number of Instruments	Total
Ground	Rock Soil	67 121	69 128	197
Structure	Brick Construction Frame Structure Dam Bridge Port Nuclear Power Plant	9 15 33 2 1 2	14 28 81 6 1 2	132
Others	Intensity Reporting Site Effect Topography Effect	152 2 1	152 5 2	159
Total			488	488

The Figure 1 shows the distribution of the stations and arrays in China mainland (from www.iem.net.cn/1d/gcgdfbt.htm).

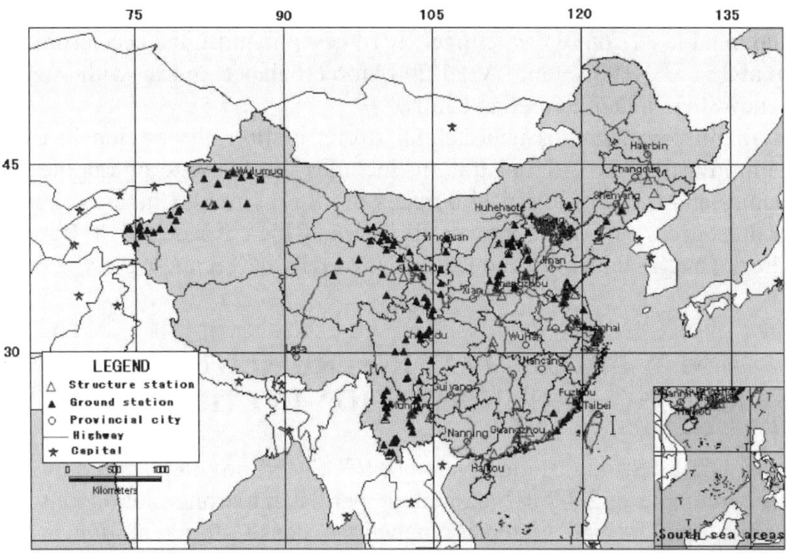

Figure.1 Distribution of strong motion stations (arrays) in China Mainland

Table 2 The accelegraphs in the network of China

Recording Medium	Type	Country of Production	Number of Instrument	Total
Photographic Film Analog	RDZ-1	China	47	201
	GQ□	China	9	
	GQ□-A	China	76	
	SMA-1	USA	69	
Magnetic Tape Analog / Digital	SCQ-1	China	13	57
	EDS-A	China	24	
	EDS-B	China	10	
	PDR-1	USA	5	
	SMA-2	USA	5	
Solid-State Digital	GDQJ-1A	China	118	270
	GSMA-2400	China	72	
	Etna	USA	34	
	K2	USA	25	
	SSR	USA	5	
	SSA2	USA	5	
	EDAS-3	USA	5	
	CA901VR	Japan	6	
Other			15	15
Total				488

The information of the accelerographs in strong motion network in China is listed in table 2 (from www.iem.net.cn/1d/qzyqylb.htm); the accelerometers are mainly SLJ-100B and WLJ-200 force-balance ones with natural frequency 80Hz manufactured in China.

The main organizations undertaking strong motion observation in China are China Earthquake Administration, includes the Institute of Engineering Mechanics and the seismological bureaus of 18 provinces, China Institute of Water Resources and Hydropower Research, China Academy of Building Research, Tongji University and Dalian University of Technology.

3. A NEW PROJECT ON INSTRUMENTATION OF STRONG GROUND MOTION IN CHINA [3]

China Digital Strong Motion Network (CDSMN) program started in 2002 and will complete in 2007, which include permanent strong motion network, metropolis rapid seismic intensity networks, dense strong motion arrays, mobile strong motion network, state and regional strong motion data centers, state strong motion instrument testing and calibrating center. It is in layout to build state-of-the-art digital strong motion networks in some seismic zones and rapid seismic intensity systems in some metropolis, to strengthen mobile strong motion observation ability, and to improve the strong motion station density and ability for obtaining near-field record in future strong earthquake. In the first purchase for the project, 228 accelerographs manufactured by US Kinemetrics Inc., Swiss GeoSIG Ltd. and Chinese IEM are ordered in this spring.

3.1 Permanent Strong Motion Network

The permanent network consists of 1,160 free-field stations, on bedrock and soil ground surface. All stations will have near-real-time dial-up digital telemetry and will be equipped with three-component force-balance accelerometers and digital recorders with resolution greater than 19-bit. Because China is a vast territory country, the number of these instruments is still limited, 21 seismic zones are chosen as the regions that may occur destructive earthquakes in the next decade [4], include 8 first rank areas as red in Fig.2 with density of 1 station per 600km^2, i.e. interval between stations is about 25 km, and 13 second rank areas as baby pink in Fig.2 with 1 station per 1800 km^2, i.e. interval is about 42 km. State strong motion observation center and 3 new-built regional strong motion observation centers will be in charge of data acquisition and data process. It is believed that some station(s) can record obtain near-field ground motion, if a strong earthquake with magnitude 5 or greater occurs in one of the above areas, after the CDSMN completed.

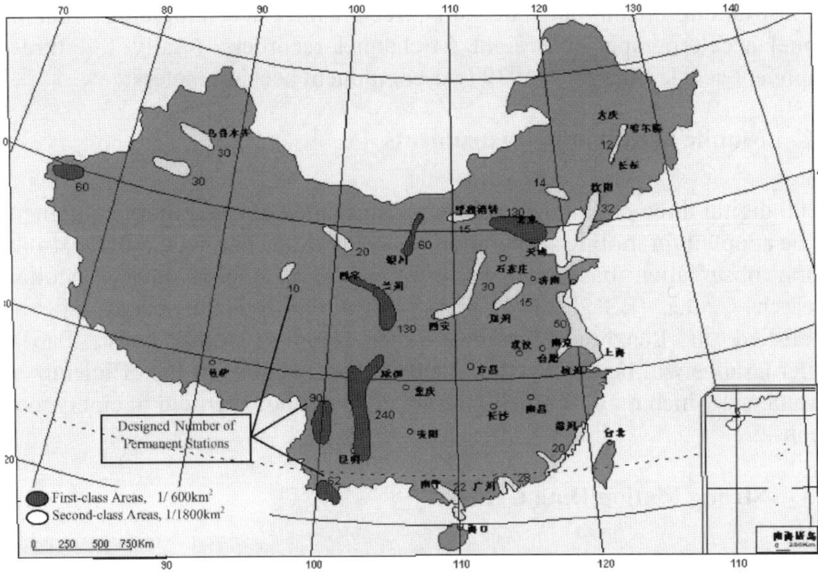

Fig.2 Seismic zones for CDSMN perminent networks

3.2 Metropolis Rapid Seismic Intensity Networks

The networks consist of 310 stations with the same configuration of the permanent station. As the centers of the networks, the candidate cities are Beijing, Tianjin, Lanzhou, Urumchi and Kunming. The station density is designed as 1 station per 50km^2, i.e. the interval between stations is about 7km. The network is expected to report distribution of ground motion and seismic intensity for emergency response in 10 minutes after a near earthquake occurs. The communication between the center and station will implement by telephone line with the autodial function.

3.2.1 Dense Arrays

There will be totally 12 dense arrays, one for fault effect with deploying 30 stations, in Yunnan Province, two for ground motion attenuation with 50 stations in eastern and western China respectively, one for topography effect with 8 stations in western China, two for site effect with 4 ground stations and 4 underground stations in eastern and western China respectively, six for structural response with 126 channels for a skyscraper, 60 channels for a high-rise building, 126 channels for a large span gymnasium, 63 channels for an isolation building, 63 channels for a large bridge, 63 channels for a large dam.

The instruments mainly include 146 three-channel accelerographs, four 6-channel accelerographs and eight 64-channel recorders, totally 134 three-component accelerometers and 39 two-component accelerometers.

3.2.2 Mobile Observation Instruments

200 digital three-component accelerographs and some auxiliary equipment will be adopted for mobile strong motion observation. The state mobile strong motion observation base will be established in Beijing Strong Motion Research Center (BSMRC) of IEM and three regional bases will be established in Lanzhou, Kunming and Nanjing, respectively. Partial accelerographs will be deployed in deposited arrays to exert the efficiency of instruments, which may be quickly removed to earthquake field in emergency response.

3.2.3 Strong Motion Data Centers

One state strong motion data center and three regional strong motion data centers will be built or rebuilt during the project period. The main functions of state strong motion data center are receiving data from regional strong motion centers, processing strong motion data and building strong motion databank, serving for the society with strong motion data, exchanging data internationally or regionally and so on. The center is in BSMRC. The main functions of regional strong motion data centers at Lanzhou, Kunming and Nanjing are acquiring strong motion data from regional stations and transferring them to state strong motion data center through seismic appropriative lines. The strong motion data obtained in northern and northeastern areas of China will be acquired directly by state strong motion data center.

3.2.4. State Strong Motion Instrument Testing and Calibrating Center

The center in BSMRC is pursuing the new advancements in digital technique of the international strong motion observation, testing and calibrating strong motion instruments, providing technical service of instrument for provincial seismological bureaus.

State strong motion data center, state mobile strong motion observation base, state strong motion observation instrument testing and calibrating center are as a whole called state strong motion observation center. Regional centers are similarly called as regional strong motion observation centers.

The planning investment of this project is about 300 million RMB yuan, and all instruments include 1808 three-component accelerographs, eight six-component accelerographs and eight 64-channel, totally are equivalent to

1882 three-component accelerographs, and other equipments for data processing, instrument testing and calibrating.

Meanwhile, some provincial seismological bureaus will build additional regional digital strong motion networks by the resources from the local governments, in company with this project.

4. STRONG GROUND MOTION DATA ACCUMULATED IN CHINA

In China mainland, strong motion databank were built in 1980s. Up to 2002, totally 3255 accelerograms had been accumulated in China mainland, in which, 206 with peak values larger than 100 gal, 59 larger than 200 gal, as shown in Figure 1. The maximum horizontal peak ground acceleration is 541.7 gal, and the maximum vertical peak ground acceleration is 528.2 gal. All of the records have been processed by a set of routine software in IEM, Strong Motion Data Processing Center. Now the databank is in a new version managed by a GIS based system, and starts the strong motion data service on Internet (Web site www.iem.net). It also consists of more than 15 thousand records over the world (mainly in the western U.S.A. and in Japan and its surroundings), which was obtained through an international data-exchanging channel. Some major strong motion records of China mainland are listed in table 3[2] (mainly from www.iem.net.cn/1d/qzzhyml.htm).

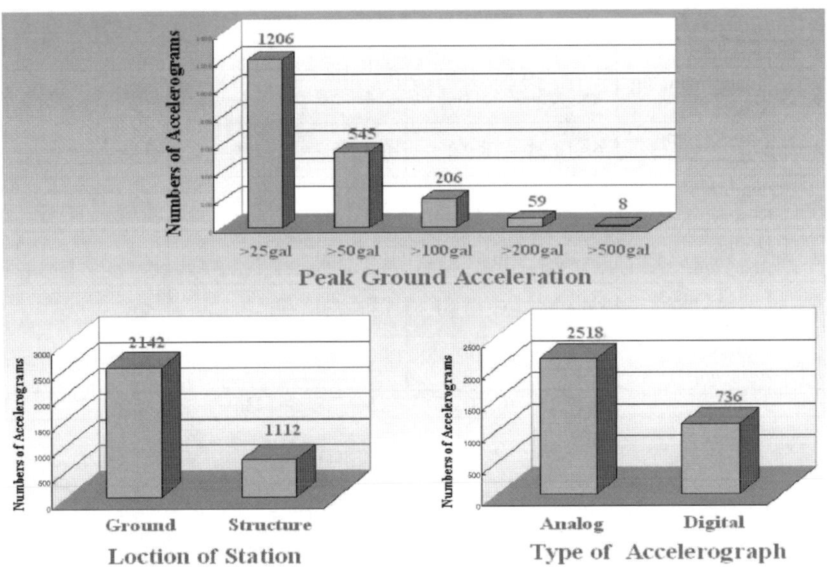

Fig.3 Statistics of the accelerograms accumulated in China mainland

Table 3 Some major strong motion accelerograms in China mainland

No.	Earthquake (array) name	Date (time period) of earthquake(s)	Magnitude (range)	Number of events	Number of accelerograms
1	Tangshan	76.7.28 —76.11.15	3.5—7.8	38	267
2	Tangshan	82.7—84.12	2.3—5.7	52	555
3	Guye	95.10.6	5.9	1	24
4	Zhangbei	98.1.10	6.4	1	3
5	Haicheng	76.2.8—76.2.28	3.0—5.4	24	366
6	Tonghai	70.1.14—70.9.4	2.6—5.7	11	87
7	Longling	76.6.5—76.6.29	2.3—6.2	11	87
8	Gengma	88.11.6—88.12.6	2.6—7.6	92	406
9	Wuding	95.10.25—95.10.27	4.1—5.3	10	24
10	Lijiang	95.1.31—96.2.9	7.0	13	54
11	Yaoan	00.1.15—00.1.31	2.7—6.5	9	38
12	Shidian	01.4.10—01.5.19	1.5—5.9	41	132
13	YongSeng	01.10.27	4.1—6.0	2	6
14	Wuqia	85.8.28—85.9.20	3.0—6.3	18	52
15	Wushi	87.1.24—90.10.25	1.9—6.4	23	55
16	Wusu	95.5.2—96.3.22	3.3—6.9	7	24
17	Atushi	95.5.2—96.3.22	3.3—6.9	7	24
18	Jiashi	96.3.19—98.8.27	3.6—6.9	14	420
19	Songping	76.8.16—76.9.2	4.2—7.2	9	36
20	ZhangYe	87.9.12	4.5	1	3
21	Sunan	88.11.22—88.12.26	2.7—5.7	5	33
22	JiayuGuan	92.1.12	5.9	1	6
23	Linze	87.10.25—88.12.26	5.0—5.1	2	6
24	Gonghe	94.2.16—94.9.24	5.2—5.6	3	8
25	Datong	89.10.18	5.7	1	20
26	Huangbizhuang Reservoir	68.7.25—74.6.6	2.4—5.2	6	70
27	Xinfengjiang Reservoir	68.10.21—78.9.2	2.6—5.3	24	345
28	Guanting Reservoir	94.12.23	3.8	1	10
29	Douhe Reservoir	79.8.25—80.4.16	3.6—4.0	3	25
30	Longyangxia Reservoir	90.4.26—94.10.10	1.9—6.9	3	50
31	Liujiaxia Reservoir	95.7.22	5.8	1	19
	Total	68.7.25—01.10.27	1.5—7.8	434	3255

Fig.4 The Web site of the databank of strong ground motion in IEM

In addition, China institute of water resources and hydropower research has edited and published one volume of *Important Data on Strong Motion and Analysis in China Hydraulic Structures.*

Strong motion records of China mainland are adopted in draw up various seismic design codes, ground motion attenuation relationship, and as the input seismic wave in structural response analysis and vibration table experiment.

5. ENGINEERING PREDICTION OF STRONG GROUND MOTION IN CHINA

In China, specific seismic hazard assessment is required for significant engineering sites, while the earthquake action is characterized by response spectra in seismic design codes for popular buildings. The term of prediction of ground motion mentioned in this paper is just for the former. In practice, the strong ground motion input for the seismic safety analysis of engineering project is mainly predicted by attenuation relationship from one or more scenario earthquake(s) which characterize(s) the seismic hazard of the site. In 1980s, studies on strong ground motion were initiated in China mainland[6]-[9]. Unfortunately, the ground motion records are insufficient, since China is a country with a huge territory, the attenuation characteristics should be deferent in regions of different geological backgrounds. Hu et al. suggested a procedure to modify available relations in other region, e.g. California, by comparing the local intensity attenuation with those in that region and assuming that the relation between intensity and ground motion parameter is one-to-one only when the magnitudes are the same or the epicenter distances are the same [10]-[11].

At the beginning of this century, the near-fault effects, such as rupture directivity effect and hanging wall effect, are emphasized for ground motion prediction in China. It is generally believed that the widely adopted empirical attenuation relationships could not describe these significant engineering features of the motion in the near field. The finite fault model (FFM in brief) is adopted, which incorporates the fact that ground motion at a site near the fault is affected mainly from the nearby, local and finite portion of the rupture plane, the effects of more distant parts of the fault are less important, and it allows a realistic inhomogeneous distribution of the dislocation on the rupture surface. In the model, the faulting plane is divided into many sub-sources, and the specified dislocation energy is assigned to each one of them, as point source. The simplest method to generate sub-event seismograms, the random synthesis procedure was introduced [12]-[13]. The following figure shows the comparison of the response spectra (left-hand) and time histories (right-hand) of simulated ground motions from a source model of Northridge earthquake 1994 (after Beresnev, 1998) at three near fault stations MCN, LV3 and PCD, all rock sites with hypocenter distance 27.1, 54.5 and 27.2 km respectively, with those recorded during the shock, for validation of the developed procedure [14]. One can find the fact that the spectra are quite consistent, while the envelopes of the time histories are yet with some different. A Numerical Green Function method (NGFM) is developed to take into account the regional realistic wave velocity structure of the Earth crust, which typically includes sedimentary basins beneath urban centers. The NGFM has a greater chance of computing realistic results at low frequencies. Since the source size of a large earthquake is tens to more than a hundred km, the computational region must be quite large. In general, it may be several hundred km; the size of the sub-sources also must be large from the computer resources restriction, e.g. several hundred meters. Thus the high frequency component could not be simulated well by NGFM. One key idea is to merge the ground motion of random synthesis at high frequency with the NGFM results at low frequencies, in time domain.

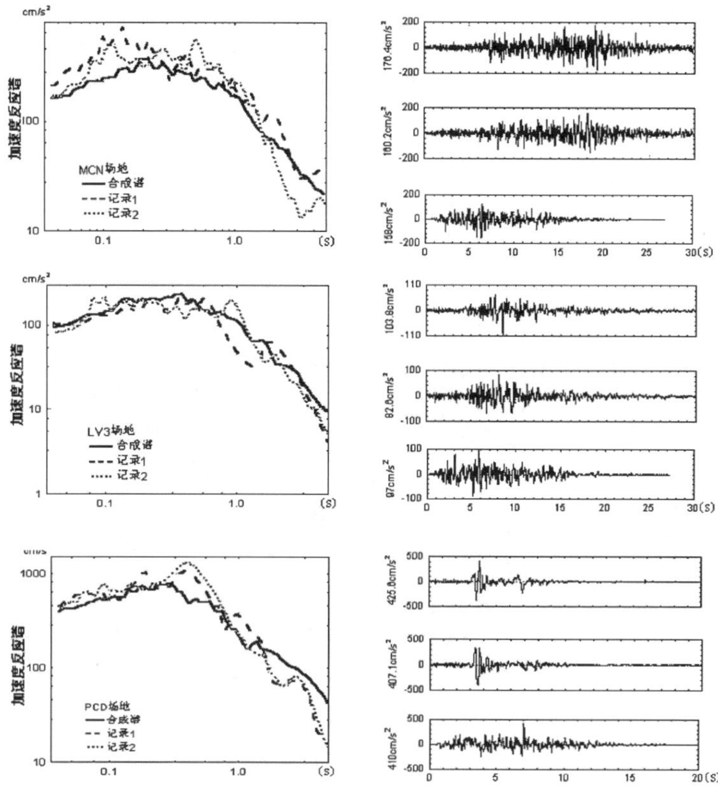

Fig.5 Comparison of the simulated ground motions with the records

6. A STOCHASTIC SOURCE MODEL FOR PREDICTION OF STRONG GROUND MOTION

As well known, the geometrical characteristics, such as the fault location, size (fault length and width), attitude (strike and dip), and average displacement, can be obtained from geological, geophysical investigations, thickness of the seismogenic zone, paleoseismology and scaling laws of the region. The latter provides the relationships between earthquake moment magnitude and rupture area, length, width, and average slip [15]. Detailed studies of the spatial distribution of slip on the fault plane for earthquakes in tectonically active regions, derived from strong motion recordings and other data, have shown that the slip distribution is highly variable. At present, the most difficult problem concerning with prediction of strong ground motion, is how to estimate the source parameters that describe the slip heterogeneity or roughness on a specific fault plane and the rupture process before an earthquake. A random source model is developed recently [16]-[17], which emphasized both parts of the slip distribution, the deterministic and random. Firstly, the former is characterized by significant asperity(s), the length, width, average slip of maximum asperity, relative coordinate values of the maximum asperity center, size and average slip of the other asperities, are derived respectively from earthquake moment magnitude by statistical relationships for the given fault slip type. The slip values then are assigned to each sub-source of FFM as weighting factor according to the asperity layout on FFM. The k square model is adopted for the random part.

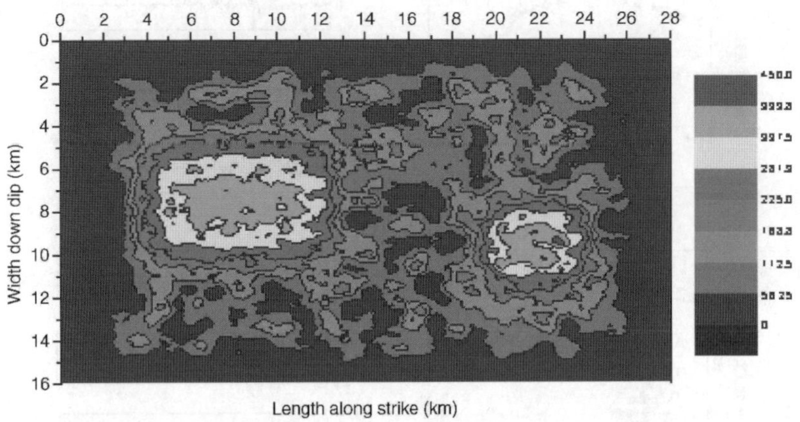

Fig.6 A random source model example

The final dislocation distribution can be worked out by superpose those of the two parts and normalized by the constraint that the total moment in all sub-sources must be the same of the moment of the earthquake, i.e. the product of the area of the fault, the shear modulus and average slip on the whole fault which estimated at the first step. Figure 6 shows a source model generated for an earthquake with magnitude 6.7 on dip slip fault.

The rupture velocity could be taken as depth independent at a value of 0.8 times the shear-wave velocity generally at the half-depth of the fault plane. The rupture start point(s) in the source model is/are located by a statistical relation with the center of the maximum asperity. The predicted acceleration time histories (below) and response spectra (up) for the same earthquake at the same sites in Figure 5 are compared with those from records in Figure 7.

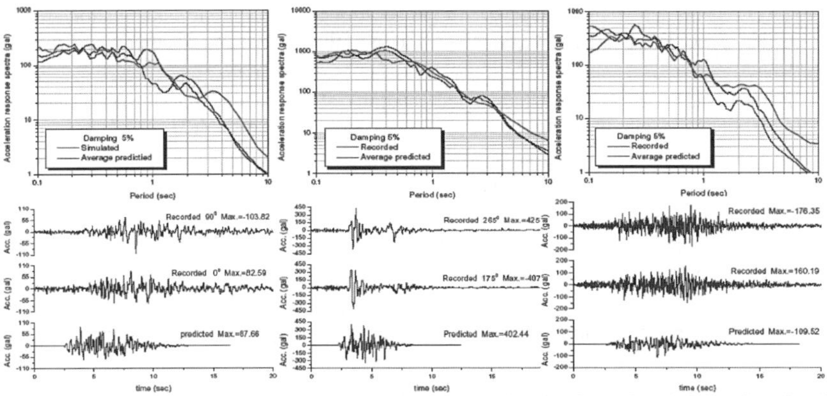

Figure.7 The simulated response spectra and time histories for site MCN (left-hand), PCD (in middle) and LV3 (right-hand)

7. CONCLUSION

The evolution and the state-of-the-art of instrumentation of strong ground motion in China are summarized in this paper. Furthermore a new project on instrumentation of strong ground motion in China is briefly introduced. The prediction of strong ground motion in practice is reviewed and finite fault model is emphasized. A random source model is introduced, after a talk on synthesis procedures of ground motion.

ACKNOWLEDGEMENT

The authors appreciate the help of their co-worker, Ms. Zhang Dongli, in preparing this paper. Some results of this paper are from the research work sponsored partly by the Earthquake Science Foundation under Contract No.201009 and by the National Social Commonweal Research Foundation under Grant No. 2002DIB30076.

REFERENCES

1. GAO Guangyi, YU Haiying, LI Shanyou (2001). Strong motion observation in China Mainland (in Chinese). *World Information on Earthquake Engineering* 17(4)
2. Yu Haiying, Strong motion database and analysis method in China mainland, Proc. of ANCEER workshop, Hong Kong, China
3. China Earthquake Administration (2002), The construction program of China Digital Strong Motion Network (in Chinese).
4. Li Shanyou, Yu Haiying and Zhou Zhenghua (2002), Development of strong ground motion instrumentation in China mainland, Proc. of ANCEER whorkshop, Harbin, China

5. TAO Xiaxin and ZHENG Guangfen (1994), Strong ground motion attenuation study in China, Proc. of the GSHAP regional workshop of the Eastern and Southern Asia, Beijing, China
6. HU Yuxian et al. (1982), Functional relations of bedrock motion parameters with earthquake magnitude and distance, ACTA Seismologica Sinica, 4(2)
7. WANG Yayong and LI Hong (1986), Site dependent attenuation statistics of strong ground motion parameters, Earthquake Engineering and Engineering Vibration, 6(3)
8. GUO Yuxue and WANG Guoxin (1989), Determinate method of parameters of earthquake ground motion in the North China, World Information on Earthquake Engineering, (3)
9. HU Yuxian and ZHANG Minzheng (1983), Attenuation of ground motion for regions with no ground motion data, Proc. of the 4th Canadian Conference on Earthquake Engineering, Canada
10. HUO Junrong and HU Yuxian (1992), Study on attenuation laws of ground motion parameters, Earthquake Engineering and Engineering Vibration, 12(2)
11. Wang Guoxin and TAO Xiaxin (2001). Strong ground motion of the 1994 M6.7 Northridge earthquake simulated by a stochastic approach. Proc. of IAGA-IASPEI joint scientific Assembly, Hanoi, Vietnam
12. TAO Xiaxin and G. J. Anderson (2002), Near field strong ground motion simulation, Proc. of ANCEER, Harbin, China
13. TAO Xiaxin and WANG Guoxin(2003), Rupture directivity and hanging wall effect in near filed strong ground motion simulation, ACTA Seismologica Sinica, 16(2)
14. WANG Haiyun and TAO Xiaxin (2003), Relationships between moment magnitude and fault parameters: theoretical and semi-empirical relations, Earthquake Engineering and Engineering Vibration, 2(2)
15. WANG Haiyun and TAO Xiaxin (2004), Global source parameters of finite fault model for strong ground motion predictions, Proc.of 13WCEE, Canada, Vancouver
16. TAO Xiaxin and WANG Haiyun (2004), A random source model for near filed strong ground motion prediction,Proc.of 13WCEE, Canada, Vancouver
17. Somerville, Irikula *et al* (1999), Characterizing Crustal Earthquake Slip Models for the Prediction of Strong Ground Motion, *Seismological Research Letters*, 70(1)
18. Kamae, K. and K. Irikura(2000), Source characterization for an inland large earthquake and near-source strong ground motion simulation, *Proc. 12th WCEE*, New Zealand, Auckland

STRONG MOTION INSTRUMENTATION PROGRAMS IN TAIWAN

- Past and Present -

Yi-Ben Tsai[1] and Chien-Ping Lee[1]

[1]*Institute of Geophysics, National Central University, Chung-Li 320, Taiwan, ROC*

Abstract: Earthquakes occur frequently in Taiwan. A modern seismological program was initiated there in early 1970's, with strong motion instrumentation as an integral component. We present here an overview of the strong motion networks and arrays deployed in past thirty years by two major operators, the Institute of Earth Sciences of Academia Sinica and the Seismology Center of Central Weather Bureau. The strong motion networks and arrays deployed by the former include SMA1 network, SMART1 array, LLSST array, SMART2 array, HLLST array, CMSMA network, and Taipei basin downhole array. The strong motion networks deployed by the latter include TSMIP networks that include free-field stations, structural arrays, the RTD network, and Taiwan railroad system. These dense networks and special arrays have recorded large quantity of valuable data that greatly accelerated development of seismology in Taiwan. We also highlight some examples of data applications, including studies of ground motion attenuation by the TSMIP free-field data, of nonlinear soil amplification effects by the LLSST downhole array data, of permanent changes in structural response of a damaged building due to the Chi-Chi earthquake by the TSMIP structural array data, and of the Chelungpu fault rupture propagation by the SMART2 data.

Key words: strong motion; network; array; Taiwan.

1. INTRODUCTION

Taiwan is located in the collision boundary zone of the Phillipine Sea and Eurasian plates. It experiences very active seismicity with frequent damaging earthquakes, as shown in Fig. 1. Some of them brought great disasters upon people in Taiwan. The 1999 Chi-Chi earthquake was the latest reminder of what destruction an earthquake can cause. It was recognized since early 1930s

P. Gülkan and J.G. Anderson (eds.), Directions in Strong Motion Instrumentation, 255–278.

that instrumentation observation of strong ground motion is a key to more effective earthquake hazard mitigation. This idea was taken seriously form the beginning when a modern seismological program was initiated in Taiwan in early 1970s.

In past thirty years, a number of strong motion networks and arrays were deployed in Taiwan by two major operators, the Institute of Earth Sciences (IES), Academia Sinica and the Seismology Center, Central Weather Bureau (CWB). In this paper, we will describe several past and present strong motion networks operated by these two institutions and listed in Table 1. We will also highlight some of the diverse applications of the strong motion data acquired by these networks.

2. STRONG MOTION NETWORKS DEPLOYED BY THE INSTITUTE OF EARTH SCIENCES (IES)

2.1 The SMA1 Network

The SMA1 network was installed and operated by the Institute of Earth Sciences of Academia Sinica since the mid - 1970s. This network was consisted of nearly 200 analog accelerographs and recorded more than 800 events (Fig. 2). Most of the stations were placed in the basement of civil structures or buildings. In 1992 the CWB began to install a dense network of digital strong-motion seismographs under the Taiwan Strong-Motion Instrumentation Program (TSMIP). This new network consists of 650 stations covering the urban areas of entire Taiwan. Consequently, the IES relocated the SMA network to the Cross-island highways in the Central Mountain Range which suffered a lack of strong motion instruments. Since 2001 the SMA network was renamed Central Mountain Strong-Motion Array (CMSMA). Currently, the network consists about 40 digital accelerographs, including Kinemetrics Model SSA-1 and Model SSR-1; and Terra Technology Model IDS-3602.

2.2. The SMART1 Array

In order to provide earthquake ground motion data for a wide range of research topics and hazard-reduction applications, the IES and the University of California at Berkeley cooperated to build a dense digital array of strong-motion seismographs in 1980. This array was called the Strong Motion Array in Taiwan, Phase 1, i.e., SMART1. The array was located in northeastern Taiwan near the city of Lotung on the Lanyang Plain (Fig. 3). It was designed to record near-field ground motion, hopefully for a very large earthquake ($M \geqq 8$) and definitively for some damaging earthquakes ($M>6.5$).

The original SMART1 array consisted of thirty-seven stations with a central station, and configured in three concentric rings with a radius of 200 m, 1000 m and 2000 m, respectively, each ring with twelve approximately equally spaced stations. Later, two additional stations were added to the south in 1983, and two more to the north in 1987, respectively. The addition of these four new stations was designed to study response of ground motion at various site conditions, as well as to study seismic wave propagation across the array. The SMART1 array was discontinued in 1991 after succession by a more advanced SMART2 array in Hualien area about 80 km to the south of SMART1.

The equipment at each SMART1 station included the following:

- A three-component Model SA-3000 force-balance accelerometer which was capable of recording at 2g full scale.
- A Model DR-100 digital recorder with 12-bit resolution and 100 sps. It has an internal crystal clock and time code generator which provides for synchronization with highly precise time signals generated by a portable time comparator.
- The accelerometer, recorder and battery are all tied down to a 10 cm thick concrete pad poured in place over a steel wire mesh and are protected by a prefabricated light-weight fiberglass hut.

During its operational period (1980~1991), SMART1 recorded sixty earthquakes with different magnitudes, focal depths and epicentral distances (Fig. 3). Numerous studies on ground motion coherence, site response, wave propagation, fault ruptures have been made using these data. Fig. 4 shows sample records of the array from the November 14, 1986 earthquake of M_L6.5.

2.3. The LLSST Array

The Lotung Large Scale Seismic Test (LLSST) Array was a joint project of the Taiwan Power Company (Taipower) and the U. S. Electric Power Research Institute (EPRI). The IES was the array operator responsible for instrument installation and maintenance, data collection, reduction and analysis. This array lasted from 1985 to 1991.

The LLSST array was sited inside the SMART1 array very close to the outer ring O08 station (Fig. 5). Two scaled down reinforced concrete PWR containments (1/4 and 1/12 scale) were constructed and instrumented in the Taipower's Lotung Substation. It consisted of thirty-seven stations. They included fifteen surface accelerometers, eight downhole accelerometers (Fig. 6), and fourteen containment structure response accelerometers (Fig. 7). Each station was instrumented with a three-component Kinemetrics FBA accelerometer and a DSA-3 digital cassette recorder for data acquisition. Two types of accelerometers were used: FBA-13 and FBA-13DH.

During its operational period (1085~1991), the LLSST array recorded thirty earthquakes. Sample records by the LLSST array for the November 14, 1986 earthquake of M_L6.5 are shown in Fig. 7. Data collected by the LLSST have been used very effectively for analyses of containment structural response, to earthquake ground motion and soil-structure interaction. These outstanding results attracted worldwide attention and encouraged EPRI and Taipower to continue for the experiment second-phase at Hualien in 1993.

2.4. SMART2 Array

After ten years of operation of the SMART1 array, the Second Workshop on Strong-Motion Array was held from January 14 to 16, 1988 in Taipei. This workshop recommended a second-phase deployment of the strong-motion dense array. It was suggested that a new strong-motion array on a firm-soil site in the Hualien area be installed to address some of the problems which had remained unsolved by the SMART1 array and to make a comparison of these results with those form SMART1. After the workshop, the IES undertook a plan to deploy the SMART2 array near the city of Hualien. Instrumentation of this array was completed in late 1991.

The SMART2 array consisted of forty-five stations (Fig. 8). Later, two sets of downhole accelerographs were installed in 1992 and 1993, respectively. The first set composed of three downhole accelerometers at a depth of 50, 100 and 200 meters, respectively. The second set had only one downhole accelerograph at a depth of 100 meters.

Each SMART2 station was instrumented with the following:

- A three-component Model FBA-23 accelerometer which is capable of recording at 2g full scale.
- A Model SSR-1 digital recorder with 16-bit resolution and 200 sps. Each recorder was equipped with two megabyte memory and an Omega clock which provides a relative timing accuracy of about one millisecond between two strong-motion stations.
- The accelerometer, recorder and battery were all tied down to a 120 cm X 120 cm X 10 cm concrete pad poured in place over a steel wire mesh and are protected by a prefabricated light-weight fiberglass hut.
- The SSR-1 recorder provided 10 seconds pre-event memory and 30 seconds post-event memory. The pre-event memory ensured recording of first P-arrivals.

A total of 663 events were recorded by SMART2 from 1990 to 2002. Fig. 9 shows the array records of the 1999 Chi-Chi earthquake. The peak ground accelerations across the array vary greatly from 33.4 gals to 113.9 gals.

2.5. The HLSST Array

After the successful LLSST was closed in 1991, Taipower and EPRI actively searched for a suitable place to conduct the second phase of Large Scale Seismic Test. In order to record near-field ground motion of earthquakes and to study response of structures of engineering significance, a site was selected at the Hualien Veteran's Marble Plant for deployment of the HLSST array. Meanwhile, the EPRI and Taipower assembled an impressive international consortium to cosponsor the project. The consortium members included many governmental agencies and power companies of the U.S., Japan, Korea, and France.

The HLSST Array was sited in the northern part of the SMART2 array (Fig. 10). One scaled down reinforced concrete cylindrical containment model (1/4 scale) were constructed and instrumented in the yard of Hualien Veteran's Marble Plant. The instrumentation of this array consisted of forty-two stations, including fifteen surface accelerometers, twelve downhole accelerometers (Fig. 11) and fifteen containment structural response accelerometers (Fig. 12).

By December 2000, HLSST recorded 112 earthquakes (Fig. 10). Sample records by this array of the 1999 Chi-Chi earthquake are shown in Fig. 12. The reinforced concrete PWR containment model was removed in 2002. The IES has rearranged the instrumentation of surface and downhole accelerographs to continue observations since then.

2.6. The CMSMA Network

In 1992 the CWB began to install a new digital strong-motion network (TSMIP) which consists 650 stations in urban areas of Taiwan. Concurrently, the IES began to deploy the CMSMA strong motion network in the Central Mountain areas which suffered severe lack of strong motion data. The program was designed to collect records for understanding the ground motion characteristics of the mountainous areas.

The CMSMA instruments were installed in 1993. All stations were deployed along highways in the mountainous region (Fig. 13, left). It took special efforts to deploy and maintain the network during the past few years. Even with extra efforts, there are still too few stations to attain a high probability of recording strong motion of earthquakes. In 2003 the IES decided an action plan to increase the density of strong-motion stations and to upgrade the instrument. In the initial phase, 20 stations will be upgraded and 20 more new stations will be installed in the next few years.

2.7. Taipei Basin Downhole Array

In 1991 the Central Geological Survey (CGS) undertook a major research program on Taipei basin. One of the research tasks was to study the basin seismic effects and seismic wave propagation by IES. Under the project about ten downhole sites in Taipei basin were selected to install downhole accelerometers from surface down to different depths (Fig. 13, right). By the end of 2001, three downhole sites were still operational. In 2002, the CGS finished the research project and transferred the array to IES. After few months of rearrangement, installation of four additional downhole sites was under consideration.

3. STRONG MOTION NETWORKS DEPLOYED BY THE CENTRAL WEATHER BUREAU (CWB)

3.1. TSMIP Networks

The CWB in Taiwan has the official duty to monitor and report earthquakes in and near Taiwan. It was remarkable for CWB to have completed an extensive strong motion instrumentation program in 1996 at a cost of about US$40 million. The network, called TSMIP, has a total of 650 digital accelerographs deployed at the free-field sites (Fig. 14, left). Four academic groups in Taiwan are sponsored by the CWB, under a grant support for supervising monitoring and data collection of the CWB free-field accelerographs. About 85 of these free-field stations are connected to send real-time data to CWB headquarters for rapid earthquake reporting (Fig. 15, left). There are also 55 strong-motion arrays (typically of 30 seismic channels) in operation in buildings and bridges (Fig. 14, right). Each strong-motion array also has a three-component free-field accelerometer. Recently, a real-time network of accelerographs is installed across the Taiwan Railway System by CWB for safe train operations (Fig. 15, right).

The TSMIP uses several types of digital accelerographs. Most of them have ±2 g full scale, 200 or higher samples per second, and 16-bit or better resolution with up to 20-sec pre-event recording. All waveform data are stored as integers in unit of digital counts. Conversion to a physical unit, such as g, depends on the accelerograph model as shown in Table2.

The A-800 is a 12 bit digital data recorder and contains three Geotech Model S-110 accelerometers that are mounted orthogonally inside a cast aluminum base. Data is stored in CMOS static RAM using a non-distorting format without data compression.

The A-900 has all the features of the A-800 while being upgraded with a 16-bit ADC, new Geotech DC Force Balanced Accelerometers (FBA) and the capability to use a GPS receiver for its time synchronization source.

The A-900A is an improved version of the A-900 with modifications that provide for submersion to a limited depth, realtime serial data output and accelerometers with improved performance.

All three units have an internal timing accuracy of ±26 milliseconds per 24 hours over the operating temperature range when not connected to an external timing reference. The time can be initially set by hand via a PC connected to the unit. On the A-900 and A-900A, a GPS-1 receiver can be used to synchronize the internal timing to absolute UTC time to within ±1 millisecond.

Finally, the DEMUX program can be used to take a quick look at the data files and to convert them to binary files or to PC-SUDS format.

The Models IDS and IDSA accelerographs are 16-bit digital instruments with 250 samples per second sampling. They are manufactured by Terra Technology Corp. Recently, Terra Tech was acquired by Refraction Technology, Inc.

The Etna is a digital accelerograph with 18-bit resolution at 200 samples per second.

3.2. TSMIP Data Set for the 1999 Chi-Chi, Taiwan Earthquake

After the 1999 Chi-Chi earthquake, it was clear that CWB has recorded the most extensive set of strong-motion data for a major earthquake since strong-motion seismology began in the 1930s. An intensive effort was made to perform the first-order quality control and to make the data available as soon as possible. For practical reason, the free-field strong-motion data from the Chi-Chi earthquake was placed as computer readable files on a CD-ROM with explanatory notes. This pre-publication CD with a brochure (Lee et al., 1999) was distributed free-of-charge on December 9, 1999 to any requester for academic research. Later on an improved version of a total of 441 records were selected to represent the main shock strong motion.

In brief, the refined strong-motion records for the Chi-Chi earthquake contained the following significant improvements over the data in the pre-publication CD (Lee et al, 1999) released earlier.

- About 10,000 strong-motion data files recorded on September 20, 1999 were examined and over 200 additional data files were used to complete and extend the available strong-motion records for the Chi-Chi main shock.
- With cooperation from the manufacturers all known errors in the data files were corrected.
- The coordinates for 85% of the stations were "verified" or "confirmed".
- The site classification was added in the station list according to Lee et al. (2001).

- All stations equipped with a GPS timing device were examined. It was found that 39 accelerographs had absolute timing during Chi-Chi main shock.
- The first P-arrival time was picked (whenever possible) and the DC offset was removed by using the pre-event data, if available.
- Tables were provided on the CD so that user can select data by various criteria.
- A complete set of 441 accelerograms was stored in the CD so that users can view them conveniently.

3.3. The Taiwan Rapid Reporting and Warning System for Earthquakes

The desire for seismological observation in real time has long been recognized, and significant advances have been made during the past decade in many countries (Kanamori et al., 1997). The idea of an islandwide rapid earthquake reporting and warning system using existing telemetry in Taiwan was first proposed by T. L. Teng in early 1990s. The Taiwan Rapid Earthquake Information Release System (RTD) is based on a simple hardware/software design first introduced by Lee et al. (1989), and was subsequently improved and refined (Lee, 1994; Lee et al., 1996; Shin et al., 1996; Teng et al., 1997; Wu et al., 1997, 1998, 1999). The RTD system consists of 85 telemetered strong-motion accelerographs in Taiwan (Fig. 15, left). Digital signals are continuously telemetered to the headquarters of the CWB in Taipei via 4800-baud leased telephone lines. Each telemetered signal contains three-component seismic data digitized at 50 samples per second and at 16-bit resolution. The full recording range is ±2 g. The incoming digital data streams are processed by a computer program called XRTPDB (Tottingham and Mayle, 1994). Whenever the prespecified trigger criteria are met, the digital waveforms are stored in memory and are automatically analyzed by a series of programs (Wu et al., 1998). The results are immediately disseminated to emergency response agencies and interested individuals electronically in four ways, namely, by e-mail, World Wide Web, fax, and a pager system. The system is being upgraded to become an earthquake warning system which will provide about 10 to 20 seconds warning for cities 100 km beyond the epicenter of a major earthquake.

4. HIGHLIGHTS OF DIVERSE APPLICATIONS OF STRONG MOTION DATA

Usually rich and diverse sets of strong motion data from Taiwan earthquakes have become available for research and application in a wide range of topics and fields. Here we present highlights of a few applications of strong motion data obtained by the SMART1, LLSST, SMART2, HLSST, TSMIP free-field network and structural arrays. These include studies of ground motion attenuation by the TSMIP free-field data (Fig. 16)(Chao, 2001), nonlinear soil amplification effects by the LLSST downhole array data (Fig. 17)(Lee, 1998), permanent changes in structural response of a damaged building due to the Chi-Chi earthquake by the TSMIP structural array data (Fig. 18)(Wen, 2002), and tracking of the Chelungpu fault rupture propagation by the SMART2 data (Fig. 19)(Huang, 2001).

ACKNOWLEDGMENTS

This research was supported by National Science Council under Grant NSC92-2625-Z-008-011, and by the Ministry of Education under Grant 91-N-FA07-7-4.

REFERENCES

1. Central Weather Bureau, 2004, Disastrous earthquakes in Taiwan in twentith century (in Chinese) (July 31, 2004); http://www.cwb.gov.tw/V4/index.htm.
2. Chao, H. L., 2001, Attenuation relations for strong ground motion in Taiwan and rapid mapping of spatial distributions of ground motion response spectra from the Chi-Chi earthquake sequence, Master thesis, National Central University, Taiwan (in Cineses with abstract in English).
3. Huang, B. S., 2001, Evidence for azimuthal and temporal variations of the rupture propagation of the 1999 Chi-Chi, Taiwan earthquake from dense seismic array observations, *Geophys. Res. Lett.* **28:**3377-3380.
4. Institute of Earth Sciences of Academia Sinica, 2004, Strong motion Networks in Taiwan (July 31, 2004); http://www.earth.sinica.edu.tw/~smdmc.
5. Kanamori, H., Hauksson, E., and Heaton, T., 1997, Real-time seismology and earthquake hazard mitigation, *Nature* **390:**461-464.
6. Lee, C. P., 1998, Analysis of nonlinear site response using LSST down-hole accelerograph array data, Master thesis, National Central University, Taiwan (in Cineses with abstract in English).
7. Lee, C. T., Cheng, C. T., Liao, C. W., and Tsai, Y. B., 2001, Site classification of Taiwan free-field strong-motion stations, *Bull. Seism. Soc. Am.* **91:**1283-1297.
8. Lee, W. H. K., (Editor) 1994, *Realtime Seismic Data Acquisition and Processing, IASPEI Software Library, Volume 1 (2nd Edition),* Seismological Society of America, El Cerrito, CA, (285 pp. and 3 diskettes).
9. Lee, W. H. K., Tottingham, D. M., and Ellis, J. O., 1989, Design and implementation of a PC-based seismic data acquisition, processing, and analysis system, *IASPEI Software Library* **1:**21-46.
10. Lee, W. H. K., Shin, T. C., and Teng, T. L., 1996, Design and implementation of earthquake early warning systems in Taiwan, *Proc. 11th World Conf. Earthq. Eng., Paper No. 2133.*
11. Lee, W. H. K., Shin, T. C., Kuo, K. W., and Chen, K. C., 1999, *CWB Free-Field Strong-Motion Data from the 921 Chi-Chi Earthquake: Volume 1. Digital Acceleration Data, Pre-Publication CD,* Central Weather Bureau, Taipei, Taiwan.

12. Shin, T. C., Tsai, Y. B., and Wu, Y. M., 1996, Rapid response of large earthquake in Taiwan using a realtime telemetered network of digital accelerographs, *Proc. 11th World Conf. Earthq. Eng., Paper No. 2137*.
13. Teng, T. L., Wu, Y. M., Shin, T. C., Tsai, Y. B., and Lee, W. H. K., 1997, One minute after: strong motion map, effective epicenter, and effective magnitude, *Bull. Seism. Soc. Am.* **87:**1209-1219.
14. Tottingham, D. M., and Mayle, A. J., 1994, User manual for XRTPDB. *IASPEI Software Library* **1** (2nd Edition)**:**255-263.
15. Wen, C. L. (2002). Study of the structure oscillation when earthquakes occur by using response spectral ratio, Master thesis, National Central University, Taiwan (in Chinese).
16. Wu, Y. M., Shin, T. C., Chen, C. C., Tsai, Y. B., Lee, W. H. K., and Teng, T. L., 1997, Taiwan rapid earthquake information release system, *Seism. Res. Lett.* **68:**931-943.
17. Wu, Y. M., Chen, C. C., Chung, J. K., and Shin, T. C., 1998, An automatic phase picker of the real-time acceleration seismic network, *Meteorol. Bull., Central Weather Bureau, Taipei* **42:**103-117 (in Chinese).
18. Wu, Y. M., Chung, J. K., Shin, T. C., Hsiao, N. C., Tsai, Y. B., Lee, W. H. K., and Teng, T. L., 1999, Development of an integrated seismic early warning system in Taiwan – case for Hualien area earthquakes, *Terres. Atmos. Ocean. Sci.* **10:**719-736.

Table 1. Strong motion instrumentation networks in Taiwan.

IES Networks and Arrays	CWB Networks and Arrays
- SMA1 (1976-1999)	- TSMIP (1991-)
- SMART1 (1980-1991)	• Free field statons
- LLSST (1985-1990)	• Structural arrays
- SMART2 (1990-)	• RTD network
- HLSST (1993-)	• Taiwan railroad system
- CMSMA (1993-)	
- Taipei basin downhole array (1993-)	

Table 2. Conversion to a physical unit in different accelerograph models.

Model	Numbers deployed	Full-scale	Data Stored as	Digital Counts
A800	46	±1 g	12-bit integer	±2048
A900 & A900A	393	±2 g	16-bit integer	±32768
IDS & IDSA	163	±2 g	16-bit integer	±32768
ETNA & K2	38	±2 g	24-bit integer	±8388608

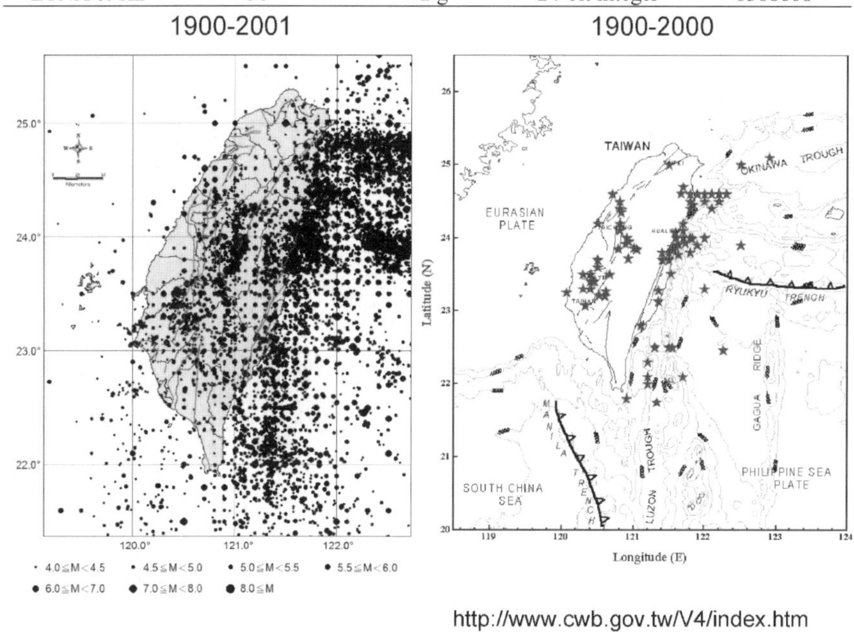

Figure 1. Seismicity in Taiwan from 1900 to 2001 (left). Large earthquakes in Taiwan from 1900 to 2000 (right).

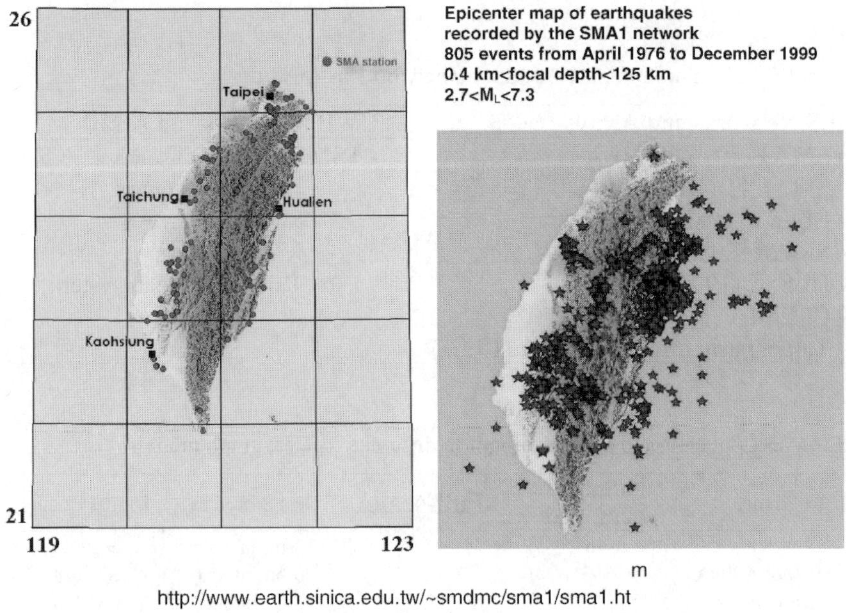

Figure 2. The SMA1 network and the recorded earthquakes.

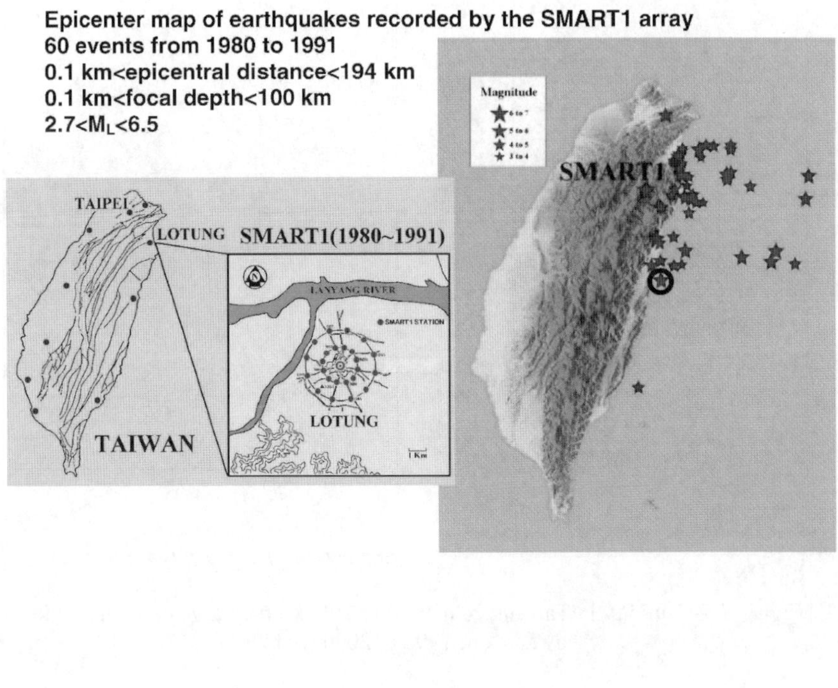

Figure 3. The SMART1 array and the recorded earthquakes.

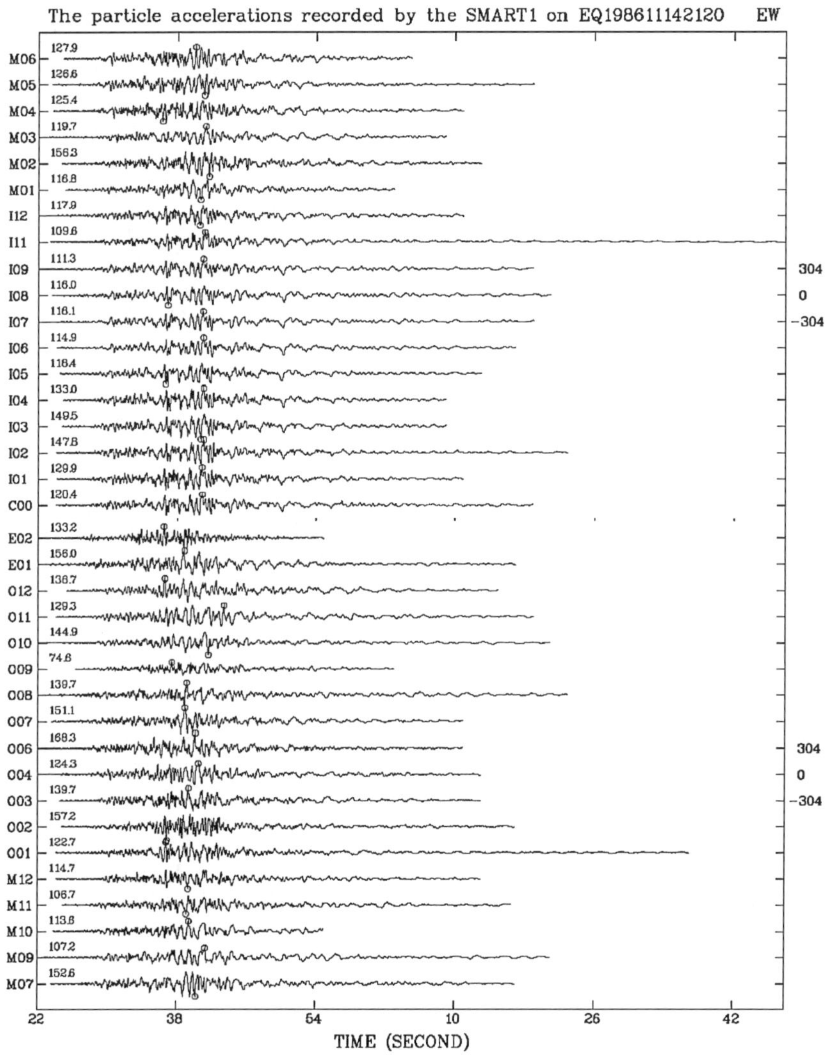

Figure 4. Sample records of SMART1 array from the M_L6.5 earthquake of November 14, 1986.

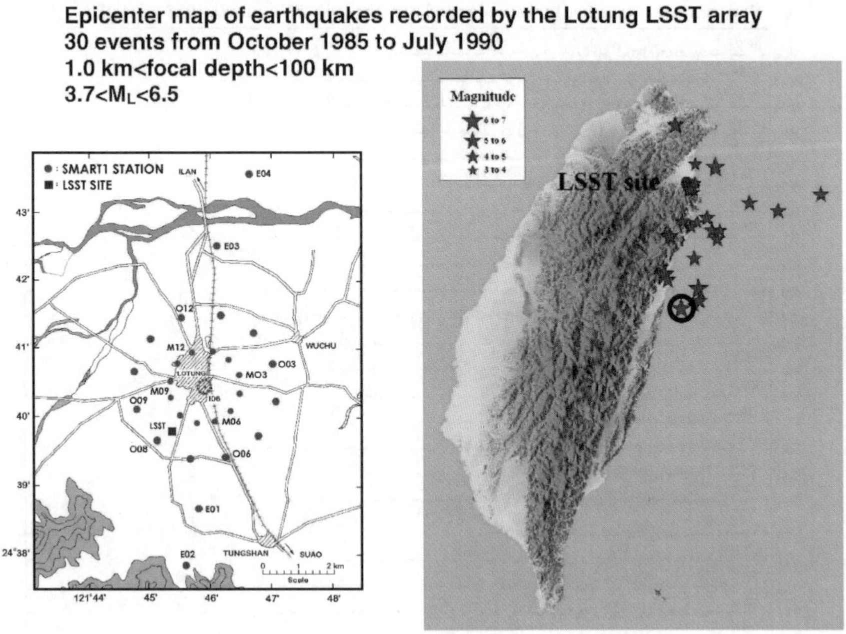

http://www.earth.sinica.edu.tw/~smdmc/llsst/llsst.htm

Figure 5. The Lotung LSST array and the recorded earthquakes.

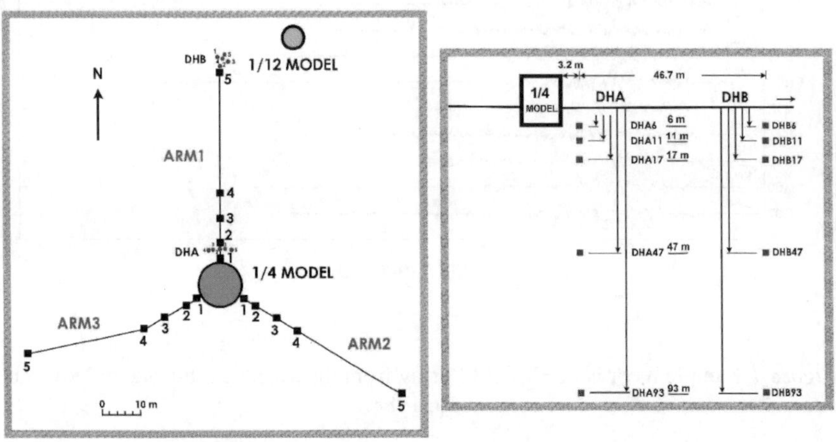

http://www.earth.sinica.edu.tw/~smdmc/llsst/llsst.htm

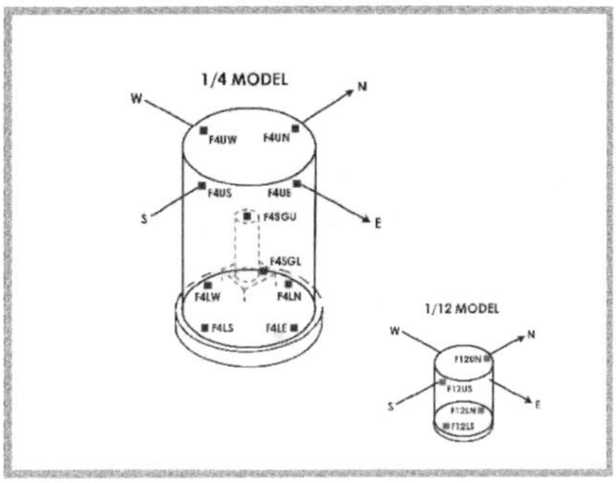

http://www.earth.sinica.edu.tw/~smdmc/llsst/llsst.htm

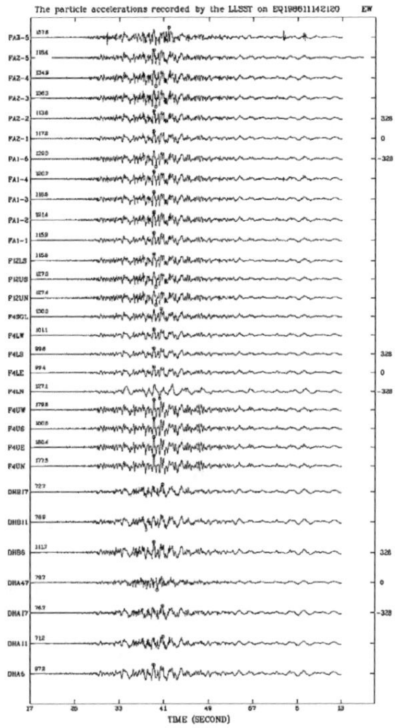

Figure 7. Structure response accelerometers of the LLSST array (left) and sample records of the M_L6.5 earthquake of November 14, 1986 (right).

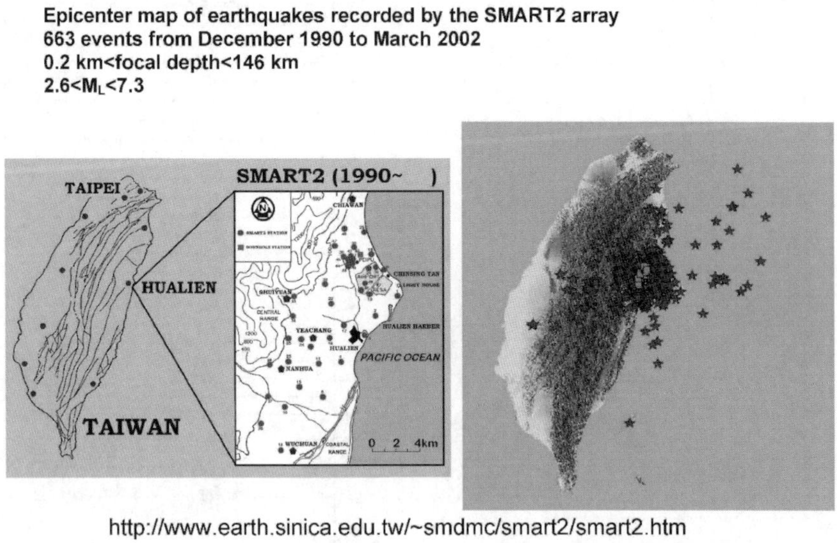

Figure 8 The SMART2 array and the recorded earthquakes.

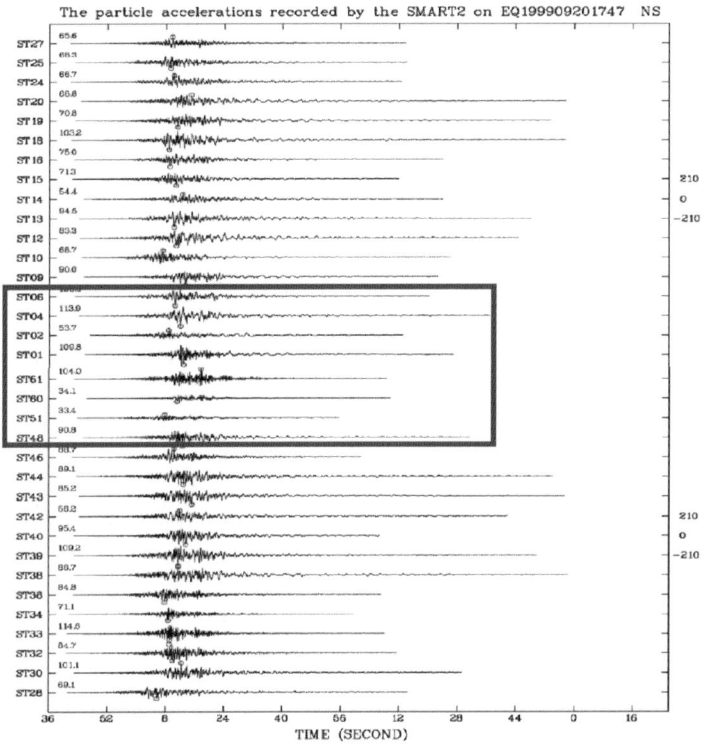

Figure 9. The records of the 1999 Chi-Chi earthquake by SMART2 array.

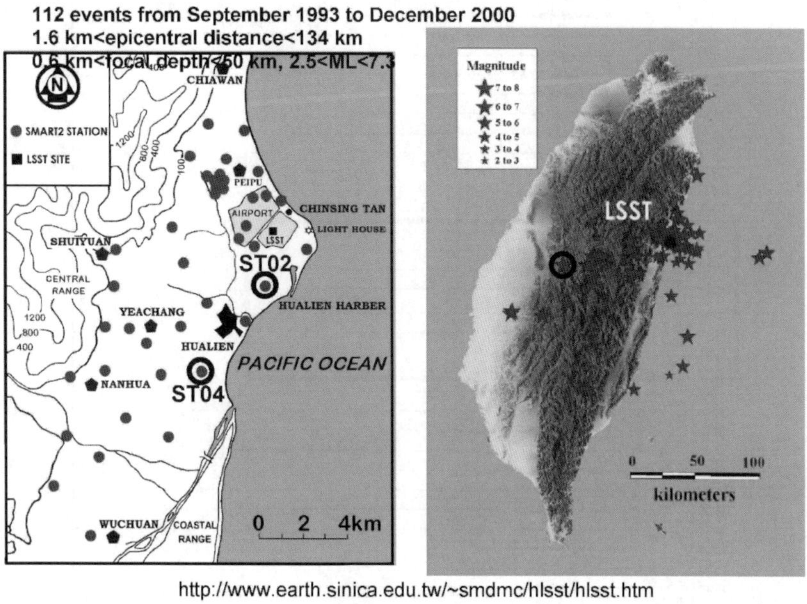

Figure 10. The HLSST array and the recorded earthquakes.

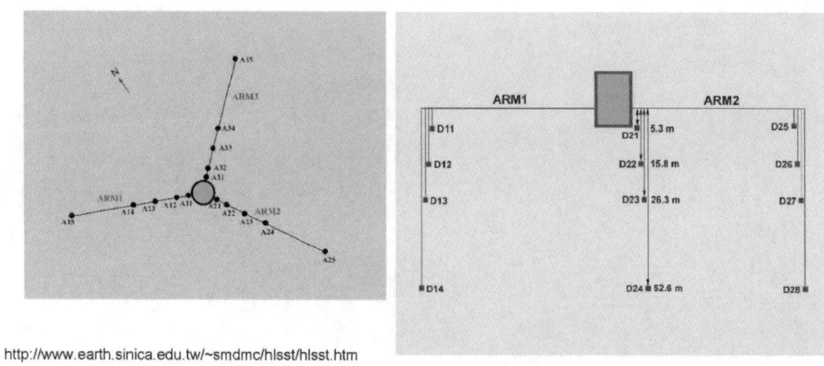

Figure 11. Surface (left) and downhole (right) accelerometers of the HLSST array.

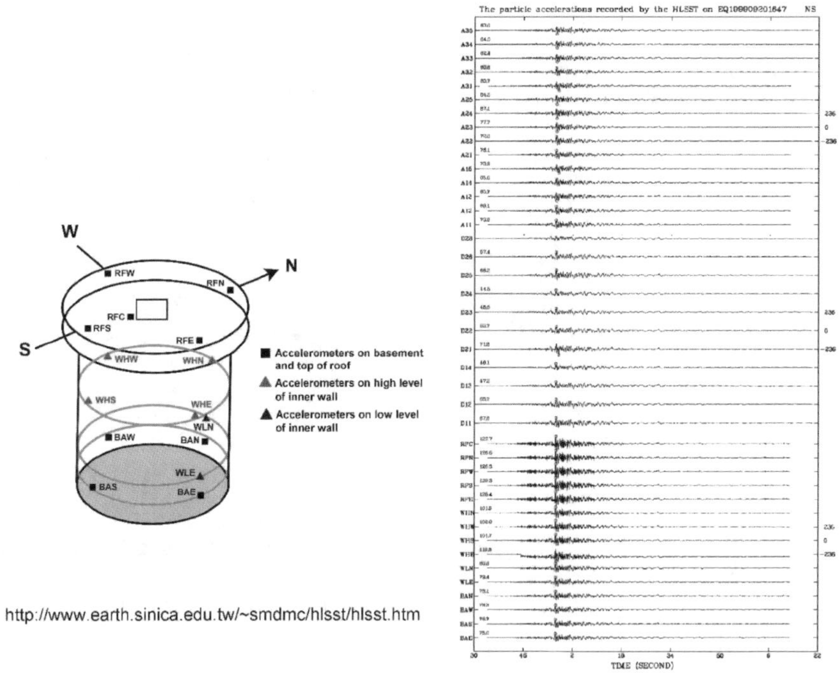

Figure 12. Structure response accelerometers of the HLSST array (left) and sample records of the Chi-Chi earthquake (right).

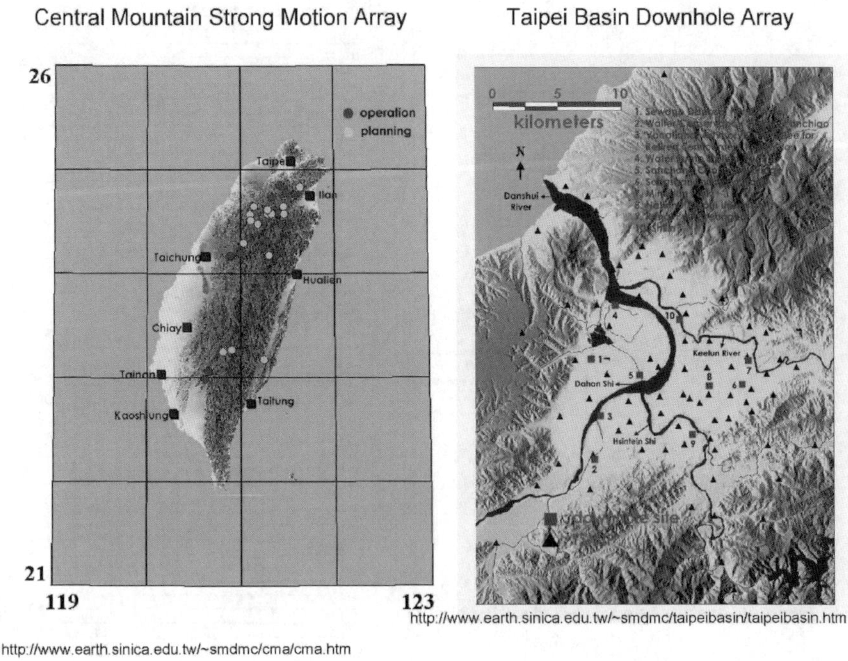

Figure 13. The central mountain strong motion array (left) and Taipei basin downhole array (right).

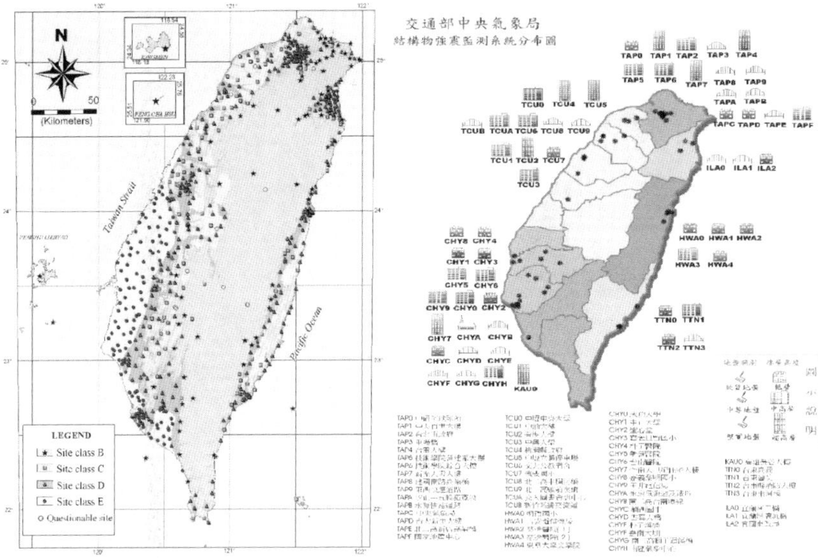

Figure 14. The free-field stations (left) (Lee et al., 2001) and structural arrays of TSMIP.

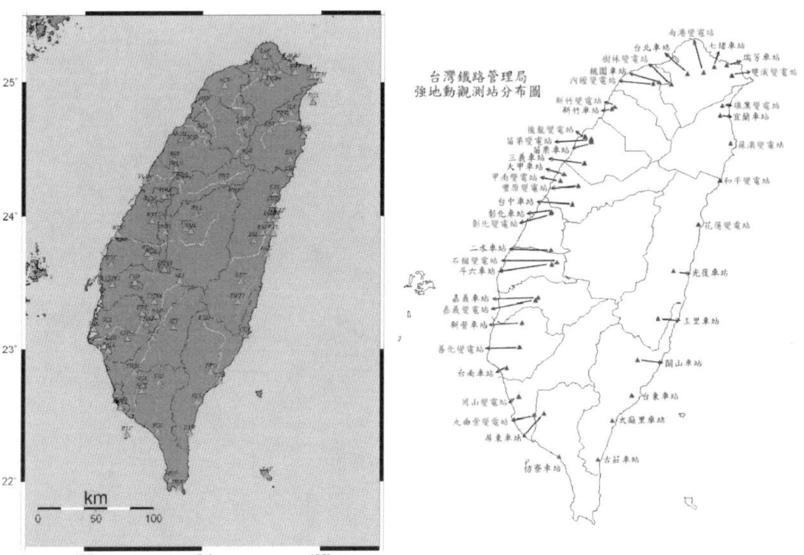

Figure 15. The RTD network (left) and Taiwan railroad system (right).

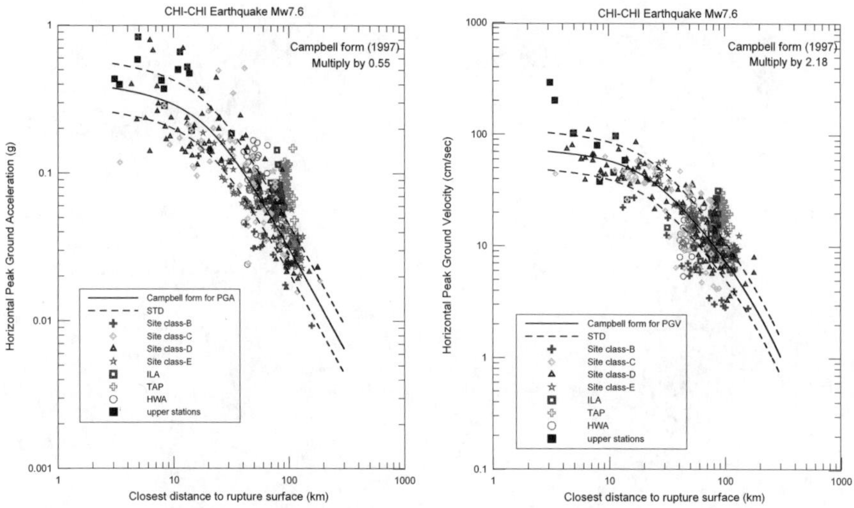

Figure 16. The ground motion attenuation relationship of PGA (left) and PGV (right) by the TSMIP free-field data for the Chi-Chi earthquake (Chao, 2001).

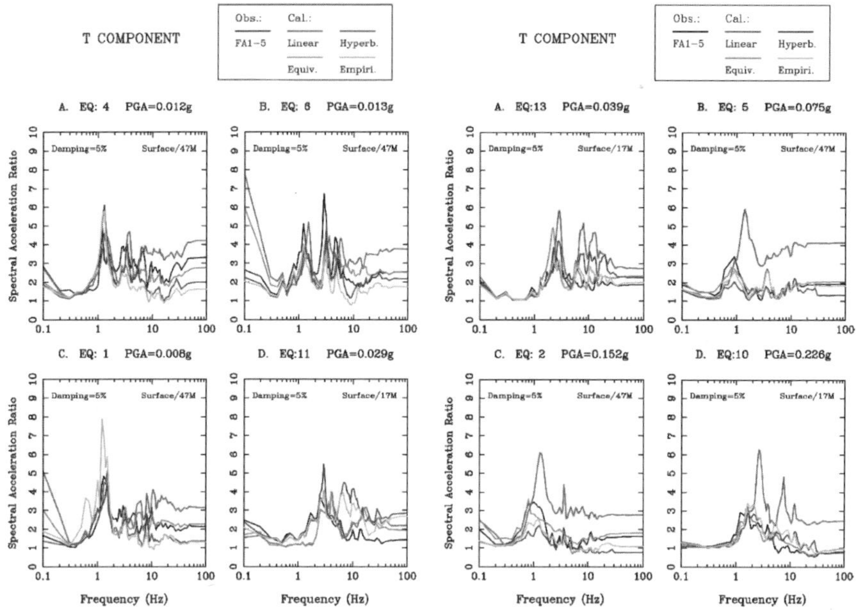

Figure 17. Nonlinear soil amplification effects by the LLSST downhole array data (Lee, 1998).

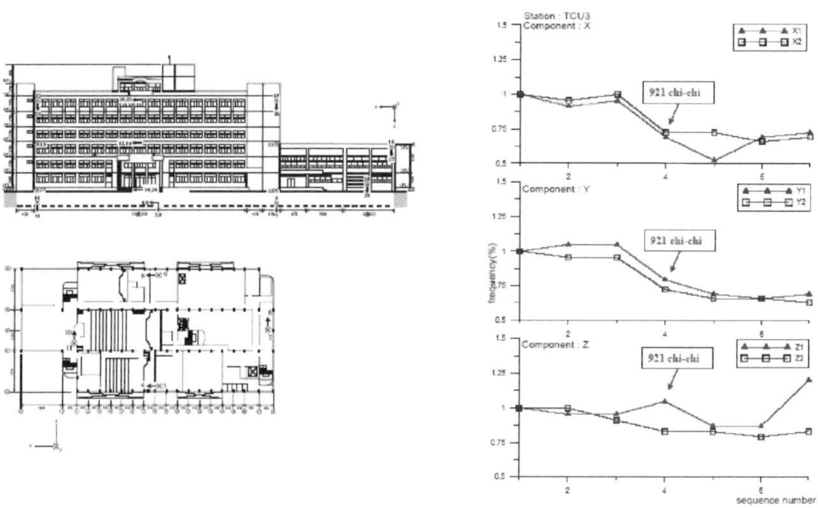

Figure 18. Permanent changes in structural response of a damaged building due to the Chi-Chi earthquake using the TSMIP structural array data (Wen, 2002).

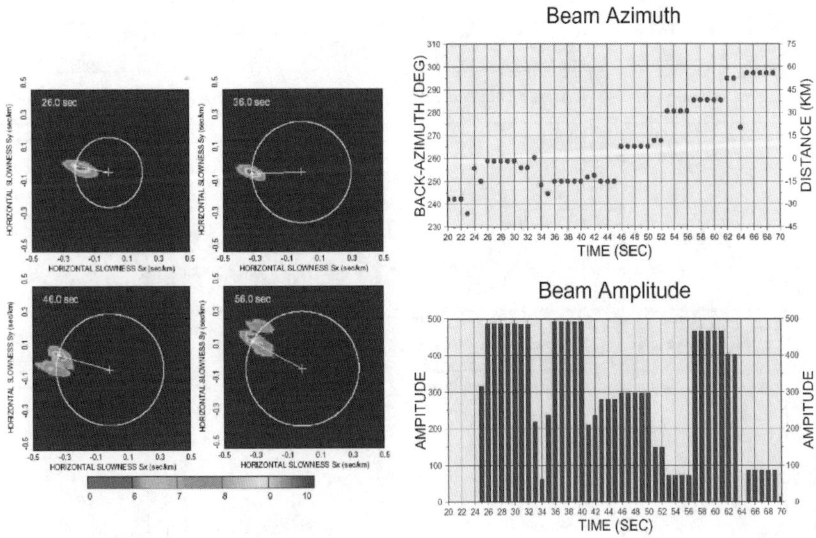

Figure 19. Tracking of the Chelungpu fault rupture propagation by the SMART2 data (Huang, 2001).

STRONG-MOTION DATA ACQUISITION, PROCESSING AND UTILIZATION WITH APPLICATIONS TO ISTANBUL STRONG MOTION NETWORK

M. Erdik[1], Y. Fahjan[1] and E. Durukal[1]
[1] *Boğaziçi University, Istanbul, Turkey*

Abstract: The strong ground motion network operated by Boğazici University in and around İstanbul has been described, including the rapid response and early warning system, structural instrumentation, instrument calibration and, processing, availability and utilization of the data.

Key words: ground motion; Istanbul; rapid response; early warning; structural instrumentation; calibration

1. INTRODUCTION

There exist some 361 strong motion accelerographs in Turkey. About 200 digital strong motion accelerographs are owned and operated by Kandilli Observatory and Earthquake Research Institute (KOERI) of Bogazici University as dense urban network in and around Istanbul (Rapid response and early warning system, structural instrumentations) and North Cyprus.

2. RAPID RESPONSE AND EARLY WARNING SYSTEM

Potential impact of large earthquakes on urban societies can be reduced by timely and correct action after a disastrous earthquake. The rapid response systems implemented in California (USGS, Caltech and CDMG TriNet ShakeMaps - http://www.trinet.org/shake/, Caltech-USGS Broadcast of Earthquakes (CUBE) System http://www.gps.caltech.edu/~bryant/cube.html,

P. Gülkan and J.G. Anderson (eds.), Directions in Strong Motion Instrumentation, 279–294.

UC Berkeley Seismological Laboratory and USGS Menlo Park (REDI) www.seismo.berkeley.edu/seismo/redi), Taiwan (Earthquake Rapid Reporting System of CWB http://www.earth.sinica.edu.tw/cdr/IASPEI/data/cwb/rapid.html) and Japan (Real-time Earthquake Assessment Disaster System in Yokohama -READY - http://www.city.yokohama.jp/me/bousai/dppc/handout-e.html) are current examples. Earthquake Early Warning in urban and industrial areas allows for clean emergency shutdown of systems susceptible to damage such as power stations, transportation, computer centers and telephone systems. Early Warning Systems, currently in operation in Bucharest (http://www.infp.ro/publications/ews.htm) and Mexico (http://www.gfz-potsdam.de/ewc98/abstract/espinosa.html), have the capability of issuing an earthquake early warning in advance, by relying on significant distances between the source and the populated urban regions. A general treatment of the earthquake early warning and rapid response systems exist in Kanamori[1]. Espinosa-Aranda[2], Gee[3] and Hauksson[4] provides respectively the descriptions of the Seismic Alert System in Mexico City, REDI and TriNet Projects.

In 2002 a dense urban strong motion recording network that encompasses 110 stations for rapid response and early warning purposes were installed by KOERI. This network and its functions are called Istanbul Earthquake Rapid Response and Early Warning System[5] (IERREWS). One hundred (100) of the strong motion recorders are stationed in dense settlements in the Metropolitan area of Istanbul in dial-up mode for Rapid Response information generation (Figure 1). Ten (10) of the strong motion stations are sited at locations as close as possible to the Great Marmara Fault in on-line data transmission mode to enable Earthquake Early Warning (Figure 2). The remaining 40 strong motion recorder units will be placed on critical engineering structures in addition to the already instrumented structures in Istanbul (http://www.koeri.boun.edu.tr/ depremmuh/stronmotion.htm). All together this network and its functions is called Istanbul Earthquake Rapid Response and Early Warning System (IERREWS). The system is designed and operated by Boğazici University with the logistical support of the Governorate of İstanbul, First Army Headquarters and İstanbul Metropolitan Municipality. The construction of the system is realized by the GeoSig Inc. (http://www.geosig.com) and Electrowatt-Ekono (www.ewe.ch) consortium. Communications are provided by AVEA GSM service provider.

IERREWS consists of the following components:

1. Monitoring system composed of various sensors,
2. Communication link (off-line for the Rapid Response and on-line for the Early Warning) that transmits data from the sensors to computers,
3. Data processing facilities that converts data to information, and
4. System that issues and communicates the rapid response information and early warning.

2.1. Rapid Response System

The Rapid Response part of the IERREWS has the objective of providing:

Reliable information for accurate, effective characterization of the shaking and damage by other rapid post-earthquake maps (Shake, Damage and Casualty maps) for rapid response;

Recorded motion for post-earthquake performance analysis of structures;

Empirical basis for long-term improvements in seismic microzonation, seismic provisions of building codes and construction guidelines; and

Seismological data to improve the understanding of earthquake generation at the source and seismic wave propagation.

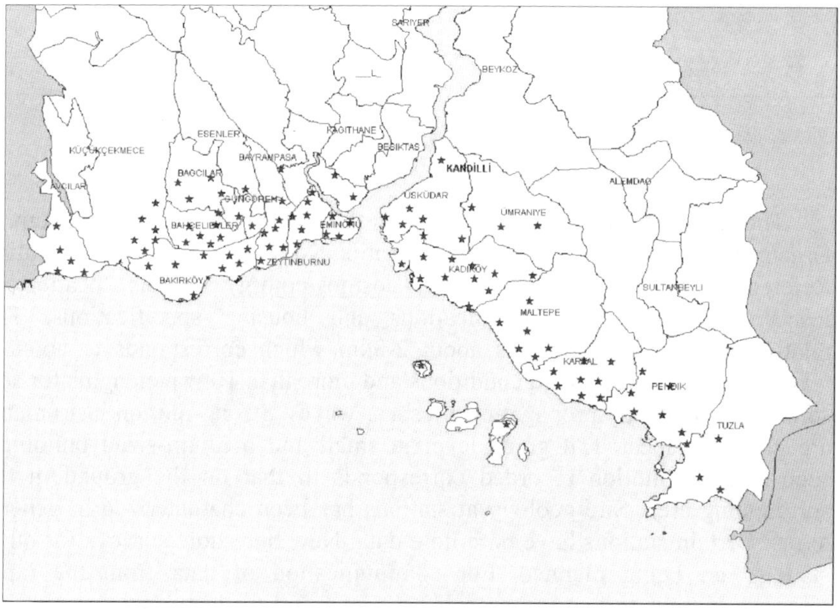

Figure 1. Rapid Response Stations

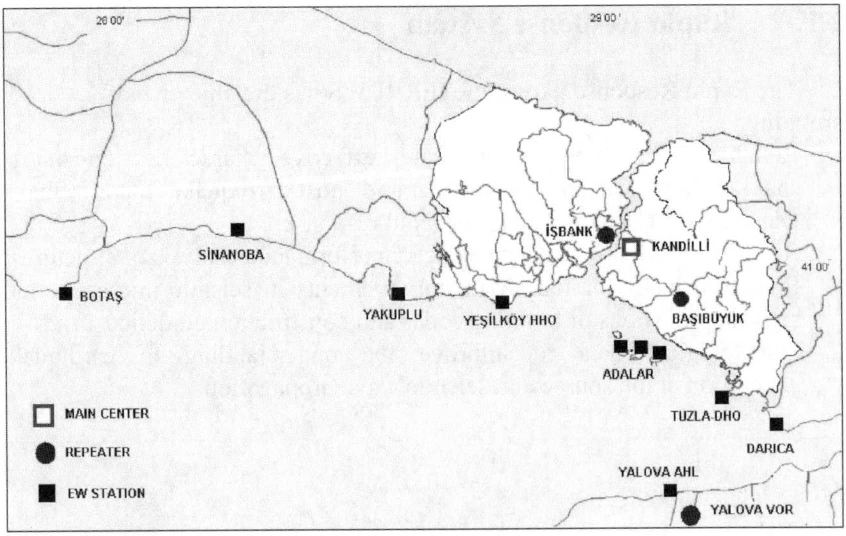

Figure 2. Early Warning Stations

The Rapid Response System satisfies the COSMOS (The Consortium of Organizations for Strong-Motion Observation Systems) Urban Strong-Motion Reference Station Guidelines (www.cosmos-eq.org) for the location of instruments, instrument specifications and housing specifications. The relative instrument spacing is about 2-3km which corresponds to about 3 wavelengths in firm ground conditions and more than 10 wavelengths for soft soils for horizontally propagating 1 s shear waves. Strong-motion instruments are generally located at grade level in small and medium-sized buildings, such that the motion recorded corresponds to that on the ground in the surrounding area. Site geology at stations has been characterized in general terms. Certain stations have bore-hole data. New bore-hole surveys for other stations are being planned. For communication of data from the rapid response stations to the data processing center and for instrument monitoring a reliable and redundant GSM communication system (backed up by dedicated landlines and a microwave system) is used, on the basis of a protocol agreement with the AVEA GSM Service provider. In normal times the rapid response stations are be interrogated (for health monitoring and instrument monitoring) on regular basis. After triggered by an earthquake, each station will process the streaming three-channel strong motion data to yield the spectral accelerations at specific periods, 12Hz filtered peak ground acceleration and peak ground velocity and will send these parameters in the form of SMS messages at every 20s directly to the main data center through the GSM communication system.

The main data processing center is located at the Department of Earthquake Engineering - Kandilli Observatory and Earthquake Research

Institute of Boğazici University (KOERI-BU). A secondary center located at the Seismological Laboratory of the same Institute serves as a redundant secondary center that can function in case of failure in the main center. Shake, damage and casualty distribution maps will be automatically generated at the data centers after the earthquake and communicated to the end users within 5 minutes. Full-recorded waveforms at each station can be retrieved using GSM and GPRS modems subsequent to an earthquake.

For the generation of Rapid Response information the methodologies essentially rely on the Istanbul Earthquake Risk Assessment Study (http://www.koeri.boun.edu.tr/depremmuh/EXEC_ENG.pdf) conducted bu the Department of Earthquake Engineering of Bogazici University . For the computation of input ground motion parameters, spectral displacements obtained from the SMS messages sent from stations will are interpolated to determine the spectral displacement values at the center of each geo-cell using two-dimensional splines. The earthquake demand at the center of each geo-cell is computed using these spectral displacements. The instrumental intensity at each the center of each geo-cell is computed as a function of short-period spectral acceleration. Using the response spectra and the instrumental intensities the building damage and the casualties are computed separately by using the spectral-displacement based and intensity based fragility curves. The computations are conducted at the centers of a 0.01 x 0.01 grid system comprised of geo-cells (1120 m x 830 m) size. The building inventories (in 24 groups) for each geocell together with their spectral displacement and intensity based fragility curves are incorporated in the software. The casualties are estimated on the basis of the number of collapsed buildings and degree of damage. Example of building damage map that results from a randomly simulated strong motion data is provided in Figure 3.

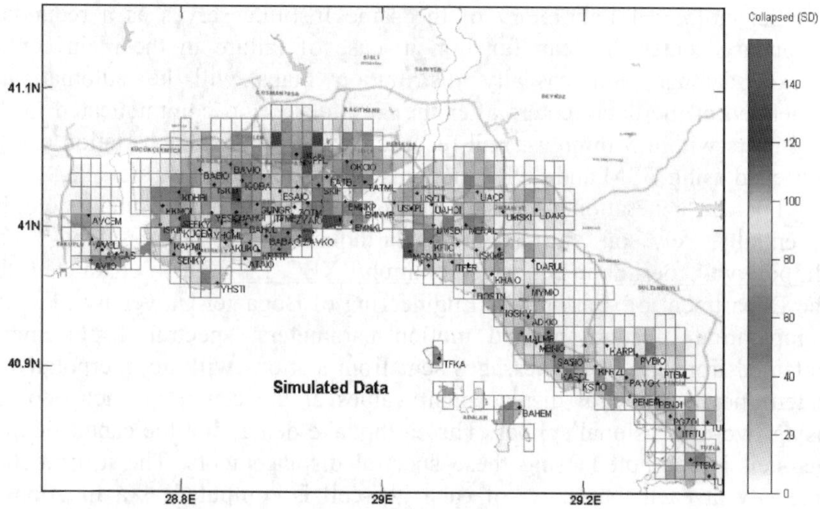

Figure 3. Distribution of buildings damaged beyond repair based on randomly simulated earthquake ground motion

For transmission of the Rapid Response information to the concerned agencies (Istanbul Governorate, İstanbul Municipality and First Army Headquarters) digital radio modem and GPRS communication systems are used. The strong motion data, shake maps and the building damage maps will be made available on the Internet immediately after an earthquake (http://www.koeri.boun.edu.tr/depremmuh/EWRR/EWRRMain.htm). The waveforms will also be catalogued for a "Green's Function" database to be used in earthquake simulations. So far (July, 2004) the system has been exposed to only small magnitude earthquakes. Figure 4 illustrates the accelerations traces associated with May 16, 2004 M4.5 earthquake in eastern Marmara Sea (Figure 8).

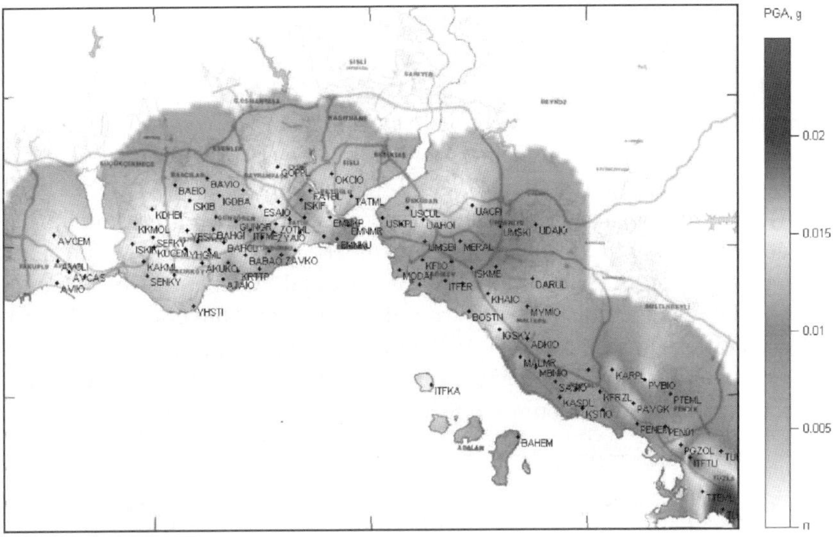

Figure 4. Distribution of accelerations associated with May 16, 2004 M4.5 earthquake in eastern Marmara Sea (Earthquake location in Figure 8)

2.2. Early Warning System

The Early Warning part of the IERREWS ten strong motion stations were located as close as possible to the Great Marmara Fault in "on-line" mode. Continuous telemetry of data between these stations and the main data center is realized with digital spread spectrum radio modem system involving repeater stations selected in the region. Considering the complexity of fault rupture and the short fault distances involved, a simple and robust Early Warning algorithm, based on the exceedance of specified threshold time domain amplitude levels is implemented. The band-pass filtered accelerations and the cumulative absolute velocity (CAV-time integral of the absolute acceleration) are compared with specified threshold levels. When any acceleration or CAV (on any channel) in a given station exceeds specific selectable threshold values it is considered a vote. Whenever we have 2 or 3 (selectable) station votes within selectable time interval, after the first vote, the first alarm is declared. The early warning information (consisting three alarm levels) will be communicated to the appropriate servo shut-down systems of the recipient facilities, which will automatically decide proper action based on the alarm level. Depending on the location of the earthquake (initiation of fault rupture) and the recipient facility the alarm time can be as high as about 8s.

3. STRUCTURAL MONITORING

Fatih Sultan Mehmet Suspension Bridge: Immediately following the Marmara and Duzce 1999 earthquakes, Boğaziçi University Earthquake Engineering Department in connection with General Directorate of State Highways decided to install strong motion instrumentation on the FSM suspension bridge in Istanbul. The monitoring system encompassing 12 tri-axial accelerometers is installed in November 2001 (Figure 5). The data, digitized at the source, is monitored by two parallel PC units operating under the SEISLOG data acquisition software, developed by University of Bergen. The acceleration data obtained is transmitted on-line to the analysis center at Department of Earthquake Engineering of Boğaziçi University. The modal vibration parameters are monitored using the measured response.

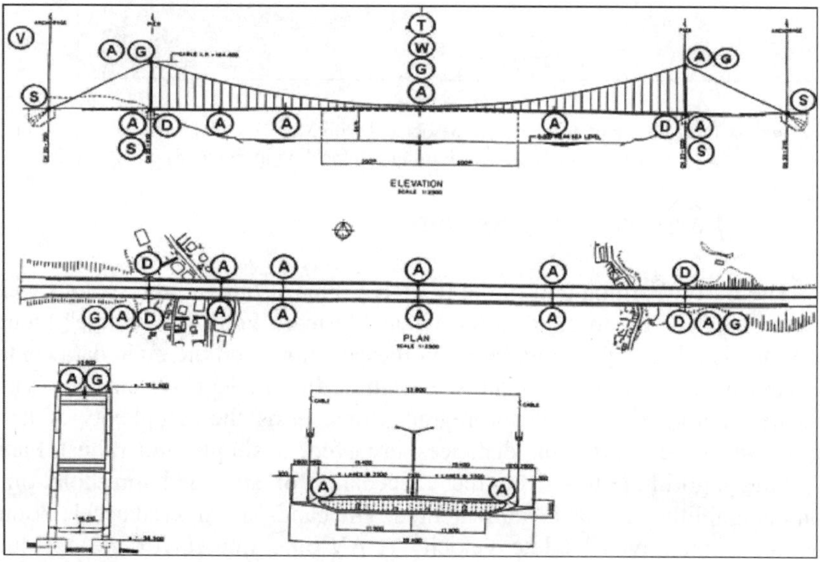

Figure 5. Strong Motion Instrumentation on Fatih Sultan Mehmet Bridge

Boğazici Suspension Bridge: In 1983 a structural health monitoring system was installed by General Directorate of State Highways, consisting of four vertical, three transverse and two longitudinal acceleration transducers at the mid-span, one-fourth of the span and at the top of east tower; four seismometers at the foundation of towers and cable anchorages; two relative displacement transducers at the western expansion joint and; one optical displacement transducer at the mid-span. The response of these transducers, digitized at the source is transferred to central recording/monitoring system. This system is currently non-functional. 10 temporary strong motion recorders installed in the deck temporarily by Department of Earthquake Engineering of

Boğazici University will be operated pending the overhaul (or replacement) of the existing monitoring system.

Hagia Sophia Museum: A network of strong motion accelerometers consisting of nine interconnected Kinemetrics SSA-2 units has been operational in the structure since August 1991 as part of the efforts for the determination of its earthquake worthiness. The configuration of the accelerometers is shown in Figure 6. The instruments cover three levels. One instrument is at the ground level; four of them are located at the cornice level next to the springing points of the four main arches; and four instruments are at the dome base level on the crowns of the main arches[6].

Suleymaniye Mosque: An accelerometer system consisting of eight individual Kinemetrics SSA-2 units operated in the Süleymaniye mosque between 1994 and 1999. The system was replaced by an interconnected array of eight GeoSig recorders in 1999. The locations of the accelerometer units in Süleymaniye can be seen in Figure 7. The positioning of the instruments is similar to the Hagia Sophia array except that there is no instrument at the ground level[6].

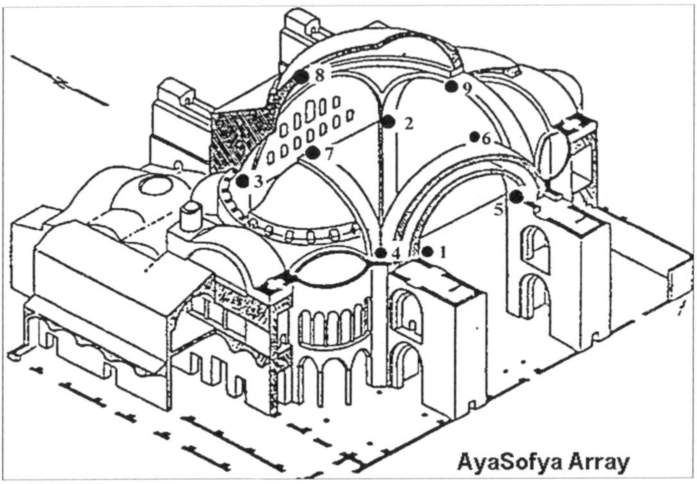

Figure 6. Strong Motion Instrumentation on Hagia Sophia Museum

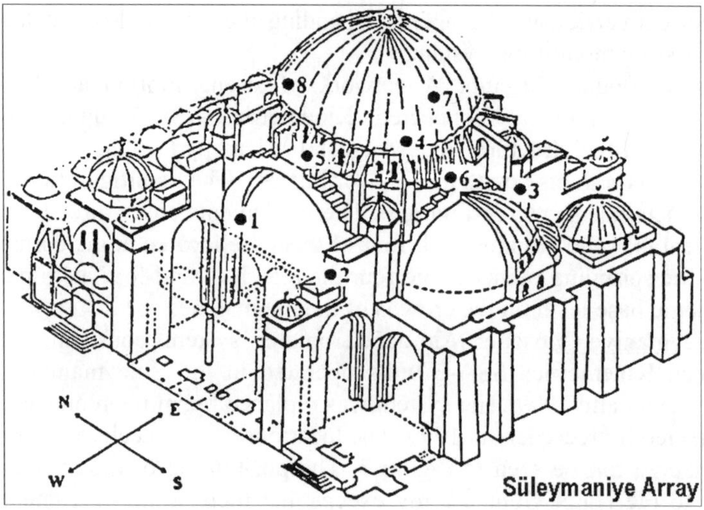

Figure 7. Strong Motion Instrumentation on Suleymaniye Mosque

Enron Power Plant: Earthquake monitoring network system for *Enron-Trakya Elektrik Power Plant* facilities includes eight on-line strong ground motions recorders distributed at critical locations within the industrial facility. The recorders are connected on-line to the main processing center (PC) to send near real-time data packets. GeoDAS software is used for the data acquisition and analysis. To synchronize the timing in the system, Garmin type GPS is utilized. The network monitors the earthquake response and could also give warning before the strong motion from a large earthquake arrives. The thresholds and alarm levels are selected according to operational design levels of the facility and the emergency measures.

Is Bank Tower: Earthquake monitoring and alert systems for ISKULE building had been designed and installed to monitor the structural response and to generate earthquake alarm signal for the building facilities and emergency system. For this purpose, four GSR-18 strong ground motion stations are utilized. The operation system is similar to that employed for the Enron Power Plant.

4. INSTRUMENT TESTING AND CALIBRATION

To increase the system reliability all of accelerometers and the strong motion recorder units used in KOERI instrumentation were tested by using tilt, pulse, noise and dynamic test procedures. In tilt tests each component of the sensor is tilted 90 degrees and the measured acceleration is compared with the correct 1g value. Accelerometers were also excited by an electronic test pulse, consisting of a 1g square wave signal. Test measurements are compared with the signal shape. The electronic noise levels of the accelerographs are measured under zero level excitation at an (almost) ambient vibration free location and the results are compared with instrument specifications.

For calibration purposes the accelerometers and accelerographs are tested under prescribed dynamic excitation using special electro-dynamic shaker. The shaker is a long stroke, electrodynamic actuator coupled to a horizontal table with bearing to ensure very low noise and cross-axis motion. A specially calibrated piezo-electric acceleration sensor is mounted on the shake table to provide the reference measurement. The driver unit and guide bar assembly are mounted on a common rigid base, insuring correct alignment of all moving parts. For dynamic excitation a sine wave test signal with 0.5 g amplitude and a sweep frequency between 2 to 50 Hz is employed. In the first stage of dynamic testing, the response of each component of the accelerometer is compared with the reference accelerometer. The measurement data obtained from the units was compared with the recording from reference accelerometer and calibration status was assessed. For 160 strong motion accelerographs tested we have found problems in about 4% of the seismometers and 16% of the recorders.

5. DATA AVAILIBILITY

After 1999 Kocaeli and Duzce earthquakes KOERI has compiled and disseminated all strong motion data[7]. Strong-motion records included in this data set are obtained from the stations operated by Earthquake Engineering Department of Bogazici University, Kandilli Observatory and Earthquake Research Institute (KOERI), Istanbul Technical University (ITU), Earthquake Research Department of General Directorate of Disaster Affairs (ERD) and U.S. Geological Survey (USGS). KOERI contributed 1161 records from 392 earthquakes with a magnitude range of Mw7.4 - Md2.6 recorded at 42 strong-motion stations between years 1994 - 2001. 13 of these stations are permanent stations installed in the Marmara Region. 25 stations were installed after the 17 August 1999 earthquake in Adapazari, Duzce, and Yalova as part of aftershock studies. KOERI and ITU records were obtained through the use of GeoSig GSR-16 and Kinemetrics SSA-12 accelerographs with no GPS timing. Array microtremer and SASW methods were used to determine the soil conditions (NEHRP site classifications) at most of the KOERI stations. ITU contributed 27 records from 12 earthquakes with a magnitude range of Mw7.4 - Md3.7 recorded at five strong-motion stations. The first stage of processing for the other records was the elimination of non-earthquake signals caused by voltage fluctuation, quarry explosions, vehicle-generated waves, etc. The second stage was the record-event (earthquake) matching process. The inaccurate timing on the recorders made this stage very difficult. After matching the main shocks, the time shifts of all the stations were corrected in order to facilitate the event identification. Since no other processing has been applied to these records, they qualify as an uncorrected data set. The resulting database has been disseminated in CD format by KOERI.

Celebi[8] also provided a collection of all available strong ground motion records from main shock and aftershocks of August 17, 1999 (Mw=7.4) Kocaeli and November 12, 1999 Düzce (Mw=7.2) earthquakes in Turkey (http://geopubs.wr.usgs.gov/open-file/of01-163). The records contained in this compilation were obtained from permanent stations of Earthquake Research Institute- Ministry of Public Works and Resettlement, Kandilli Observatory and Earthquake Research Institute - Boğazici University and Istanbul Technical University and the temporary stations of USGS, LDEO and other institutions. All the data are presented in ASCII format with the original header information as provided by the contributing agency. There is no event association and information for the aftershocks.

Most of the database is also included in the CD volume of European Strong Motion Database (http://www.isesd.cv.ic.ac.uk). Selected events from this database can be reached in PEER Strong Motion Database (http://peer.berkeley.edu/smcat/) and COSMOS Virtual Data Center (http://db.cosmos-eq.org). The strong motion data recorded by the Istanbul Earthquake Rapid Response Network can be obtained from http://www.koeri.boun.edu.tr/depremmuh/EWRR/EWRRMain.htm.

6. PROCESSING OF STRONG MOTION DATA

The usable frequency band of a digitized strong motion accelerogram is inherently restricted by the combined noise of transduction, digitizing and processing, and thus for proper processing and utilization of accelerograms various corrections have to be incorporated and the limits of this band have to be assessed. Strong motion data obtained from Turkey has been processed for engineering applications and for research in numerous studies [9-17].

Selected strong motion records from KOERI stations (about 10% of the total data set) were incorporated in the CD-ROM Volume 2 of the European Strong-Motion Database[18]. This CD ROM is an updated and improved version of the previous compilation[19] and contains corrected acceleration, velocity and displacement time-histories of 462 triaxial strong-motion records from 110 earthquakes and 261 stations in Europe and the Middle East. The data set contains 112 records (about 25% of all data) from Turkey. KOERI contributes 48 records. 47 records are from ERD and 17 records are from Lamont Doherty Earth Observatory of Columbia University. The correction technique implemented in the Basic Strong-Motion Accelerogram Processing (BAP) software was used for the correction of all of time-histories. This method consists of high-cut filtering, with a cosine transition from the roll-off frequency to the cut-off frequency; followed by low-cut bidirectional Butterworth filtering of the acceleration after padding the time-history with zeros. A method based on the estimated signal-to-noise ratio of each record was used for the selection of appropriate cut-off frequencies. In addition to corrected time history data and peaks, linear and non-linear response spectra, Fourier amplitude spectra, code-based spectra, Arias intensity, normalized energy density, Housner spectral intensity with user-specified limits, acceleration spectral intensity with user-specified limits, durations and root-mean-square acceleration are provided.

The strong ground motion progressing procedure that has been used on several scientific and technical papers originating from KOERI essentially use the following procedure that has its roots in Erdik[20].

1. Low - pass filter the uncorrected accelerogram (roll-off of frequency of 50 Hz and a cut-off of 100 Hz).
2. Least square fit a straight line to the acceleration trace and adjust accordingly.
3. Integrate the acceleration to obtain velocity trace assuming zero initial velocity.
4. High - pass filter the velocity trace using Butterworth type recursive filter with a cut - off frequency determined from the shape of the Fourier amplitude spectrum.
5. Least square fit a straight line to the filtered velocity trace and adjust accordingly.
6. Differentiate velocity trace to obtain the corrected acceleration trace.
7. Integrate velocity trace to obtain the displacement trace assuming zero initial displacement.
8. Least square fit a straight line to this displacement and adjust accordingly to obtain the corrected displacement trace.
9. Add the slope of the straight line of step 8 to the velocity of step 5 to obtain the corrected velocity trace.

This routine preserves the phase of the accelerogram and the correspondence of acceleration, velocity and displacement traces. However, after filtering the corrected accelerogram can not represent the true ground motion outside the pass band. Also the initial motion estimates for velocity and displacement, based on least square fitting of a straight line and high - pass filtering, may influence the shape of the response spectra.

In near field records, where the absolute permanent displacements are substantial (i.e. more that 10's of cm), attempts has been made for the estimation of the DC component of the displacements. Few digital accelerometer data can be doubly integrated directly to yield the permanent displacements (DC Component) without any base line adjustment. (Sakarya). For other near-field records special techniques have been used.

For the correction of the Kocaeli earthquake data we used the methodology presented in [21]. It involves application of a least-mean-square linear fit to the uncorrected accelerogram, integration of acceleration to obtain velocity time history, segmented polynomial baseline-fit to the velocity, integration of velocity to obtain corrected displacements and differentiation of velocity to get corrected accelerations.

7. UTILIZATION OF STRONG MOTION DATA

The strong motion data obtained from KOERI networks were (and will be) used in the development of attenuation relationships in studies for the assessment of source and propagation path characteristics, structural system identification and health monitoring. Some of these studies are summarized below.

Özbey[22] have developed attenuation relationship for the geometric mean of horizontal peak ground acceleration (PGA) and 5% damped spectral acceleration (Sa) for periods up to 4 s using Turkish strong motion data. The overall database[7] consists of 1188 records from 392 earthquakes recorded at 47 strong motion stations between 1994 and 2001. A subset of this comprising 195 records from 17 earthquakes in the Marmara Region of Turkey, with Mw>5.0 are used in the regression analyses. Substantial part of the records are associated with 1999 Kocaeli (Mw 7.4) and the Düzce (Mw 7.1) earthquakes and their aftershocks. The site classes used in this study are defined according to the average shear wave velocity over the top 30 meters using NEHRP Classification. A nonlinear mixed effects model, defined to account for both inter-event and intra-event variability, is employed for the regression analysis. This model accounts for inter- and intra-event variability and leads to smaller standard error compared to a fixed effects model. Over all distances less than about 100 km, the Western U.S. model based attenuation relationships predict higher PGA and spectral acceleration levels than the proposed attenuation model. For short distances this difference may reach one standard deviation. It should be noted that this study is earthquake and region-specific as only recordings from earthquakes that have occurred in Northwestern Turkey have been used in the analyses.

The earthquake response of Hagia Sophia Museum and Süleymaniye Mosque in Istanbul were monitored since 1990. Strong motion networks installed in the two monuments, recorded the mainshock and aftershocks 1999 Kocaeli and Düzce earthquakes. The data recorded were used for system identification and structural health monitoring. The time- and frequency-domain analysis of data presented in Durukal[6] yielded the earthquake specific variation of modal frequencies and damping ratios. Local structural problems were identified in Hagia Sophia in the southwest main pier and at the crowns of the east and west main arches in the form of excessive vibratory motion and, possibly, in Süleymaniye in the west main arch.

REFERENCES

1. H. Kanamori, "Earthquake Prediction an Overview", in *International Handbook of Earthquake and Engineering Seismology*, Ed. By W.H.K. Lee, H. Kanamori, P.C. Jennings and C. Kisslinger, Academic Press, pp. 1205-1216, 2003.

2. J. Espinoza-Aranda, and F.H. Rodriquez-Cayeros, "The seismic Alert System of Mexico City", in *International Handbook of Earthquake and Engineering Seismology*, Ed. By W.H.K. Lee, H. Kanamori, P.C. Jennings and C. Kisslinger, Academic Press, pp. 1253-1260, 2003.

3. L. Gee, D. Neuhauser, D. Dreger, M. Pasyanos, R. Uhrhammer, and B. Romanowicz, "The Rapid Earthquake Data Integration Project", in *International Handbook of Earthquake and Engineering Seismology*, Ed. By W.H.K. Lee, H. Kanamori, P.C. Jennings and C. Kisslinger, Academic Press, pp. 1261-1274, 2003.

4. E. Hauksson, L.M. Jones, and A.F. Shakal, "TriNet: Modern Ground-Motion Seismic Network", in *International handbook of Earthquake and Engineering Seismology*, Ed. By W.H.K. Lee, H. Kanamori, P.C. Jennings and C. Kisslinger, Academic Press, pp. 1275-1288, 2003.

5. Erdik, M., Fahjan Y., Ozel O., Alcik H., Mert A., and Gul M. (2003), Istanbul Earthquake Rapid Response and the Early Warning System. *Bull. of Earthquake Engineering*, V.1, Issue 1, pp. 157-163.

6. Durukal, E., S.Cimilli and M.Erdik (2003), Dynamic Response of two Historical Monuments in İstanbul Deduced from the Recordings of Kocaeli and Düzce Earthquakes, *Bull.Seism.Soc.Am.* V.93-2, pp.694-712.

7. Özbey, C. and Y. Fahjan (2000) Strong Ground Motion Data Base for 18 August, 1999 Kocaeli and 12 November, 1999 Düzce Earthquakes Database, prepared by under the coordination of M. Erdik, Boğaziçi University and E.Safak, USGS, Bogazici University-Istanbul, Turkey.

8. Celebi, M (Scientific Editor), S.Akkar, U.Gulerce, A.Sanli, H.Bundock, A.Salkin (2001), 1999 Turkey Earthquakes Data Base: Main Shock And Aftershock Records of The 1999 Izmit & Duzce, Turkey Earthquakes, USGS Open-File Report 01-163

9. Ates, R., Bayulke N. (1982) The 19 August 1976 Denizli, Turkey, earthquake: evaluation of the strong motion accelerograph record. *Bull Seismol Soc Am* 1982; 72(5):1635–49.

10. Erdik, M. (1984): Report on the Turkish earthquake of October 30, 1983. *Earthquake Spectra*, vol. 1, no. 1, pp. 151-172.

11. Ates, R. (1985): Turkish strong ground motion data acquisition and analysis. *Physics of the Earth and Planetary Interiors*, vol. 38, pp. 123-133.

12. Celebi M, Erdik M, Avci J. (1995), Observations on the effects of surface geology on ground motion: Erzincan (Turkey) Earthquake of March 13, 1992. In: Duma G, editor. *Proceedings of the Tenth European Conference on Earthquake Engineering*, Vol. 4. Amsterdam: A.A. Balkema; 1995. p. 2599–605.

13. Safak, E., M. Erdik (Chapter Coordinators); (in alphabetical order) : K. Beyen, D. Carver, E. Cranswick, M. Celebi, E. Durukal, W. Ellsworth, T. Holzer, M. Meremonte, C. Mueller, C. Mueller, O. Ozel, E. Safak ve S. Toprak (2000), Recorded Main Shock and Aftershock Motions, *Earthquake Spectra*, Supplement A to Volume 16, 97-112, 2000.

14. Anderson, J., H. Sucuoglu, A. Erberik, T.Yilmaz, E.Inan, E.Durukal, M.Erdik, R.Anooshehpoor, J. Brune ve S. Ni, (2000), Strong Ground Motions from the Kocaeli and

Düzce, Turkey, Earthquakes and Possible Implications for Seismic Hazard Analysis, *Earthquake Spectra*, Supplement A to Volume 16, 113-137, 2000.

15. Anderson, J. G., Zeng, Y. and Sucuoglu, H. (2001): Analysis of accelerations from the 1 October 1995 Dinar, Turkey, earthquake. Bull. Seism. Soc. Am., 91(6), 1433-1445.
16. Durukal E, Erdik M, Avci J, Yuzugulu O, Alpay Y, Avar B, Zulfikar C, Biro T, Mert A. Analysis of the strong motion data of the 1995 Dinar, Turkey earthquake. *J Soil Dyn Earthquake Engng* 1998;7,8: 557–78.
17. Zare,M. and P-Y. Bard (2002), Strong motion dataset of Turkey: data processing and site classification, *Soil Dynamics and Earthquake Engineering* 22 (2002) 703–718.
18. Ambraseys, N. N., J. Douglas, R. Sigbjornsson, C. Berge-Thierry, P. Suhadolc, G. Costa and P. M. Smit (2004), Dissemination of European Strong-Motion Data, Vol. 2
19. Ambraseys, N., Smit, P., Berardi, R., Rinaldis, D., Cotton, F., & Berge, C. 2000. Dissemination of European Strong-Motion Data. CD-ROM collection. European Commission, Directorate-General XII, Environmental and Climate Programme, ENV4-CT97-0397, Brussels, Belgium.
20. Erdik, M. (1979), Processing Of Strong Motion Accelerograms In Earthquake Engineering: Corrected Accelerogram, Velocity And Displacement Traces, Middle East Technical University, *Journal of Applied Sciences*.
21. Iwan W.D. and Chen, X. (1995), Important near-field ground data from the Landers earthquake, *Proc. 10th European Conference on Earthquake Engineering*: 229-234.
22. Özbey, C., A. Sari, L. Manuel, M. Erdik, Y. Fahjan (2004), An Empirical Attenuation Relationship For Northwestern Turkey Ground Motion Using A Random Effects Approach, Soil Dynamics and Earthquake Engineering

Addresses of Principal Contributors

Sinan Akkar
Department of Civil Engineering
Middle East Technical University
Ankara 06531, Turkey
sakkar@metu.edu.tr

John G. Anderson
Nevada Seismological Laboratory
University of Nevada, MS 174
Reno, NV 89557, USA
jga@unr.edu

Ralph J. Archuleta
Dept. of Geological Sciences and
Institute for Crustal Studies
University of Calif., Santa Barbara
Santa Barbara, CA 93106, USA
archuleta@geol.ucsb.edu

Julian J. Bommer
Department of Civil & Env. Eng.
Imperial College
London, SW7 2AZ, UK
j.bommer@imperial.ac.uk

David M. Boore
US Geological Survey
345 Middlefield Road, MS 977
Menlo Park, CA 94025, USA
boore@usgs.gov

Roger D. Borcherdt
US Geological Survey
345 Middlefield Road, MS 977
Menlo Park, CA 94025, USA
borcherdt@usgs.gov

Mehmet Çelebi
US Geological Survey
345 Middlefield Road, MS 977
Menlo Park, CA 94025, USA
celebi@usgs.gov

Mustafa Erdik
Kandilli Obs. & Earthq. Res. Inst.
Boğaziçi University
81220 Çengelköy, İstanbul, Turkey
erdik@boun.edu.tr

Polat Gülkan
Department of Civil Engineering &
Disaster Management Research Ctr.
Middle East Technical University
Ankara 06531, Turkey
a03516@metu.edu.tr

Thomas C. Hanks
US Geological Survey
345 Midlefield Road, MS 977
Menlo Park, CA 94025, USA
thanks@usgs.gov

Shigeo Kinoshita
Yokohama City University
Seto 22-2, Kanazawa-ku
Yokohama-shi, 236-0027, Japan
kkk001@yokohama-cu.ac.jp

Helmut Krawinkler
Department of Civil & Env. Eng.
Stanford University
Stanford, CA 94305, USA
krawinkler@stanford.edu

Saburoh Midorikawa
Department of Built Environment
Tokyo Institute of Technology
4259-G3-3, Nagatsuta, Midori-ku
Yokohama 226-8502, Japan
smidorik@enveng.titech.ac.jp

Eduardo Miranda
Department of Civil & Env. Eng
Stanford University
Stanford, CA 94305, USA
emiranda@stanford.edu

Jochen Schwarz
Bauhaus Universitaet Weimar
Institut für Konstr. Ingenieurbau
Erdbebenzentrum
Marienstrasse 7
D-99421, Weimar, Germany
schwarz@bauing.uni-weimar.de

Haluk Sucuoğlu
Department of Civil Engineering
Middle East Technical University
Ankara 06531, Turkey
a06074@metu.edu.tr

Tao Xiaxin
Harbin Institute of Technology &
Institute of Engineering Mechanics
China Earthquake Administration
29 Xuefu Road
Harbin, 150080 China
taoxx@public.hr.hl.cn

Yi-Ben Tsai
Institute of Geophysics
National Central University
Jung-Da Rd. 300, Chung-Li 320
Taiwan, ROC
ybtsai@geps.gep.ncu.edu.tw

INDEX